D0623255

A TO Z OF COMPUTER SCIENTISTS

NOTABLE SCIENTISTS

A TO Z OF COMPUTER SCIENTISTS

HARRY HENDERSON

Facts On File, Inc.

For Lisa
From whom I learned how to live in a story.

A TO Z OF COMPUTER SCIENTISTS

Notable Scientists

Copyright © 2003 by Harry Henderson

All rights reserved. No part of this book may be reproduced or utilized in any form or by any means, electronic or mechanical, including photocopying, recording, or by any information storage or retrieval systems, without permission in writing from the publisher. For information contact:

Facts On File, Inc.
132 West 31st Street
New York NY 10001

Library of Congress Cataloging-in-Publication Data

Henderson, Harry, 1951–
A to Z of computer scientists / Harry Henderson.
p. cm.—(Notable scientists)
Includes bibliographical references and index.
ISBN 0-8160-4531-3
1. Computer scientists—Biography. I. Title. II. Series.
QA76.2.A2 H46 2003
004'.092'2—dc21
2002153880

Facts On File books are available at special discounts when purchased in bulk quantities for businesses, associations, institutions, or sales promotions. Please call our Special Sales Department in New York at (212) 967-8800 or (800) 322-8755.

You can find Facts On File on the World Wide Web at http://www.factsonfile.com

Text design by Joan M. Toro
Cover design by Cathy Rincon

Printed in the United States of America

VB TECHBOOKS 10 9 8 7 6 5 4 3 2 1

This book is printed on acid-free paper.

Contents

LIST OF ENTRIES

Acknowledgments

I wish to thank the computer pioneers and their institutions with which I have corresponded in my search for information and photos. Without their kind help the writing of this book would have been a much more arduous task. I also want especially to thank Frank K. Darmstadt, Executive Editor at Facts On File, for his sorely-tried patience and tireless good nature, which helped sustain me throughout the process of writing. Finally, I want as always to express my appreciation to my wife, Lisa Yount, for her inspirational creativity and daily love and support.

INTRODUCTION

The Industrial Revolution demonstrated the growing ability to organize and control the production of farms, mines, and factories. What had been relatively simple tasks became increasingly precise and elaborate systems. However with the growing complexity of economic and social life came a growing need to organize and manipulate the vast torrent of information produced by society.

In the 19th century, pioneers such as Charles Babbage and Herman Hollerith designed and to some extent implemented mechanical systems for data processing. However, it was only with the advent of electricity and particularly electronics that a practical automatic digital computer became possible.

In the beginning were the inventors. To build early computers, the inventors drew mainly upon two disciplines: the abstract world of mathematics and the more practical discipline of electrical and electronic engineering. The pressing needs of World War II then created a crucible in which the first computers—such as ENIAC, designed by J. Presper Eckert and John Mauchly—were built.

As the decades passed, a specific body of knowledge and methods called computer science gradually came into being. It drew on a variety of fascinating, powerful minds such as those of Alan Turing, Claude Shannon, and Marvin Minsky. Computer scientists created evolving paradigms (such as the structured programming movement with Edsger Dijkstra and Niklaus Wirth) and implemented them in new programming languages such as Pascal, C, Smalltalk, C++, and Java. Meanwhile, these researchers continually extended the boundaries of computer capability into such realms as artificial intelligence.

To inventors and computer scientists must be added a crucial third party: the entrepreneur. Such visionary business leaders as Thomas Watson Sr. and Thomas Watson Jr. of IBM, the two Steves (Jobs and Wozniak) of Apple, and Kenneth Olsen of Digital Equipment Corporation who turned the inventors' ideas into products that changed the industry. In the realm of software, other entrepreneurs such as Mitchell Kapor and An Wang popularized new applications such as spreadsheets and word processing, while others, among them Bill Gates, cut a swath across the entire industry.

In the 1990s, the growth of the Internet, the World Wide Web, and e-commerce created a new generation of innovators and entrepreneurs such as Jeff Bezos of Amazon.com, Pierre Omidyar of eBay, and Jerry Yang of Yahoo! They are joined by other young pioneers, including virtual reality researcher Jaron Lanier and Linux developer Linus Torvalds.

A science and an industry need a variety of voices with which to debate its values and explore its future. Some of these voices, such as those of Michael Dertouzos and Raymond

Kurzweil, are essentially optimistic, believing that technology will empower humanity in unimaginable ways. Others, such as Clifford Stoll, offer a critical and cautionary view. Difficult legal and social questions are addressed by still other voices, such as Richard Stallman and Eric Raymond (advocates for software free from corporate control) and Howard Rheingold and Sherry Turkle, chroniclers of virtual communities and explorers of their psychology and sociology. For all these reasons we have not limited our selections to pure computer scientists.

The persons selected for inclusion in this book thus offer a varied array of inventors, computer scientists, and entrepreneurs as well as some people from other backgrounds who have had a major impact on computer science and technology.

The Entries

A to Z of Computer Scientists presents the stories of more than 100 people. Rather than being characterized by research area or specialty, the entrants are characterized by background (engineer, mathematician, or other scientist) and by role (computer scientist, programmer, entrepreneur, and inventor). For a few people, writing played an important role in their careers, so they are given the category "writer." Many people fit into two or occasionally three categories. See the appendix "Entries by Field."

It should be clear that hundreds of people have played important roles in computer science and technology. The decision was made to explore a somewhat smaller number of people in greater depth, in part because brief biographical entries are already easy to come by. Each entry is an essay that not only recounts achievements but places them in context and explains their significance.

Format

Entries are arranged alphabetically by surname. The heading for each entry provides the entrant's complete name, birth and death dates, countries where the subject was born and (if different) where the subject lived at the time of his or her chief scientific achievement, and field of work.

The text of the entries ranges generally from about 900 to about 1,750 words, with most around 1,250. They include the usual biographical information: date and place of birth and death, family and childhood (where known), educational background, places worked and positions held, prizes awarded, and so on.

Each entry focuses on one or more major achievements of the person, such as inventions, ideas, or enterprises. Technical terms are generally briefly explained in the text, but the glossary at the back of the book provides somewhat more extensive definitions. Quotations from the subject or by other persons commenting on the subject are often provided. Longer quotations are set off in an indented paragraph and placed in italics.

Names in small capital letters within the essays indicate cross-references to other persons who have entries in the book. For those who wish to learn more about a particular person and his or her work, a short list of further reading, including both print and Internet resources, is provided at the end of each entry.

The book concludes with several appendixes that may aid readers seeking particular types of information. In addition to the glossary, the appendixes include a chronological chart showing the chief achievements of the persons in the book arranged by approximate date. This chronology provides a sort of capsule history of highlights, focusing mainly on the period from 1940 to date. There is also a general bibliography listing books that explore the history of the computer, other biographies, and concepts of computer science.

The computer field is certainly one of the most fascinating areas of human activity, as well as a vital part of the economy and our daily lives. As fascinating as the machines and ideas are, the people are equally fascinating in their variety. Later editions will likely include a larger number of women and people of different nationalities and ethnic backgrounds as the computer field becomes truly global.

A

Aiken, Howard
(1900–1973)
American
Inventor

Howard Hathaway Aiken was a pioneer in the development of automatic calculating machines. Born on March 8, 1900, in Hoboken, New Jersey, he grew up in Indianapolis, Indiana. He pursued his interest in electrical engineering by working at a utility company while in high school. Aiken then earned a B.A. degree in electrical engineering in 1923 at the University of Wisconsin.

By 1935, Aiken was working on the physics of how electric charges were conducted in vacuum tubes—an important question for the new technology of electronics. This work required tedious, error-prone hand calculation. Aiken therefore began to investigate the possibility of building a large-scale, programmable, automatic computing device. As a doctoral student at Harvard, Aiken aroused considerable interest in his ideas, particularly from THOMAS J. WATSON SR., head of International Business Machines (IBM). In 1939, IBM agreed to underwrite the building of Aiken's first calculator, the Automatic Sequence Controlled Calculator (ASCC), which became known as the Harvard Mark I.

Mechanical and electromechanical calculators were nothing new: indeed, machines from IBM, Burroughs, and others were being increasingly used in business settings. However, ordinary calculators required that operators manually set up and run each operation step by step in the complete sequence needed to solve a problem. Aiken wanted a calculator that could be programmed to carry out the sequence automatically, storing the results of each calculation for use by the next. He wanted a general-purpose programmable machine rather than an assembly of special-purpose arithmetic units.

Earlier complex calculators (such as the Analytical Engine which CHARLES BABBAGE had proposed a century earlier) were very difficult to implement because of the precise tolerances needed for the intricate assembly of mechanical parts. Aiken, however, had access to a variety of tested, reliable components, including card punches, readers, and electric typewriters from IBM and the mechanical electromagnetic relays used for automatic switching in the telephone industry.

Aiken's Mark I calculator used decimal numbers (23 digits and a sign) rather than the binary numbers of the majority of later computers. Sixty registers held whatever constant data numbers were needed to solve a particular problem. The operator turned a rotary dial to enter each digit of each constant number required for the calculation. Variable data and program instructions were entered from punched paper tape. Calculations had to be broken down into specific instruction codes similar to those in later low-level programming languages such as "store this number in this register" or "add this number to the number in that register." The results (usually tables of mathematical function values) could be printed by an electric typewriter or output on punched cards.

The Mark I was built at IBM's factory in Endicott, New York. It underwent its first full-scale test on Christmas Day 1943, illustrating the urgency of work under wartime conditions. The bus-sized machine (about eight feet high by 51 feet long) was then painstakingly disassembled and shipped to Harvard University, where it was up and running by March 1944. Relatively slow by comparison with the vacuum tube-based computers that would soon be designed, the Mark I was a very reliable machine. A *New York Times* article enthused, "At the dictation of a mathematician, it will solve in a matter of hours equations never before solved because of their intricacy and the enormous time and personnel which would be required to work them out on ordinary office calculators."

Aiken then went to work for the U.S. Navy (and was given the rank of commander), where his team included another famous computer pioneer, the future admiral GRACE MURRAY HOPPER. The Mark I worked 24 hours a day on a variety of problems, ranging from solving equations used in lens design and radar to the ultrasecret design for the implosive core of the atomic bomb. Unlike many engineers, Aiken was comfortable managing fast-paced projects. He once quipped, "Don't worry about people stealing an idea. If it's original, you'll have to ram it down their throats."

Aiken completed an improved model, the Mark II, in 1947. The Mark III of 1950 and Mark IV of 1952 were electronic rather than electromechanical, replacing relays with vacuum tubes. The Mark III used a magnetic core memory (analogous to modern RAM, or random-access memory) that could store and retrieve numbers relatively quickly, as well as a magnetic drum that served the function of a modern hard disk.

Compared to slightly later digital computers such as ENIAC and Univac, the sequential calculator, as its name suggests, could only perform operations in the order specified, rather than, for example, being able to loop repeatedly. (After all, the program as a whole was not stored in any sort of memory, and so previous instructions could not be reaccessed.) Yet although Aiken's machines soon slipped out of the mainstream of computer development, they did include the modern feature of parallel processing, because different calculation units could work on different instructions at the same time. Further, Aiken recognized the value of maintaining a library of frequently needed routines that could be reused in new programs—another fundamental of modern software engineering.

Aiken's work demonstrated the value of large-scale automatic computation and the use of reliable, available technology. Computer pioneers from around the world came to Aiken's Harvard computation lab to debate many issues that would become staples of the new discipline of computer science. By the early 1950s Aiken had retired from computer work and became a Florida business entrepreneur, enjoying the challenge of rescuing ailing businesses.

The recipient of many awards, including the Edison Medal of the Institute of Electrical and Electronics Engineers and the Franklin Institute's John Price Award, Howard Aiken died on March 14, 1973, in St. Louis, Missouri.

Further Reading

Cohen, I. B. *Howard Aiken: Portrait of a Computer Pioneer.* Cambridge, Mass.: MIT Press, 1999.

Cohen, I. B., R. V. D. Campbell, and G. Welch, eds. *Makin' Numbers: Howard Aiken and the Computer.* Cambridge, Mass.: MIT Press, 1999.

Ferguson, Cassie. "Howard Aiken: Makin' a Computer Wonder." *Harvard University Gazette*. Available on-line. URL: http://www.news.harvard.edu/gazette/1998/04.09/HowardAikenMaki.html. Posted on April 9, 1998.

Amdahl, Gene M.

(1922–)
American
Inventor, Entrepreneur

In a long and fruitful career as a computer designer Gene Myron Amdahl created many innovations and refinements in the design of mainframe computers, the hefty workhorses of the data processing industry from the 1950s through the 1970s. Amdahl was born on November 16, 1922, in Flandreau, South Dakota. Amdahl did his college work in electrical engineering and physics. When his studies were interrupted by World War II, he served as a physics instructor for an army special training program and then joined the navy, where he taught electronics until 1946. He then returned to school, receiving his B.S. degree from South Dakota State University in 1948 and his doctorate in physics at the University of Wisconsin in 1952.

As a graduate student, Amdahl worked on a problem involving the forces binding together parts of a simple atomic nucleus. He and two fellow students spent a month performing the necessary computations with calculators and slide rules. Amdahl realized that if physicists were going to be able to move on to more complex problems they would need greater computing resources. He therefore designed a computer called the WISC (Wisconsin Integrally Synchronized Computer). This computer used a sophisticated procedure to break calculations into parts that could be carried out on separate processors, making it one of the earliest examples of the parallel computing techniques found in today's computer processors.

In 1952, Amdahl went to work for IBM, which was beginning the effort that would lead to its dominating the business computer industry by the end of the decade. Amdahl worked with the team that designed the IBM 704. The 704 improved upon the 701, the company's first successful mainframe, by adding many new internal programming instructions, including the ability to perform floating point calculations (involving numbers that have decimal points). The machine also included a fast, high-capacity magnetic core memory that let the machine retrieve data more quickly during calculations. In November 1953, Amdahl became the chief project engineer for the 704.

On the heels of that accomplishment, new opportunities seemed to be just around the corner. Although IBM had made its reputation in business machines, it was also interested in the market for specialized computers for scientists. Amdahl helped design the IBM 709, an extension of the 704 designed for scientific applications. When IBM proposed extending the technology by building a powerful new scientific computer called STRETCH, Amdahl eagerly applied to head the new project. However he ended up on the losing side of a corporate power struggle, and did not receive the post. He left IBM at the end of 1955.

Amdahl then worked for several small data processing companies. He helped design the RW440, a minicomputer used for industrial process control. This period gave Amdahl some experience in dealing with the problems of startup businesses, experience he would call upon later when he started his own company.

In 1960, Amdahl rejoined IBM and soon was involved in several design projects. The one with the most lasting importance was the IBM System/360, which would become the most ubiquitous and successful mainframe computer of all time. In this project, Amdahl further refined his ideas about making a computer's central processing unit more efficient. He designed logic circuits that enabled the processor to analyze the instructions waiting to be executed (the "pipeline") and determine which instructions could be executed immediately and which would have to wait for the results of other instructions. He also used a cache, or special memory area in which the instructions that would be needed next could be stored ahead of time so they could be retrieved quickly from high-speed storage. Today's desktop personal computers (PCs) use these same ideas to get the most out of their chips' capabilities.

The problem of parallel computing is partly a problem of designing appropriate hardware and partly a problem of writing (or rewriting) software so its instructions can be executed simultaneously. It is often difficult to predict how much a parallel computing arrangement will improve upon using a single processor and conventional software. Amdahl created a formula called Amdahl's law, which attempts to answer that question. In simple terms, Amdahl's law says that the advantage gained from using more processors gradually declines as more processors are added. The amount of improvement is also proportional to how much of the calculation can be broken down into parts that can be run in parallel. As a result, some kinds of programs can run much faster with several processors being used simultaneously, while other programs may show little improvement.

As a designer, Amdahl coupled hard work with the ability to respond to sudden bursts of intuition. "Sometimes," he recalled to author Robert Slater, "I wake up in the middle of the night and I'll be going 60 miles an hour on the way to a solution. I see a mental picture of what is going on and I dynamically operate that in my mind."

In 1965, Amdahl was awarded a five-year IBM fellowship that allowed him to study whatever problems interested him. He also helped establish IBM's Advanced Computing Systems Laboratory in Menlo Park, California, which he directed. However, Amdahl became increasingly frustrated with what he thought was IBM's too-rigid approach to designing and marketing computers. IBM insisted on basing the price of a new computer not on how much it cost to produce, but on how fast it could calculate.

Amdahl wanted to build much more powerful computers—what would soon be called "supercomputers." But if IBM's policy were followed, these machines would be so expensive that virtually no one would be able to afford them. Thus at a time when increasing miniaturization was leading to the possibility of much more powerful machines, IBM did not seem to be interested in building them. The computer giant seemed to be content to gradually build upon its financially successful 360 line (which would become the IBM 370 in the 1970s). Amdahl therefore left IBM in 1970, later recalling to Slater that he left IBM that second time "because I wanted to work in large computers. . . . I'd have had to change my career if I stayed at IBM—for I wanted personal satisfaction."

To that end, in 1970 he founded the Amdahl Corporation. Amdahl resolved to make computers that were more powerful than IBM's machines, but would be "plug compatible" with them, allowing them to use existing hardware and software. Business users who had already invested heavily in IBM equipment could thus buy Amdahl's machines without fear of incompatibility. Since IBM was known as "Big Blue," Amdahl decided to become "Big Red," painting his machines accordingly.

Amdahl would later recall his great satisfaction in "getting those first computers built and really making a difference, seeing it completely shattering the control of the market that IBM had, causing pricing to come back to realistic levels." Amdahl's critics sometimes accused him of having unfairly used the techniques and knowledge that he had developed at IBM, but he has responded by pointing to his later technical innovations. In particular, he was able to take advantage of the early developments in integrated electronics to put more circuits on a chip without making the chips too small, and thus too crowded for placing the transistors.

After it was introduced in 1975, the Amdahl 470 series of machines, doubled in sales in each of its first three years. Thanks to the use of larger-scale circuit integration, Amdahl could sell machines with superior technology to that of the IBM 360 or even the new IBM 370, and at a lower price. IBM responded belatedly to the competition, making more compact and faster processors, but Amdahl met each new IBM product with a faster, cheaper alternative. However, IBM also countered by using a sales technique that opponents called FUD—fear, uncertainty, and doubt. IBM salespersons promised customers that IBM would soon be coming out with much more powerful and economical alternatives to Amdahl's machines. As a result, many potential customers were persuaded to postpone purchasing decisions and stay with IBM. Amdahl Corporation began to falter, and Gene Amdahl gradually sold his stock and left the company in 1980.

Amdahl then tried to repeat his early success by starting a new company called Trilogy. The company promised to build much faster and cheaper computers than those offered by IBM or Amdahl. He believed he could accomplish this by using the new, very-large-scale integrated silicon wafer technology, in which circuits were deposited in layers on a single chip rather than being distributed on separate chips on a printed circuit board. However, the problem of dealing with the electrical characteristics of such dense circuitry, as well as some design errors, somewhat crippled the new computer design. Amdahl also found that the aluminum substrate that connected the wafers on the circuit board was causing short circuits. Even weather, in the form of a torrential rainstorm, conspired to add to Amdahl's problems by flooding a chip-building plant and contaminating millions of dollars' worth of chips. Amdahl was forced to repeatedly delay the introduction of the new machine, from 1984 to 1985 to 1987. He attempted to infuse new technology into his company by buying Elxsi, a minicomputer company, but Trilogy never recovered.

After the failure of Trilogy, Amdahl undertook new ventures in the late 1980s and 1990s, including Andor International, an unsuccessful developer of minicomputers, and Commercial Data Servers (CDS), which is trying to compete with IBM in the low-priced end of the mainframe market.

Amdahl has received many industry awards, including "Data Processing Man of the Year" from the Data Processing Management Association (1976) and the Harry Goode Memorial Award from the American Federation of Information Processing Societies.

Further Reading

The History of Computing Foundation. "Gene Amdahl." Available on-line. URL: http://www.thocp.net/biographies/amdahl_gene.htm. Downloaded on November 25, 2002.

Slater, Robert. *Portraits in Silicon*. Cambridge, Mass.: MIT Press, 1987.

⊠ Andreessen, Marc

(1971–)
American
Entrepreneur, Programmer

Marc Andreessen brought the World Wide Web and its wealth of information, graphics, and services to the desktop, setting the stage for the "e-commerce" revolution of the later 1990s. As founder of Netscape, Andreessen also created the first big "dot-com," as companies doing business on the Internet came to be called.

By the early 1990s, the World Wide Web (created by TIM BERNERS-LEE) was poised to change the way information and services were delivered to users. However, early Web browsers ran mainly on machines using UNIX, a somewhat esoteric operating system used primarily by

Marc Andreessen made the World Wide Web consumer friendly with his graphical Netscape Web browser. In the mid-1990s, he went on to challenge Microsoft for dominance on the new frontier of e-commerce. *(CORBIS SABA)*

students and scientists on college campuses and at research institutes (Berners-Lee had been working at CERN, the European nuclear physics laboratory.) The early Web generally consisted only of linked pages of text, without the graphics and interactive features that adorn webpages today. Besides looking boring, early webpages were hard for inexperienced people to navigate. Marc Andreessen would change all that.

Marc Andreessen was born on July 9, 1971, in New Lisbon, Wisconsin. That made him part of a generation that would grow up with personal computers, computer games, and computer graphics. Indeed, when Marc was only nine years old he learned the BASIC computer language from a book in his school's library, and then proceeded to write a program to help him with his math homework. Unfortunately, he did not have a floppy disk to save the program on, so it disappeared when the school's janitor turned off the machine.

Marc got his own personal computer in seventh grade, and tinkered on many sorts of programs through high school. He then studied computer science at the University of Illinois at Urbana-Champaign. Despite his devotion to programming, he impressed his fellow students as a Renaissance man. One of them recalled in an interview that "A conversation with Andreessen jumps across a whole range of ungeekish subjects, including classical music, history, philosophy, the media, and business strategy. It's as if he has a hypertext brain."

Andreessen encountered the Web shortly after it was introduced in 1991 by Tim Berners-Lee. He was impressed by the power of the new medium, which enabled many kinds of information to be accessed using the existing Internet, but became determined to make it more accessible to ordinary people. In 1993, while still an undergraduate, he won an internship at the National Center for Supercomputing Applications (NCSA). Given the opportunity to write a better Web browser, Andreessen, together with colleague Eric Bina and other helpers, set to work on what became known as the Mosaic web browser. Since their work was paid for by the government, Mosaic was offered free to users over the Internet. Mosaic could show pictures as well as text, and users could follow Web links simply by clicking on them with the mouse. The user-friendly program became immensely popular, with more than 10 million users by 1995.

After earning his B.S. degree in computer science, Andreessen left Mosaic, having battled with its managers over the future of Web browsing software. He went to the area south of San Francisco Bay, a hotbed of startup companies known as Silicon Valley, which had become a magnet for venture capital in the 1990s. There he met Jim Clark, an older entrepreneur who had been chief executive officer (CEO) of Silicon Graphics. Clark liked Andreessen and agreed to help him build a business based on the Web. They founded Netscape Corporation in 1994, using $4 million seed capital provided by Clark.

Andreessen recruited many of his former colleagues at NCSA to help him write a new Web browser, which became known as Netscape Navigator. Navigator was faster and more graphically attractive than Mosaic. Most important, Netscape added a secure encrypted facility that people could use to send their credit card numbers to online merchants. This was part of a two-pronged strategy: First, attract the lion's share of Web users to the new browser, then sell businesses the software they would need to create effective Web pages for selling products and services to users.

By the end of 1994, Navigator had gained 70 percent of the Web browser market. *Time* magazine named the browser one of the 10 best products of the year, and Netscape was soon selling custom software to companies that wanted a presence on the Web. The e-commerce boom of the later 1990s had begun, and Marc Andreessen was one of its brightest stars. When Netscape offered its stock to the public in summer 1995, the company gained a total worth of $2.3 billion, more than that of many traditional blue-chip industrial companies. Andreessen's own shares were worth $55 million.

Microsoft under BILL GATES had been slow to recognize the growing importance of the Web. However, as users began to spend more and more time interacting with the Netscape window, Microsoft began to worry that its dominance of the desktop market was in jeopardy. Navigator could run not only on Microsoft Windows PCs, but also on Macintoshes and even on machines running versions of UNIX. Further, a new programming language called Java, developed by JAMES GOSLING made it possible to write programs that users could run from Web pages without being limited to Windows or any other operating system. If such applications became ubiquitous, then the combination of Navigator (and other Netscape software) plus Java could in effect replace the Windows desktop.

Microsoft responded by creating its own Web browser, called Internet Explorer. Although technical reviewers generally considered the Microsoft product to be inferior to Netscape, it gradually improved. Most significantly, Microsoft included Explorer with its new Windows 95 operating system. This "bundling" meant that PC makers and consumers had little interest in paying for Navigator when they already had a "free" browser from Microsoft. In response to this move, Netscape and other Microsoft competitors helped promote the antitrust case against Microsoft that would result in 2001 in some of the company's practices being declared an unlawful use of monopoly power.

Andreessen also responded to Microsoft by focusing on the added value of software for Web servers, while making Navigator "open source," meaning that anyone was allowed to access and modify the program's code. He hoped that a vigorous community of programmers might help keep Navigator technically superior to Internet Explorer. However, Netscape's revenues began to decline steadily. In 1999 America Online (AOL) bought Netscape, seeking to add its technical assets and Webcenter online portal to its own offerings.

After a brief stint with AOL as its "principal technical visionary," Andreessen decided to start his own company, called LoudCloud. The company provided website development, management and custom software (including e-commerce "shopping basket" systems) for corporations that have large, complex websites. Through 2001, Andreessen vigorously promoted the company, seeking to raise enough operating capital to continue after the crash of the Internet industry. However, after a continuing decline in profitability Andreessen sold LoudCloud's Web management business to Texas-based Electronic Data Systems (EDS), retaining the smaller (software) side of the business under a new name, Opsware.

While the future of his recent ventures remains uncertain, Marc Andreessen's place as one of the key pioneers of the Web and e-commerce is assured. His inventiveness, technical insight, and business acumen made him a model for a new generation of Internet entrepreneurs. Andreessen was named one of the Top 50 People Under the Age of 40 by *Time* magazine (1994) and has received the *Computerworld*/Smithsonian Award for Leadership (1995) and the W. Wallace McDowell Award of the Institute of Electrical and Electronic Engineers Computer Society (1997).

Further Reading

"Andreessen, Marc," *Current Biography Yearbook*. New York: H. W. Wilson, 1997, p. 12.

Clark, Jim, with Owen Edwards. *Netscape Time: The Making of the Billion-Dollar Startup That Took on Microsoft*. New York: St. Martin's Press, 1999.

Quittner, Joshua, and Michelle Slatalla. *Speeding the Net: the Inside Story of Netscape and How It Challenged Microsoft*. New York: Atlantic Monthly Press, 1998.

Sprout, Alison. "The Rise of Netscape," *Fortune*, July 10, 1995, pp. 140ff.

According to a federal court, it was John Atanasoff, not John Mauchly and J. Presper Eckert, who built the first digital computer. At any rate, the "ABC" or Atanasoff-Berry Computer, represented a pioneering achievement in the use of binary logic circuits for computation. *(Photo courtesy of Iowa State University)*

Atanasoff, John Vincent
(1903–1995)
American
Inventor

It is difficult to credit the invention of the first digital computer to any single individual. Although the 1944 ENIAC, designed by J. PRESPER ECKERT and JOHN MAUCHLY, is widely considered to be the first fully functional electronic digital computer, John Vincent Atanasoff and his graduate assistant Clifford Berry created a machine five years earlier that had many of the features of modern computers. Indeed, a court would eventually declare that the Atanasoff-Berry Computer (ABC) was the first true electronic computer.

Atanasoff was born October 4, 1903, in Hamilton, New York. His father was an electrical engineer, his mother a teacher, and both parents encouraged him in his scientific interests. In particular, the young boy was fascinated by his father's slide rule. He learned about logarithms so he could understand how the slide rule worked. He also showed his father's aptitude for electrical matters: When he was nine years old, he discovered that some wiring in their house was faulty, and fixed it.

John blazed through high school in only two years, making straight A's. By then, he had decided to become a theoretical physicist. When he entered the University of Florida, however, he majored in engineering and mathematics because the school lacked a physics major. Offered a number of graduate fellowships, Atanasoff opted for Iowa State College because he liked its programs in physics and engineering. He earned his master's degree in mathematics in 1926.

Atanasoff continued on to the University of Wisconsin, where he earned his doctorate in physics in 1930. He would remain there as a professor of physics for the next decade.

Like HOWARD AIKEN, Atanasoff discovered that modern physics was encountering an increasing burden of calculation that was becoming harder and harder to meet using manual methods, the slide rule, or even the electromechanical calculators being used by business. One alternative in development at the time was the analog computer, which used the changing relationships between gears, cams, and other mechanical components to represent quantities manipulated in equations. While analog computers such as the differential analyzer built by VANNEVAR BUSH achieved success in tackling some problems, they tended to break down or produce errors because of the very exacting mechanical tolerances and alignments they required. Also, these machines were specialized and hard to adapt to different kinds of problems.

Atanasoff made a bold decision. He would build an automatic, digital electronic calculator. Instead of the decimal numbers used by ordinary calculators, he decided to use binary numbers, which could be represented by different amounts of electrical current or charge. The binary logic first developed by GEORGE BOOLE could also be manipulated to perform arithmetic directly.

Equally important, at a time when electric motors and switches drove mechanical calculators, Atanasoff decided to design a machine that would be electronic rather than merely electrical. It would use the direct manipulation of electrons in vacuum tubes, which is thousands of times faster than electromechanical switching.

Atanasoff obtained a modest $650 grant from Iowa State and hired Clifford Berry, a talented graduate student, to help him. In December 1939, they introduced a working model of the Atanasoff-Berry Computer (ABC). The machine used vacuum tubes for all logical and arithmetic operations. Numbers were input from punched cards, while the working storage (equivalent to today's random-access memory, or RAM), consisted of two rotating drums that stored the numbers as electrical charges on tiny capacitors. The ABC, however, was not a truly general purpose computer: It was designed to solve sets of equations by systematically eliminating unknown quantities. Because of problems with the capacitor-charge memory system, Atanasoff and Berry were never able to solve more than five equations at a time.

As the United States entered World War II, Atanasoff had to increasingly divide his time between working on the ABC and his duties at the National Ordnance Laboratory in Washington, D.C., where he headed the acoustics division and worked on designing a computer for naval use. Eventually the ABC project petered out, and the machine never became fully operational.

After the war, Atanasoff gradually became disillusioned with computing. The mainstream of the new field went in a different direction, toward the general-purpose machines typified by ENIAC, a large vacuum tube computer. In 1950 Atanasoff discovered that Iowa State had dismantled and partly discarded the ABC. He spent the remainder of his career as a consultant and entrepreneur. In 1952, he and his former student David Beecher founded a defense company, Ordnance Engineering Corporation. In 1961, he became a consultant working on industrial automation, and cofounded a company called Cybernetics with his son.

In 1971, however, Atanasoff and the ABC became part of a momentous patent dispute. Mauchly and Eckert had patented many of the fundamental mechanisms of the digital computer on the strength of their 1944 ENIAC machine. Sperry Univac, which now controlled the patents, demanded high licensing fees from other computer companies. A lawyer for one of these rivals, Honeywell, had heard of Atanasoff's work and decided that he could challenge the Mauchly-Eckert patents. The heart of his case was that in June 1941 Mauchly had stayed at Atanasoff's home and had been treated to an extensive demonstration of the ABC. Honeywell claimed that Mauchly had obtained the key idea of using vacuum tubes and electronic circuits from Atanasoff. If so, the Atanasoff machine would be "prior art," and the Mauchly-Eckert patents would be invalid. In 1973, the federal court agreed, declaring that Mauchly and Eckert "did not themselves invent the automatic electronic digital computer, but instead derived that subject matter from one Dr. John Vincent Atanasoff."

The decision was not appealed. Despite the definitive legal ruling, the controversy among computer experts and historians grew. Defending his work in public for the first time, Atanasoff stressed the importance of the ideas that Mauchly and Eckert had obtained from him, including the use of vacuum tubes and binary logic circuits. Defenders of the ENIAC inventors, however, pointed out that the ABC was a specialized machine that was never a fully working general-purpose computer like ENIAC.

While the dispute may never be resolved, it did serve to give Atanasoff belated recognition for his achievements. On October 21, 1983, the University of Iowa held a special conference celebrating Atanasoff's work, and later built a working replica of the ABC. By the time Atanasoff died in 1995 at the age of 91, he had been honored with many awards, including the Computer Pioneer Medal from the Institute for Electrical and Electronics Engineers in 1984 and the National Medal of Technology in 1990.

Further Reading

Burks, A. R., and A. W. Burks. *The First Electronic Computer: The Atanasoff Story*. Ann Arbor: University of Michigan Press, 1988.

Mollenhoff, Clark. *Atanasoff: Forgotten Father of the Computer*. Ames: Iowa State University Press, 1988.

"Reconstruction of the Atanasoff-Berry Computer (ABC)." Available on-line. URL: http://www.scl.ameslab.gov/ABC/ABC.html. Updated on July 18, 2002.

B

Babbage, Charles

(1791–1871)
British
Inventor, Mathematician

More than a century before the first electronic digital computers were invented, British mathematician and inventor Charles Babbage conceived and designed a mechanical "engine" that had most of the key features of modern computers. Although Babbage's computer was never built, it remains a testament to the remarkable power of the human imagination.

Charles Babbage was born on December 26, 1791, in London, to a well-to-do banking family that was able to provide him with a first-class private education. Even as a young child Babbage was fascinated by mechanisms of all kinds, asking endless questions and often dissecting the objects in search of answers. He became a star student, particularly in math, at boarding school, where he and some intellectually inclined friends stayed up late to study.

In 1810 Babbage entered Trinity College, Cambridge, where he studied advanced calculus and helped found an organization to reform the Newtonian discipline along more modern European lines. By 1815, Babbage had made such an impact in mathematics and science that he had been elected a fellow of the prestigious British Royal Society. His reputation continued to grow, and in 1828 he was appointed Lucasian Professor of Mathematics at Cambridge, occupying the chair once held by Isaac Newton.

What was becoming a distinguished career in mathematics then took a different turn, one in keeping with the times. By the early 19th century, the role of mathematics and science in European society was beginning to change. Britain, in particular, was a leader in the Industrial Revolution, when steam power, automated weaving, and large-scale manufacturing were rapidly changing the economy and people's daily lives. In this new economy, "hard numbers"—the mathematical tables needed by engineers, bankers, and insurance companies—were becoming increasingly necessary. All such tables, however, had to be calculated slowly and painstakingly by hand, resulting in numerous errors.

One day, when poring over a table of logarithms, Babbage fell asleep and was roused by a friend. When asked what he had been dreaming about, Babbage replied that "I am thinking that all these tables might be calculated by machines."

The idea of mechanical computation was perhaps not so surprising. Already the automatic loom invented by Joseph-Marie Jacquard was being controlled by chains of punched cards containing weaving patterns. The idea of controlled, repetitive motion was at the heart of the new

If it had been built, Charles Babbage's Analytical Engine, although mechanical rather than electrical, would have had most of the essential features of modern computers. These included punched card input, a processor, a memory (store), and a printer. *(Photo courtesy of NMPTFT/Science & Society Picture Library)*

industry. Babbage was in essence applying industrial methods to the creation of the information that an industrial society increasingly required for further progress. But although his idea of industrializing mathematics was logical, Babbage was entering uncharted technological territory.

From 1820 to 1822, Babbage constructed a small calculator that he called a difference engine. The calculator exploited a mathematical method for generating a table of numbers and their squares by repeated simple subtraction and addition. When the demonstration model successfully generated numbers up to about eight decimal places, Babbage undertook to build a larger scale version, which he called Difference Engine Number One. This machine would have around 25,000 gears and other moving parts and could handle numbers with up to 20 decimal places. The machine was even to have a printer that could generate the final tables directly, avoiding transcription errors.

By 1830, work was well under way, supported by both government grants and Babbage's own funds. However, Babbage soon became bogged down with problems. Fundamentally, the parts for the Difference Engine required a tolerance and uniformity that went beyond anything found in the rough-hewn industry of the time, requiring new tools and production methods. At the same time, Babbage was a poorer manager than an inventor, and in 1833 labor disputes virtually halted the work. The big Difference Engine would never be finished.

By 1836, however, Babbage, undaunted, had developed a far bolder conception. He wrote in his notebook, "This day I had for the first time a general . . . conception of making an engine work out *algebraic* developments. . . . My notion is that the cards (Jacquards) of the calc. engine direct a series of operations and then recommence with the first, so it might be possible to cause the same cards to punch others equivalent to any number of repetitions."

As Babbage worked out the details, he decided that the new machine (which he called the Analytical Engine) would be fed by two stacks of punched cards. One stack would contain instructions that specified the operation (such as addition or multiplication), while the other would contain the data numbers or variables. In other words, the instruction cards would program the machine to carry out the operations automatically using the data. The required arithmetic would be carried out in a series of gear-driven calculation units called the mill, while temporary results and variable values would be stored in a series of mechanical registers called the store. The final results could be either printed or punched onto a set of cards for future use.

The Analytical Engine thus had many of the features of a modern computer: a central processor (the mill), a memory (the store), as well as input and output mechanisms. One feature it lacked, as revealed in Babbage's journal entry, was the ability to store programs themselves in memory. That is why a repetition (or loop) could be carried out only by repeatedly punching the required cards.

The new machine would be a massive and expensive undertaking. Babbage's own funds were far from sufficient, and the British government had become disillusioned by his failure to complete the Difference Engine. Babbage therefore began to use his contacts in the international mathematical community to try to raise support for the project. He was aided by L. F. Menebrea, an Italian mathematician, who wrote a series of articles about the Analytical Engine in France. He was further aided by ADA LOVELACE (the daughter of the poet George Gordon, Lord Byron). She not only translated the French articles into English, but greatly expanded them, including her own example programs and suggestions for applications for the device.

However, like the Difference Engine, the Analytical Engine was not to be. A contemporary wrote that Babbage was "frequently and almost notoriously incoherent when he spoke in public." His impatience and "prickliness" also made a bad impression on some of the people he had to persuade to support the new machine. Funding was not found, and Babbage was only able to construct demonstration models of a few of its components.

As he moved toward old age, Babbage continued to write incredibly detailed engineering drawings and notes for the Analytical Engine as well as plans for improved versions of the earlier Difference Engine. But he became reclusive and even more irritable. Babbage became notorious for his hatred of street musicians, as chronicled in his 1864 pamphlet *Observations of Street Nuisances*. Neighbors who supported the musicians often taunted Babbage, sometimes organizing bands to play deliberately mistuned instruments in front of his house.

After Babbage's death in October 18, 1871, his remarkable computer ideas faded into obscurity, and he was remembered mainly for his contributions to economic and social statistics, another field that was emerging into importance by the mid-19th century. Babbage therefore had little direct influence on the resurgence of interest in automatic calculation and computing that would begin in the late 1930s (many of his notes were not unearthed until the 1970s). However, computer scientists today honor Charles Babbage as their spiritual father.

Further Reading

"The Analytical Engine: the First Computer." Available on-line. URL: http://www.fourmilab.ch/babbage/. Downloaded on October 31, 2002.

Babbage, H. P., ed. *Babbage's Calculating Engines*. London, 1889. Reprinted as vol. 2 of I. Tomash, ed. Babbage Institute Reprint Series, 1984.

Campbell-Kelly, M., ed. *The Works of Charles Babbage*. 11 vols. London: Pickering and Chatto, 1989.

Henderson, Harry. *Modern Mathematicians*. New York: Facts On File, 1996.

Swade, Doron D. "Redeeming Charles Babbage's Mechanical Computer." *Scientific American*, February 1993, p. 86.

Backus, John

(1924–)
American
Computer Scientist

John Backus led the team that developed FORTRAN, one of the most popular computer languages of all time, particularly for scientific and engineering applications. Backus also made significant contributions to computer design and to the analysis of programming languages.

Born on December 3, 1924, John Backus grew up in a wealthy family in Philadelphia. Unlike many computer pioneers, John's performance in high school showed little promise: He often flunked his classes and had to go to summer school. He managed to be accepted at the University of Pennsylvania in 1942 as a chemistry major, but was expelled for failure to attend classes. He then joined the U.S. Army where he did better, becoming a corporal, and was placed in charge of an antiaircraft gun crew. After taking an aptitude test, however, Backus was revealed to have some talent for engineering, so the army assigned him to an engineering program at the University of Pittsburgh. However, he then took a medical aptitude test and was found similarly qualified for medical school, which he attended, only to drop out again.

Uncertain what to do next, Backus became interested in electronics because he wanted to build a good hi-fi sound system (something that hobbyists had to build by hand at the time). He attended a trade school to learn more about electronics, and a teacher, asking him to help with calculations for a magazine article, sparked Backus's interest in mathematics. He then went to Columbia University in New York City, where he earned a B.A. degree in mathematics in 1949.

At that time, Backus visited the new IBM Computer Center in Manhattan, where an early computer called the Selective Sequence Electronic Calculator (SSEC) was on display. A tour guide persuaded him to apply for a job, and he was hired to help improve the SSEC. This machine used hard-to-read machine codes for programming, and was also hard to debug. Backus later noted, as recounted by Dennis Shasha and Cathy Lazere, that "You had to be there because the thing would stop running every three minutes and make errors. You had to figure out how to restart it."

In trying to make the machine easier to use, Backus began his lifelong interest in programming languages. When a much more capable machine, the IBM 701, came out, Backus and some colleagues devised a system called "speedcoding" that made it easier to specify numbers in programs. With speedcoding, the significant digits were specified with a "scaling factor," similar to what are called floating point numbers. Once the number was coded, the computer could keep track of the decimal point.

As Backus moved on to the next machine, the IBM 704, he had an important insight. Why not write a program that can translate human-style arithmetic (such as T = A + 1) into the low-level instructions needed by the machine? This program, called a compiler, would revolutionize the practice of computer programming.

In 1954, Backus was given the go-ahead to develop a new high-level computer language that would use a compiler. It took three years of intense work to design the language and the program to translate it. In 1957, the first version of the language, called FORTRAN (for "FORmula TRANslating") was released for the IBM 704.

Before the development of FORTRAN, those scientists who wanted to tap into the power of computers had to spend months learning the peculiar machine language for a particular computer. With FORTRAN, scientists could describe a calculation using notation similar to that in ordinary mathematics, and let the computer translate it into machine codes. Even better, as scientists wrote programs for different applications, many of them became parts of "program libraries" that other scientists or engineers could use. Today, millions of lines of tested, reliable FORTRAN code are waiting to be used.

By the end of the 1950s, Backus was working with other computer scientists to create a new language, Algol. While FORTRAN was designed primarily for scientific computing, Algol was designed to be an all-purpose language that could more easily manipulate text as well as numbers. Algol did not become a commercial success, perhaps because of the investment in existing languages. However, Algol had features that would prove to be very influential in later languages such as Pascal and C. For example, the language provided more flexible ways to describe data structures and the ability to declare "local" variables that could not be accidentally changed by other parts of the program.

Backus's work with Algol also led him to devise a set of grammatical diagrams that could be used to describe the structure of any computer language—rather like the way English sentences can be diagrammed to show the relationships between the words and phrases. This notation, which was further developed by the Danish computer scientist Peter Naur, is now known as Backus-Naur Form, or BNF.

By the 1980s, Backus was working on a new approach to the structure of programming languages, called functional programming. Functional languages (such as LISP) use mathematical functions as their fundamental building blocks, combining them to yield the desired result.

By the time Backus retired in 1991, he had received many honorary degrees and awards, including the National Medal of Science (1975), the Association for Computing Machinery Turing Award (1977), and the Institute of Electrical and Electronic Engineers Computer Society Pioneer Award (1980). These honors reflect both his achievements in computer science and the value of FORTRAN as a key that unlocked the power of the computer for science.

Further Reading

Backus, John. "The History of FORTRAN I, II and III." *IEEE Annals of the History of Computing* 20, no. 4 (1998): 68–78.

Lee, J. A. N., and Henry Trop, eds. "25th Anniversary of Fortran," Special Issue, *IEEE Annals of the History of Computing* 6, no. 1 (1984).

Shasha, Dennis, and Cathy Lazere. *Out of Their Minds: The Lives and Discoveries of 15 Great Computer Scientists*. New York: Springer-Verlag, 1997.

Baran, Paul

(1925–)
Polish/American
Engineer, Computer Scientist

Today millions of e-mail messages travel seamlessly from computer to computer, linking people around the world. A user in Berne, Switzerland, can click on a webpage in Beijing, China, in only a few seconds. Building the infrastructure of switches, routers, and other networks to tie this worldwide network together was an engineering feat comparable to creating the U.S. interstate highway system in the 1950s. While his name is not well known to the general public, Polish-born engineer Paul Baran deserves much of the credit for building today's Internet.

Baran was born April 29, 1926, in Poland. Baran's family immigrated to the United States shortly after his birth, stayed briefly in Boston, and settled in Philadelphia, where his father ran a small grocery store. (The boy delivered groceries to neighbors with his toy wagon.)

After high school, Baran attended the Drexel Institute of Technology (later called Drexel University), where he received a degree in electrical engineering in 1949. Baran then received solid work experience in modern electronics, working at the Eckert-Mauchly Computer Corporation maintaining the vacuum tube circuits in Univac, the first commercial electronic digital computer. Later in the 1950s, Baran worked for the defense contractor Hughes Aircraft, helping design systems to process radar data and to control the Minuteman intercontinental ballistic missile (ICBM). Meanwhile, he took night classes to earn a master's degree in engineering from the University of California, Los Angeles, in 1959.

As the cold war progressed into the early 1960s, military planners were increasingly concerned with how they could maintain contact with their far-flung radar stations and missile silos. Baran went to work for the RAND Corporation, a think tank that studied such strategic problems. Looking back, Baran analyzed the problem as follows:

> Both the US and USSR were building hair-trigger nuclear ballistic missile systems. If the strategic weapons command and control systems could be more survivable, then the country's retaliatory capability could better allow it to withstand an attack and still function; a more stable position. But this was not a wholly feasible concept, because long-distance communication networks at that time were extremely vulnerable and not able to survive attack. That was the issue. Here a most dangerous situation was created by the lack of a survivable communication system.

In other words, if the communications and control systems could not survive a nuclear attack, both sides would feel that in a situation of high international tension, they had better attack first: "Use it or lose it." Further, even if the communications systems were not entirely lost and the attacked side could retaliate, communications were likely to become so degraded that negotiations to end the conflict would become impossible.

Starting in 1960, Baran studied ways to make the communications system more survivable. He realized that the existing system was too centralized and that there were no backup links between the installations that made up the nodes of the network. That meant that if a key installation were knocked out, there was no way to route messages around it to contact the surviving installations.

Baran decided that what was needed was a decentralized, or "distributed," network. In such a network, no one node is essential and each node is connected to several others. As a result, even if an attack destroyed a number of nodes, there would be enough alternate routes to allow messages to reach the surviving nodes. Baran built a further level of resiliency into the system by having it break up messages into smaller chunks, or "blocks," that could be sent over whatever links were currently working, then reassembled into a complete message after arriving at the ultimate destination. (This idea is similar to the packet switching concept, which had been developed by LEONARD KLEINROCK.)

Messages would be dispatched under the direction of minicomputers (today called "routers") that could look up possible routes in tables and quickly shift from one route to another if necessary. Baran wryly noted that "each message is regarded as a 'hot potato,' and rather than hold the 'hot potato,' the node tosses the message to its neighbor, who will now try to get rid of the message."

Baran's complete proposal was published in a lengthy RAND report with the title "On Distributed Communication." His ideas were met with considerable resistance, however. Most network communication would be over phone lines, and traditional telephone engineers thought in terms of establishing a single connection between caller and receiver and sending a complete message or series of messages. The idea of breaking messages up into little bits and sending them bouncing willy-nilly over multiple routes sounded pointless and crazy to engineers who prided themselves on maintaining communications quality.

However, the Defense Department remained interested in the advantages of distributed communication, and following a 1967 conference, began to build a system called ARPANET. The designers of the system, inspired by Baran's earlier work, employed Kleinrock's packet-switching concept.

Baran left RAND in 1968. He founded the Institute for the Future to help plan for future technological developments. He also became involved in the promotion of commercial networking systems during the 1970s and 1980s, including Cable Data Associates, Metricom (a packet radio company), and the cable networking company Com21. Baran was honored in 1987 with the Institute of Electrical and Electronics Engineers Edwin H. Armstrong Award (named for a broadcasting pioneer), and in 1993 with the Electronic Frontier Foundation Pioneer Award.

Further Reading

Baran, Paul. "Introduction to Distributed Communications Networks." RAND Corp. Publication RM-3420-PR, 1964. Available on-line. URL: http://www.rand.org/publications/RM/ RM3420/. Downloaded on November 26, 2002.

Griffin, Scott. "Internet Pioneers: Paul Baran." Available on-line. URL: http://www.ibiblio.org/pioneers/baran.html. Downloaded on October 31, 2002.

Hafner, Katie, and Matthew Lyon. *Where Wizards Stay Up Late: The Origins of the Internet*. New York: Simon and Schuster, 1996.

Bartik, Jean
(1924–)
American
Programmer, Computer Scientist

When Jean Bartik was starting out in the field in the early 1940s a "computer" was not a machine—it was a person who performed calculations by pushing buttons and pulling levers on mechanical calculators. Such clerical workers, mainly women, were necessary for carrying out the computations needed for aiming artillery, designing airplanes, or even creating the first atomic bomb. But Bartik would not end up being such a "computer"—she would learn to program one. As

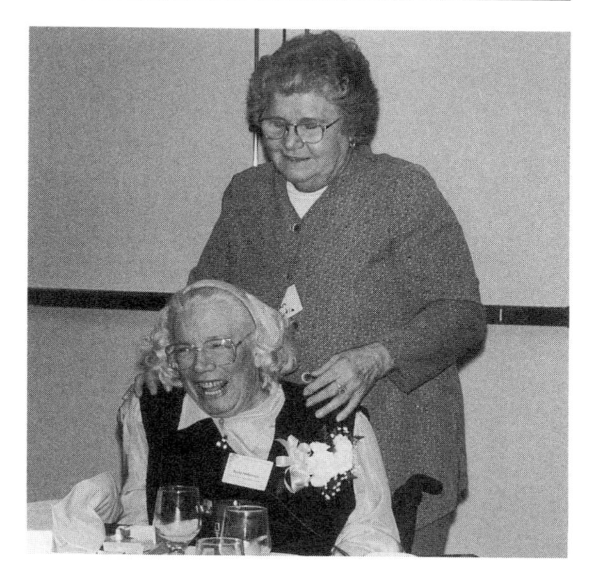

Jean Bartik (standing) and Betty Holberton answered a call for "computers" during World War II. At the time, that was the name for a clerical worker who performed calculations. But these two computer pioneers, shown here at a reunion, would go on to develop important programming techniques for the ENIAC and later machines. *(Courtesy of the Association for Women in Computing)*

one of a handful of pioneering programmers for ENIAC, the world's first general-purpose electronic digital computer, Bartik was in on the ground floor of a technology and industry that would change the world.

Born on December 27, 1924, Jean grew up on a farm in northwest Missouri. She was a good student in science and math, but her athletic talents were considered more remarkable. In particular, she was a formidable softball pitcher and always in demand in after-school games. When Bartik enrolled in the Northwest Missouri State Teachers College, its officials tried to persuade her to major in physical education. However, she majored in mathematics (as one of only two students), earning her B.S. degree in 1945.

By that time, of course, most of her male fellow students had gone off to war. Bartik seemed destined for a career as a math teacher, but her calculus professor showed her a help wanted ad for "computers" to join a top secret project at the army's Ballistics Research Laboratory in Aberdeen, Maryland. After Bartik spent a few months in this essentially clerical job, it was announced that the lab was looking for people to operate a new, highly secret computing machine—ENIAC. Bartik and five other women were accepted for the project.

When she arrived at the ENIAC facility, Bartik and her colleagues were confronted by a daunting sight. The hulking electronic monster filled the whole room, lighting it by the glow of thousands of vacuum tubes.

Today, a computer owner expects to receive a complete illustrated manual with helpful instructions for setting up and running the machine. Online help screens and technical support lines offer further assistance if a problem arises. But with ENIAC, Bartik and her five fellow "computers" had to start from scratch. They had only engineering sketches and cryptic lists of instructions to work with.

Running a program on a modern computer is as easy as slipping a disk into a drive and making a few clicks with the mouse. Running a new program on ENIAC, however, meant moving hundreds of plugs to new sockets, setting more than 3,000 switches, and arranging the input numbers in many separate "digit trays." Early programming was more like running a machine in a factory than working at a keyboard on a desk.

Thanks to Bartik and her colleagues, ENIAC became a workhorse for the war effort. After the war, ENIAC's inventors, J. PRESPER ECKERT and JOHN MAUCHLY, decided to improve the machine so that it could store programs in memory. (Previously, they had to be set up by hand.) Storing programs in memory required special instructions for putting the program commands in particular memory locations, and manipulating them through extremely detailed instructions called microcode.

In 1948, Bartik left government service and went to work at the Eckert-Mauchly Computer Company as it began to develop BINAC (a scaled-down version of ENIAC) and Univac I, which became the first commercially available computer. She wrote programs and designed a memory backup system. She then left the workforce to raise her family, but in 1967 she returned to the computer field as a technology writer and analyst.

In 1997, Bartik was inducted into the Hall of Fame of Women in Technology International. Reflecting on her experience, she says that "I was just at the right place at the right time. It was divine providence or fate that selected me to be an ENIAC programmer."

Further Reading

NASA. Quest Archives. "Female Frontiers: Jean Bartik." Available on-line. URL: http://quest.arc.nasa.gov/space/frontiers/bartik.html. Downloaded on October 31, 2002.

Women in Technology International. "Jean Bartik." Available on-line. URL: http://www.witi.com/center/witimuseum/womeninsciencet/1997/062297.shtml. Posted on June 22, 1997.

Bell, Chester Gordon

(1934–)
American
Inventor, Computer Scientist

The room-sized mainframe computers of the 1950s were amazing machines, but they were so expensive that only the largest businesses, universities, and government agencies could afford them. However, during the 1960s a new kind of computing device, the refrigerator-sized "minicomputer," would bring data processing power to many more institutions. A pioneer in the development of minicomputers, Chester Gordon Bell had a career that continued to evolve through the decades as he explored the history and future possibilities of computers.

Gordon Bell designed the first minicomputers. Using transistors and then integrated circuits, the minicomputer had shrunk to refrigerator-size by the mid-1960s. Their price tag had shrunk as well, making computing power affordable to smaller businesses and universities. *(Courtesy of Microsoft Corporation)*

Bell was born on August 19, 1934, in the small town of Kirksville, Missouri. Even as a young child he made himself useful in his father's electrical business: At age six, he installed a plug on a wire, and at an age when most kids are learning to read, Gordon was wiring and fixing appliances.

After high school, Bell went to the Massachusetts Institute of Technology (MIT), where he majored in electrical engineering. During his college years, he worked during the summers for General Electric and various manufacturing companies, as well as studying the emerging field of digital electronics. In 1957, having completed his B.S. and M.S. degrees, he went to New South Wales, Australia, on a Fullbright scholarship, avoiding a final decision about what he called "the going-to-work problem."

In Australia, Bell entered fully into the world of programming and computer design while teaching courses and helping with various projects. After returning to the United States in 1959, Bell decided to work toward his doctorate while pursuing another interest: acoustics and sound systems. His thesis adviser suggested that he might want to help develop the TX-0, a new type of computer being built by MIT. Combining his interests, Bell equipped the new machine with sound input and wrote some of the earliest software for speech recognition.

However, the problem of voice recognition proved to be very complex. It would take decades to master, and Bell's interests were turning more directly to computer design. It was a fertile time for innovation. The TX-0 used transistors rather than vacuum tubes. This meant that more computing power could be packed into a smaller space, with less power consumption and greater reliability. Bell began to work with KENNETH H. OLSEN at a new company, Digital Equipment Corporation (DEC), to develop the PDP-1, the first commercial computer to take advantage of this new technology.

Introduced in 1960, the PDP-1 was considerably smaller than mainframe computers, although it still required four six-foot-tall cabinets. Unlike mainframes, it did not need expensive air conditioning or cooling systems. The machine sold for $120,000—a stiff price by modern standards, but much less than mainframes that typically cost many hundreds of thousands or even millions of dollars.

In developing the PDP-1, they had to make tradeoffs between size requirements and computing power. The PDP had a data word length of 18 bits, rather than the 36 bits used by most mainframes. This meant that it could not easily handle the large numbers used in some scientific calculations. The machine also did not have built-in floating-point circuitry for handling decimal numbers, so such calculations would be slower than with mainframes. But DEC was betting that there was a market for more modest computing capability at a much lower price, and the market proved them right: Between 1965 and 1980, DEC sold 50,000 of the PDP-1 and its successors.

Bell continued to believe that simpler was better: the next machine, the PDP-4, had fewer programming instructions and took up half the space of the PDP-1. It sold at half the price but delivered about five-eighths the performance. The PDP-4 and the smaller PDP-5's compactness made them suitable for use in controlling industrial and chemical processes on the factory floor. The sleek new breed of minicomputers produced by DEC and other companies were taking computing out of the sequestered realm of the mainframes and bringing it to more and different users.

This proliferation would have a profound impact on the field of computer science and technology. The PDP-8, introduced in 1965, went beyond transistors to integrated circuits that packed many logic units into a small space. The entire computer now fit into a space the size of a small refrigerator. With prices eventually dropping to less than $10,000, more than 100,000 of the PDP-8 and its successors would be sold. Just about any college could afford to buy one or several of the machines, and with the development of time-sharing systems, this meant that students could gain plentiful free computer time. It was this environment that made possible the cooperative development of the UNIX operating system and its vast array of useful software.

The minicomputer architecture spearheaded by Bell would also have a great impact on later computer hardware. The PDP-11, released in 1970, pioneered the use of an integrated "bus," or data connection system, mounted on a single large board. The bus had its own controllers for routing data, freeing the central processor from managing that task. Later in the decade, the designers of the first desktop microcomputers would turn to this design for its simplicity and reliability. Meanwhile, Bell's team introduced another feature that had previously been available only on mainframes: virtual memory. The VAX (Virtual Address Extension) series of computers allowed space on disk to be used as though it were part of the computer's main memory. This meant that a VAX computer could run larger (and more) programs than a PDP.

By 1983, however, Bell had tired of the sluggish pace of work in a very large corporation, and decided to create his own company, Encore, devoted to developing multiprocessing computers (that is, machines with many separate processors that could run many programs at the same time rather than switching in turn from one program to the next). Bell studied the characteristics of successful startup ventures and developed methods for predicting their success.

After launching Encore and another startup called the Dana Group, Bell changed direction yet again. He went to Washington, D.C., and became the assistant director of the National Science Foundation, in charge of government funding for research in computer science and technology. But while finding an increasing interest in exploring the future of computing, he also found himself wanting to be a steward of its past. He cofounded the Computer Museum in Boston, an organization dedicated to studying and displaying the history of computer development. (When this organization became defunct in 1999, Bell founded the Computer Museum History Center at Moffett Field, in Mountain View, California, to carry on its collection and work.)

Awards received by Bell include the National Medal of Technology (1991), fellowships in the Association for Computing Machinery and the Institute of Electronic and Electrical Engineers, and various medals from these and other organizations. He continues to serve on a number of boards and as a senior researcher at Microsoft's Media Presence Research Group, where he is studying future uses of supercomputers and the implications of "telepresence"—the ability of people to send their images and even manipulate objects from a remote location.

Further Reading

Bell, C. Gordon, and John E. McNamara (contributor). *High-Tech Ventures: The Guide to Entrepreneurial Success*. Reading, Mass.: Addison-Wesley, 1991.

Microsoft Bay Area Research Center. "Gordon Bell's Home Page." Available on-line. URL: http://research.microsoft.com/users/GBell. Downloaded on October 31, 2002.

Slater, Robert. *Portraits in Silicon*. Cambridge, Mass.: MIT Press, 1987.

Berners-Lee, Tim

(1955–)
British
Computer Scientist

Tim Berners-Lee wanted to write a program to help nuclear scientists keep track of their projects, but he ended up creating the World Wide Web. It would change business, education, and the media during the 1990s. *(© Henry Horenstein/CORBIS)*

Tim Berners-Lee invented the World Wide Web, the interconnected realm of words, images, and sounds that now touches the daily lives of millions of people around the world.

Born on June 8, 1955, in London, Berners-Lee was the child of two mathematicians who were themselves computer pioneers who helped program the Manchester University Mark I in the early 1950s. As a boy, Berners-Lee showed considerable aptitude in solving math puzzles posed by his parents and building "pretend" computers out of cardboard and running discarded punched paper tape through them to simulate their operation.

Berners-Lee always had wide-ranging interests, however. He would later recall that as a child he had been given a Victorian children's encyclopedia with the title *Enquire Within Upon Everything*. From that time, he became intrigued with the idea that all knowledge could be collected, arranged, and connected.

During high school, in the 1970s Berners-Lee learned to work with electronics and even built a working computer using one of the first commercially available microprocessors. He then studied physics at Oxford University, receiving his B.A. degree with honors in 1976. After graduation he worked for several computer-related companies, writing printer control software and learning more about operating systems. As a result of this varied experience, Berners-Lee became familiar with scientists and their computing needs, the capabilities of current computers, and the discipline of software engineering.

By 1980, Berners-Lee was working as an independent consultant at CERN, the giant European physics research institute. It was there that his interests turned away from the nuts and bolts of computer hardware to investigating the structure of knowledge as it might be dealt with in computer networks.

In his autobiography *Weaving the Web* (1999), Berners-Lee recalls his most important insight: "Suppose all the information stored on computers everywhere were linked? Suppose I could program my computer to create a space in which anything could be linked with anything?" He proceeded to write a program called Enquire, the name harking back to his childhood encyclopedia. The basic idea of Enquire was that each bit of knowledge could potentially be connected to any other according to the needs or interests of the user. This way of embedding links from one document or piece of information to another is called hypertext, an idea introduced by the work of VANNEVAR BUSH in the 1950s and TED NELSON in the 1960s.

Accepting a fellowship at CERN in 1984, Berners-Lee struggled with organizing the dozens of incompatible computer systems and software programs that had been brought to the labs by thousands of scientists from around the world. With existing systems, each requiring a specialized access procedure, researchers had trouble sharing data with one another or learning about existing software tools that might solve their problems.

By 1989, Berners-Lee had proposed a solution to the problem of information linkage that combined the hypertext idea from his Enquire program with the growing capabilities of the Internet, which already offered a wide assortment of databases, archives, and other sources of information. Essentially, he suggested bypassing traditional database organization and treating text on all systems as "pages" that would each have a unique address, a universal document identifier (later known as a uniform resource locator, or URL). He and his assistants used highlighted text to link words and phrases on one page to another page, and adapted existing hypertext editing software to create the first World Wide Web pages. They then programmed a server to provide access to the pages and created a simple browser, a program that could be used by anyone connected to the Internet to read the pages and follow the links as desired. By December 1990, the first Web server was running within the CERN community. By the summer of 1991, Web servers had begun to appear throughout the worldwide Internet.

Between 1990 and 1993, word of the Web spread throughout the academic community as Web software was written for more computer platforms. As demand grew for a body to standardize and shape the evolution of the Web, Berners-Lee founded the World Wide Web Consortium (W3C) in 1994. Through a process of user feedback and refinement, the standards for addresses (URLs), the transmission protocol (HTTP), and the hypertext markup language (HTML) were firmly established.

Together with his colleagues, Berners-Lee has struggled to maintain a coherent vision of the Web in the face of tremendous growth and commercialization, the involvement of huge corporations with conflicting agendas, and contentious issues of censorship and privacy. His general approach has been to develop tools that empower the user to make the ultimate decision about the information he or she will see or divulge.

In the original vision for the Web, users would create Web pages as easily as they could read them, using software no more complicated than a word processor. While there are programs today that hide the details of HTML coding and allow easier Web page creation, Berners-Lee feels the Web must become even easier to use if it is to be a truly interactive, open-ended knowledge system. He believes that users should be empowered to become active participants in the creation of knowledge, not just passive recipients.

Berners-Lee is also interested in developing software that can take better advantage of the rich variety of information on the Web, creating a "semantic" Web that would not simply connect pieces of information but encode the meaning or relevance of each connection. This would allow "bots," or software agents, to aid researchers by zeroing in on the most relevant material. Ultimately, human beings and machines might be able to actively collaborate in the search for knowledge. The beginning steps in what appears to be a long process can be seen in the recent emergence of XML, an information description language, and RDF, or Resource Description Framework.

Berners-Lee has been honored with numerous awards from computer societies, including the Association for Computing Machinery and the Institute for Electrical and Electronic Engineering. In 1999 he was named by *Time* magazine as one of the 100 greatest minds of the century. As of 2003, he worked at the Massachusetts Institute of Technology Laboratory for Computer Science.

Further Reading

Berners-Lee, Tim. "Papers on Web design issues." Available on-line. URL: http://www.w3.org/DesignIssues.

———. "Proposal for the World Wide Web, 1989." Available on-line. URL: http://www.w3.org/History/1989/proposal.html. Posted in March 1989, May 1990.

Berners-Lee, Tim, and Mark Fischetti. *Weaving the Web*. San Francisco: HarperSanFrancisco, 1999.

Luh, James C. "Tim Berners-Lee: An Unsentimental Look at the Medium He Helped Propel." *Internet World*, January 1, 2000. Available on-line. URL: http://www.pathfinder.com/time/interstitials/inter.html.

The World Wide Web Consortium. "Tim Berners-Lee." Available on-line. URL: http://www.w3.org/People/Berners-Lee. Downloaded on October 31, 2002.

Bezos, Jeffrey P.

(1964–)
American
Entrepreneur

With its ability to display extensive information and interact with users, the World Wide Web of the mid-1990s clearly had commercial possibilities. But it was far from clear how traditional merchandising could be adapted to the on-line world, and how the strengths of the new medium could be translated into business advantages. In creating Amazon.com, "the world's largest bookstore," Jeff Bezos would show how the Web could be used to deliver books and other merchandise to millions of consumers.

Jeff Bezos was born on January 12, 1964, and grew up in Miami, Florida. His mother, Jacklyn Gise, remarried shortly after his birth, so he grew up with a stepfather, Miguel Bezos, a Cuban who had fled to the United States in the wake of Castro's revolution. Jeff Bezos was an intense, strong-willed boy who was fascinated by gadgets but also liked to play football and other sports. His uncle, Preston Gise, a manager for the Atomic Energy Commission, encouraged young Bezos's interest in technology by giving him

Jeff Bezos wanted a "big" name for his on-line bookstore, so he named it after the world's biggest river, calling it Amazon.com. He proved that goods could be sold quickly and efficiently on-line, and helped change the way people shopped. *(CORBIS Sygma)*

electronic equipment to dismantle and explore. Bezos also liked science fiction and became an enthusiastic advocate for space colonization.

Bezos entered Princeton University in 1982. At first he majored in physics, but decided that he was not suited to be a top-flight theorist. He switched to electrical engineering and graduated in 1986 summa cum laude. By then he had become interested in business software applications, particularly financial networks. Still only 23 years old, he led a project at Fitel, a financial communications network. He managed 12 programmers, commuting each week between the company's New York and London offices. When he became a vice president at Bankers Trust, a major Wall Street firm in 1988, Bezos seemed to be on a meteoric track in the corporate world.

Bezos became very enthusiastic about the use of computer networking and interactive software to provide timely information for managers and investors. However, he found that the "old line" Wall Street firms resisted his efforts. He recalled to author Robert Spector that their attitude was: "This was something that couldn't be done, shouldn't be done, and that the traditional way of delivering information in hard copy was better. . . . The feeling was: Why change? Why make the investment?"

In 1990, however, Bezos was working at the D. E. Shaw Company and his employer asked him to research the commercial potential of the Internet, which was starting to grow (even though the World Wide Web would not reach most consumers for another five years). Bezos ranked the top 20 possible products for Internet sales. They included computer software, office supplies, clothing, music—and books.

Analyzing the publishing industry, Bezos identified ways in which he believed it was inefficient. Even large bookstores could stock only a small portion of the available titles, while many books that were in stock stayed on the shelves for months, tying up money and space. Bezos believed that by combining a single huge warehouse with an extensive tracking database, an on-line ordering system and fast shipping, he could satisfy many more customers while keeping costs low.

Bezos pitched his idea to D. E. Shaw. When the company declined to invest in the venture, Bezos made a bold decision. He recalled to Spector, "I knew that when I was 80 there was no chance I would regret having walked away from my 1994 Wall Street bonus in the middle of the year. I wouldn't even have *remembered* that. But I did think there was a chance I might regret significantly not participating in this thing called the Internet that I believe passionately in. I also knew that if I tried and failed, I wouldn't regret that."

Looking for a place to set up shop, Bezos decided on Seattle, partly because the state of Washington had a relatively small population (the only customers who would have to pay sales tax) yet had a growing pool of technically trained workers, thanks to the growth of Microsoft and other companies in the area. After several false starts he decided to call his store Amazon, deciding that the name of the Earth's biggest river would be suited to what he intended to become Earth's biggest bookstore.

In November 1994, Bezos, his wife Mackenzie, and a handful of employees began the preliminary development of Amazon.com in a converted garage. Bezos soon decided that the existing software for mail order businesses was too limited and set a gifted programmer named Shel Kaphan to work creating a custom program that could keep track not only of each book in stock but how long it would take to get more copies from the publisher or book distributor.

By mid-1995, Amazon.com was ready go online from a new Seattle office, using $145,553 contributed by Bezos's mother from the family trust. As word about the store spread through Internet chat rooms and a listing on Yahoo!, the orders began to pour in and Bezos had to struggle to keep up. Despite the flood of orders, the business was losing money, as expenses piled up even more quickly.

Bezos went to Silicon Valley, the heart of "high tech," seeking investors to shore up Amazon.com until it could become profitable. Bezos's previous experience as a Wall Street star, together with his self-confidence, seemed to do the trick, and he raised $1 million. But Bezos believed that momentum was the key to long-term success. The company's unofficial motto became "get big fast." Revenue was poured back into the business, expanding sales into other product lines such as music, video, electronics, and software.

The other key element of Bezos's growth strategy is to take advantage of the vast database that Amazon was accumulating—not only information about books and other products, but about what products a given individual or type of customer was buying. Once a customer has bought something from Amazon, he or she is greeted by name and given recommendations for additional purchases based upon what items other customers who bought that item had also purchased. Customers are encouraged to write on-line reviews of books and other items so that each customer gets the sense of being part of a virtual peer group. Ultimately, as he told Robert Spector, Bezos believes that "In the future, when you come to Amazon.com, I don't want you just to be able to search for *kayak* and find all the books on kayaking. You should also be able to read articles on kayaking and buy subscriptions to kayaking magazines. You should be able to buy a kayaking trip to anywhere in the world you want to go kayaking, and you should be able to have a kayak delivered to your house. You should be able to discuss kayaking with other kayakers. There should be everything to do with kayaking, and the same should be true for anything."

By 1997, the year of its first public stock offering, Amazon.com seemed to be growing at an impressive rate. A year later, the stock was worth almost $100 a share, and by 1999 Jeff Bezos's personal wealth neared $7.5 billion. Bezos and Amazon.com seemed to be living proof that the "new economy" of the Internet was viable and that traditional "brick and mortar" businesses had better develop an on-line presence or suffer the consequences.

However, in 2000 and 2001 the Internet economy slumped as profits failed to meet investors' expectations. Amazon was not exempt from the new climate, but Bezos remained confident that by refining product selection and "targeting" customers the company could eventually reach sustained profitability. In January 2002, Amazon announced that it had actually made a profit during the last quarter of 2001.

Regardless of the eventual outcome, Jeff Bezos has written a new chapter in the history of retailing, making him a 21st-century counterpart to such pioneers of traditional retailing as Woolworth and Montgomery. *Time* magazine acknowledged this by making him its 1999 Person of the Year, while *Internet Magazine* put Bezos on its list of the 10 persons who have most influenced the development of the Internet.

Further Reading

Grove, Alex. "Surfing the Amazon." *Red Herring*, July 1, 1997. Available on-line. URL: http://www.redherring.com/mag/issue44/bezos.html. Downloaded on November 26, 2002.

Southwick, Karen. "Interview with Jeff Bezos of Amazon.com." Upside.com, October 1, 1996. Available on-line. URL: http://www.upside.com/texis/mvm/story?id=34712c154b. Downloaded on November 26, 2002.

Spector, Robert. *Amazon.com: Get Big Fast: Inside the Revolutionary Business Model that Changed the World*. New York: HarperBusiness, 2000.

Boole, George
(1815–1864)
British
Mathematician

The British mathematician and logician George Boole developed his algebra of logic about a century before the invention of the computer. However, his search for "the laws of thought" led to logical rules that are now embedded in the silicon of computer chips and used every day by people seeking information through Web search engines.

Boole was born on November 2, 1815, in Lincoln, England, to a family in modest circumstances. The family soon moved to London, where Boole's father worked as a shoemaker and his mother served as a maid. Although the elementary schooling available to the lower middle class was limited, Boole supplemented what he was taught at school by learning Greek and Latin at home. Boole's father, whose hobby was making scientific instruments, introduced him to mathematics and science as they worked together to make telescopes and microscopes.

The family was not able to afford a university education, so Boole went to a trade school where he studied literature and algebra. However, when Boole was only 16, his father's business failed, and the young man became responsible for supporting the family. With the careers of minister and teacher available to him, Boole chose the latter because it paid a little better. Meanwhile, Boole had bought advanced books on subjects such as differential and integral calculus and was often distracted from his teaching duties by a particularly intriguing math problem.

In 1834, when Boole was 19, he decided that he needed to make more money to support his aging family. He started his own school, using a balanced curriculum of his own design. In addition to languages and literature, his students learned science and math with an emphasis on practical applications. Boole also gave lectures at the Mechanic's Institute, an organization dedicated to providing educational opportunities to working-class people.

Boole gradually became more involved in the international world of mathematics. Together with computer pioneer CHARLES BABBAGE and astronomer William Herschel, Boole founded the Analytical Society, an organization geared toward introducing the "modern" European approach to calculus to British mathematicians, who clung to the older, less flexible Newtonian methods. At age 23, Boole wrote his first scientific paper, offering improvements on Joseph-Louis Lagrange's methods for analyzing planetary motions.

By 1844, Boole was moving toward an analysis of how mathematics itself worked. He became intrigued about how the manipulation of mathematical operators and symbols might mirror logic and the operation of the human mind. His paper "On a General Method in Analysis" was rejected at first by the prestigious Royal Society, perhaps because Boole did not have a formal university degree. However, Boole's cause was supported by some top mathematicians, and the paper was eventually not only published but awarded a gold medal.

By 1849, Boole's work had made such an impression on his colleagues that all questions about his background were dropped, and he was appointed professor of mathematics at the newly founded Queen's College in Cork, Ireland. He would hold this post for the rest of his career.

At this time Boole was working out the principles of what would later become known as Boolean logic. Essentially, Boole developed a way to represent the abstract operations of logic using symbols that could be manipulated through a form of algebra. For example, if a set of items is designated x, then everything that is not in x can be expressed as $1 - x$, since "one" is the logical symbol for the entire universe. Boole also defined operations such as AND (the items found in both of two sets) and OR (the items found in either of two sets). He introduced this work in an 1847 pamphlet titled *The Mathematical Analysis of Logic*.

Boole believed that not only logic but human thought in general might someday be summarized using a system of rules or laws, just as physicists and astronomers were developing mathematical descriptions of the behavior of planets. This belief is expressed in Boole's comprehensive work on logic, which he published in 1854 under the title *An Investigation into the Laws of Thought, on Which Are Founded the Mathematical Theories of Logic and Probabilities*.

In his last years Boole married a former student, Mary Everest, and had five daughters with her. At age 49, Boole died after walking through a cold rain and contracting pneumonia.

Boole would have been an important mathematician even if the computer had never been invented, but the computer brought his work from the realm of abstract mathematics to the practical world of engineering. In the mid-20th century, when inventors turned to designing machines that could calculate and even "think," they found that Boole's algebra of logic allowed them to create logic circuits as well as algorithms or procedures that could be followed automatically to carry out arithmetic and logical operations. Boolean logic was a perfect fit to the binary system of 1 and 0 and the use of electronic switching. And when computers began to store information in databases, the ability to specify the "Boolean operators" of AND, OR, and NOT made it possible to more precisely specify the information being sought.

Further Reading

Henderson, Harry. *Modern Mathematicians*. New York: Facts On File, 1996.

Kramer, Edna E. *The Nature and Growth of Modern Mathematics*. Princeton, N.J.: Princeton University Press, 1981.

MacHale, Desmond. *George Boole: His Life and Work*. Dublin: Boole Press, 1985.

Bricklin, Daniel

(1951–)
American
Inventor, Programmer

Whenever a business executive or manager wants "the numbers" to summarize profits and losses or to back up a proposal, the chances are very good that the data will be provided in the rows and columns of a spreadsheet. With this versatile calculating tool, a manager can easily apply formulas to explore the effects of changes in billing, shipping, or other business procedures. Investors can use spreadsheets to chart the performance of their stocks or compare the returns on different investments. Now almost as indispensable as the word processor, the spreadsheet was the invention of a young programmer named Dan Bricklin.

Bricklin was born on July 16, 1951, in Philadelphia. He attended Solomon Schechter Day School and then enrolled at the Massachusetts Institute of Technology in 1969. At first he was a mathematics major, but switched to computer science during his junior year. Working at the school's Laboratory for Computer Science, Bricklin wrote a program with which users could perform calculations on-line. He also helped implement APL, a compact, powerful computer language favored by many scientists and engineers. During this time, Bricklin developed a close working relationship with another graduate student, Bob Frankston, and the two decided that someday they would start a business together.

After receiving his B.S. degree in electrical engineering and computer science in 1973, Bricklin went to work for Digital Equipment Corporation (DEC), a leading maker of minicomputers. There he helped design video terminals and computerized typesetting systems. His most important achievement at that time, however, was designing and partly writing one of the earliest word processing programs, called WP-8.

When DEC wanted him to move to an office in New Hampshire, Bricklin decided to go back to school instead to get a degree in business administration. In 1977 he went to Harvard Business school. Bricklin continued his programming work and interest in software design, but he was also interested in gaining more knowledge of and experience in business. This combination of skills would not only help him write better software to meet business needs, but would also prepare him when it came time to start his own business.

Sitting in class at Harvard and working through complicated mock business exercises, Bricklin became increasingly frustrated with the difficulty of keeping up with different business scenarios with only pencil, paper, and calculator. As recounted by Robert Slater, he recalled that he then "started to imagine an electronic calculator, the word processor that would work with numbers." After all, he had already written word processing software that made it much easier for writers to create and revise their work. Bricklin began to imagine a sort of electronic blackboard that, when a formula was writter on it, would automatically apply it to all the numbers and calculate the results. He later decided that if he were to write such a program, "The goal [would be] that it had to be better than the back of an envelope."

Dan Bricklin was a skilled programmer but he decided to go into business administration. As he tried to keep up with calculations for a business simulation, using only pencil, paper and calculator, he decided that a computer should be able to do a better job. He invented a new kind of software—the electronic spreadsheet. *(Photo by Louis Fabian Bachrach)*

Bricklin found that his fellow students and his professors were enthusiastic about the idea of making a "word processor" for numbers. He sketched out his ideas for the program, then wrote a simple version in BASIC on an Apple II microcomputer. It was very slow because BASIC had to translate each instruction into the actual machine language used by the microcomputer. Bricklin talked to Bob Frankston, and they agreed to develop the program together. Bricklin would create the overall design and documentation for the program, while Frankston would write it in faster-running assembly language. They then met with Dan Fylstra, owner of a small software publishing company, and he agreed to market the program. In 1979, Bricklin and Frankston started their own company, Software Arts, to develop the software.

Bricklin took business classes by day while Frankston programmed all night. They rushed to complete the program, and then had to decide what to call it. After rejecting suggestions such as "Calculature" and "Electronic Blackboard," they settled on VisiCalc. As the program progressed, Bricklin astounded one of his Harvard Business School professors by providing detailed results for one of the "what-if" assignments—never revealing that VisiCalc was doing the hard work for him. Later students of software design would admire the way the program was structured to use the very limited resources of computers that typically had only about 48 kilobytes of memory.

In October 1979, VisiCalc was ready to be offered to users of Apple II and other personal computers (PCs). Software Arts began to sell about 500 copies a month, with Bricklin spending a considerable amount of time giving "demos" (demonstrations) of the product in computer stores. As word of the program spread, big accounting firms started to use it, often buying their first PCs to run the software. As investment analyst Benjamin Rosen remarked at about that time, "VisiCalc could someday become the software tail that wags (and sells) the computer dog."

Sales grew steadily, reaching 30,000 a month in late 1981 when the IBM PC appeared on the scene. Many companies that had been reluctant to buy a computer named for a fruit were reassured by the IBM name, while a new version of VisiCalc in turn became a selling point for the PC. That same year, Bricklin was honored with the Grace Murray Hopper Award of the Association for Computing Machinery, given for significant accomplishments in computing by people under 30.

But the 1980s also brought new challenges. Bricklin's development team came out with a new program called TK!Solver, designed for solving more complex equations. However, Bricklin became embroiled in a lawsuit with Fylstra over the rights to market VisiCalc. Meanwhile, a new program called Lotus 1-2-3, developed by MITCHELL KAPOR, began to take over the spreadsheet market. Unlike VisiCalc, which provided only a spreadsheet, Lotus 1-2-3 included a simple database system and the ability to create charts and graphics. Kapor eventually bought out Software Arts.

Bricklin decided to concentrate on designing programs and leave the marketing to others. His new company, Software Garden, developed a program called Dan Bricklin's Demo Program. It let designers show how a new program will work even before writing any of its code. The designer creates a storyboard, much like that used for movies, and defines what is shown when the user performs a certain action. During the 1990s, Bricklin became involved with pen computing (programs that let users draw or write with an electronic pen), but this application never became very successful. Bricklin's latest company, Trellix, designs custom Web pages for companies.

Further Reading

Bricklin, Dan. "Dan Bricklin's Web Site." Available on-line. URL: http://www.bricklin.com/default.htm. Downloaded on November 26, 2002.

Slater, Robert. *Portraits in Silicon*. Cambridge, Mass.: MIT Press, 1987.

Wylie, Margie. "The Man Who Made Computers Useful." CNET News.com. Available on-line. URL: http://news.com.com/2009-1082-233609.html?legacy=cnet. Posted on October 13, 1997.

Brooks, Frederick P.

(1931–)
American
Computer Scientist

In the early days of computing (the late 1940s and early 1950s), most programming was done in a haphazard and improvised fashion. This was not surprising: Programming was a brand-new field and there was no previous experience to draw upon. However, as computers became larger and programs more complicated, programmers began to learn more about how to organize and manage their projects. Based on his personal experience writing the operating system for the IBM 360 computer, Frederick Brooks described the obstacles and pitfalls that often wrecked software development. He did much to transform the craft of programming to the systematic discipline that became known as "software engineering."

Brooks was born on April 19, 1931, in Durham, North Carolina, but grew up in a small rural town. His father was a medical doctor and encouraged Brook's interest in science. He later recalled that "I got fascinated with computers at the age of 13 when accounts of the Harvard Mark I appeared in magazines. I read everything I could get in that field and began collecting old business machines at bankruptcy sales." In high school, Brooks plunged into electronics, joining the radio club and the electrical engineering club. However, he also worked during the

Frederick Brooks played a key role in designing the IBM System/360, perhaps the most successful computer model of all time. The difficulties he encountered in running large software projects led him to rethink management in his book *The Mythical Man-Month.* *(photo © Jerry Markatos)*

summers at hard physical labor, making sheet metal pipes for the chimneys of tobacco barns.

Brooks then attended Duke University, majoring in physics and mathematics but also maintaining his hands-on interest in electronics. For his senior project he built a closed-circuit TV system. His first love, however, remained computers. After getting his degree from Duke in 1955, he went to Harvard and did graduate work under the direction of HOWARD AIKEN, inventor of the Harvard Mark I, and received his Ph.D. in applied mathematics in 1956.

Aiken, who designed his series of computers mainly for scientific work, believed that a different kind of computer would be needed for business applications. He invited Brooks to work on designing such a system. Brooks's work impressed both Aiken and IBM, which (ironically, perhaps) gave him a job working on STRETCH, the world's first "supercomputer," a scientific machine not suitable for business. The STRETCH introduced Brooks to innovative ideas that would become basic to all modern computers, including pipelining (the efficient processing of a collection of instructions to avoid wasting time) and the use of interrupts, or special signals that allow the computer to process input (such as keystrokes) in an orderly way.

By 1960, Brooks had been placed in charge of all computer system design at IBM, responsible for creating the new generation of machines that would make the company's name a household word in data processing. Brooks set about designing what eventually would be called the IBM System/360. The "System" in the name was key: IBM decided that there would be a range of machines utilizing the same basic design, but the higher-end models would have more memory, greater processing capacity, or other features. Unlike other computer companies where software written for one machine could not be run on a more expensive model, programs written on a low-end 360 would also run on a top-of-the-line machine. This meant that users could be confident that their investment in developing software would not be lost if they chose to upgrade their hardware.

The development of the hardware for the IBM 360 proceeded remarkably smoothly, but by early 1964 the development team had run into a roadblock. Every computer needs an operating system, a master control program that runs applications programs, manages memory and disk storage, and performs all the other "housekeeping" tasks needed to allow a computer to carry out the instructions in its software. A computer with the capabilities and requirements of the System/360 would require a very complex operating system, and the team of nearly 2,000 programmers that IBM had assigned to the project had gotten hopelessly tangled up in implementing it. Brooks took charge of the project and had them start over with a "clean" design from the bottom up. In addition to compatibility and expandability, the IBM System/360 offered another key innovation. By the mid-1960s, relatively high-speed, high-capacity hard disks were beginning to replace slow tape drives for program storage. Brooks and his team designed their operating system so that if the limited amount of main random access memory (RAM) started to run out, program instructions and data could be "swapped" or temporarily stored on the disk. In effect, this allowed the computer to run programs as though it had much more memory than was physically installed. This idea of "virtual memory" is used in all of today's computers, including desktop PCs.

Brooks had created an architecture and reliable operating system that would become the standard for mainframe computing for decades to come. But in dealing with the technical and human problems of managing a large software development team, he had also gained insights that he would describe in his book *The Mythical Man-Month*. For example, he learned that simply adding more programmers to a project did not necessarily speed up the work—indeed, the need to bring new team members up to speed and coordinate them could make the situation worse. Brooks compared the situation to another human activity, noting that "the bearing of a child takes nine months, no matter how many women are assigned."

Many computer scientists believe in the "waterfall" approach to designing a program. They start with a complete but general specification of what the program is supposed to do, from which flow detailed specifications, and finally the actual program code is written and rigorously tested. While Brooks agrees that this method is necessary for programs that must be absolutely correct (such as flight control software), he believes that for most business programs, it is better to start with a limited version of the program, get it running, and then use the feedback from users to develop the more complete version.

In 1964, with the initial work on the IBM System/360 winding down, Brooks left IBM and went to the University of North Carolina to establish a department of computer science. At the time, in most universities computer science did not exist as a separate department, but instead was taught in departments of mathematics or electrical engineering. Brooks decided that the new department would tackle two "leading edge" problems: the development of computer graphics software, and research into computer understanding of human language.

In the mid-1960s, most computers had no graphics capabilities at all, other than perhaps printing pictures made of characters or dots, although some enterprising MIT hackers had created a game called Space War on a TV screen connected to a minicomputer. Brooks and his colleagues Henry Fuchs and John Poulton set out to create a system that could produce true, moving, three-dimensional graphics. He believed that such a system could help people explore and learn things in new ways. Authors Dennis Shasha and Cathy Lazere quote some examples: "Highway safety people on driving simulators, city planners designing low cost housing who wanted real-time on-line estimating . . . astronomers concerned with galactic structure, geologists concerned with underground water resources." Brooks also worked with Sun Ho Kim, a crystallographer from Duke University, to create models of molecular structure. In 1974 Brooks and Kim demonstrated the system to visitors from the National Institutes of Health.

Since the 1980s, Brooks has also led researchers into many other areas, including robotics and what is today called virtual reality—interactive scenes that users can "walk through" and manipulate. In Brooks's view, all these applications are examples of "intelligence amplification"—the ability to use technology to help humans think better and faster. Brooks believes that this approach is likely to be more successful than the traditional approach to artificial intelligence—making computers think by themselves and perhaps replace the human mind.

For his achievements in computer and operating system design, graphics, and artificial intelligence Brooks has received numerous awards, including the Institute of Electrical and Electronics Engineers (IEEE) Computer Society Pioneer Award (1970), the National Medal of Technology (1985), and the IEEE John von Neumann Medal (1993). In 1999 Frederick Brooks received the Association for Computing Machinery's A. M. Turing Award, sometimes referred to informally as the "Nobel Prize for Computing," for his contributions to the field of software engineering.

Further Reading

Blauuw, Gerrit A., and Frederick P. Brooks Jr. *Computer Architecture: Concepts and Evolution*. Reading, Mass.: Addison Wesley, 1997.

Brooks, Frederick P., Jr. *The Mythical Man-Month: Essays on Software Engineering*. 20th Anniversary Edition. Reading, Mass.: Addison-Wesley, 1995.

Shasha, Dennis, and Cathy Lazere. *Out of Their Minds: The Lives and Discoveries of 15 Great Computer Scientists*. New York: Copernicus, 1995.

University of North Carolina. "Frederick P. Brooks Home Page." Available on-line. URL: http://www.cs.unc.edu/~brooks. Downloaded on October 31, 2002.

Burroughs, William S.

(1855–1898)
American
Inventor

Computers were not the first machines to change the way businesses were run. By the 1880s, the typewriter was starting to be used for preparing letters and reports. However, for bookkeeping, accounting, and other tasks involving numbers, the only tools available were ledgers, pens, and the ability to perform arithmetic by hand. But the late 19th century was an age of invention, and William Seward Burroughs would invent the first practical mechanical calculator to come into widespread use.

Burroughs was born on January 28, 1855, in Auburn, New York. Like most young men of his day, he had no formal schooling past eighth grade. Starting at age 15, he tried his hand at many different jobs. These included working in a store and as a bank clerk, but office work seemed to put a strain on his health. He then worked in his father's shop making models for metal casting and then at a woodworking firm called Future Great Manufacturing Company. Burroughs's mix of clerical experience and hands-on mechanical knowledge would prove to be ideal for the inventor of an office machine.

From his business experience Burroughs knew that that bookkeeping by hand was a tedious and eye-straining activity and that bookkeepers had to spend about half their time rechecking each calculation for possible errors. From his work in manufacturing and machine shops Burroughs believed that it should be possible to make a reliable calculating machine on which numbers could be added by pressing keys and pulling levers.

Mechanical calculators were not new: The French scientist and philosopher Blaise Pascal, for example, had hand-built primitive calculators in the 17th century. But early calculators jammed easily and usually could only add one column of figures at a time, requiring the operator to manually carry the total from one column to the next. In 1886, Burroughs and three investors formed the American Arithmometer Company to develop and market his design for a reliable calculator.

Burroughs's machine had a column of numbered keys for each column in the numbers to be input. The operator simply punched the appropriate keys and pulled a lever to load the number into the machine and add it to the previously loaded numbers. The keyboard had a special locking feature that prevented accidentally striking an additional key before entering the number.

Their first model sold for $475—equivalent to several thousand dollars today. The first batch of machines had mechanical problems and tended to register the wrong numbers if the clerk pulled too hard. This problem was fixed by a feature called the "dash pot" which produced the same amount of force on the handle regardless of the clerk's efforts.

The improved version began to sell in increasing numbers. By 1895, the American Arithmometer Company was selling almost 300 machines a year from a factory employing 65 clerks and manufacturing workers and three salespersons. By 1900, the sales and staff had roughly tripled, but Burroughs did not live to see it. He died in 1898. Just before his death, the Franklin Institute gave him an award for his invention.

In 1905, the company was renamed the Burroughs Adding Machine Company. Improved machines now had the ability to subtract and to print both subtotals and totals, making them more of a true calculator. A motorized version appeared in 1918. By the 1920s, more elaborate machines could print text (for billing) as well as numbers, and the keys were electrically activated so the operator could simply touch rather than strike them.

In the 1950s, under the name Burroughs Corporation, the firm competed with office machine giant IBM in the emerging computer market. Although IBM would dominate the mainframe market, Burroughs became a respected competitor, the Avis to IBM's Hertz. In 1986 the Burroughs and Sperry Corporations merged.

In creating a reliable calculator, Burroughs helped offices become used to the idea of automation and thus helped pave the way for the computer revolution.

Further Reading

Charles Babbage Institute. "Burroughs Corporation History." Available on-line. URL: http://www.cbi.umn.edu/collections/inv/burros/burhist.htm. Updated on April 2, 2002.

Morgan, B. *The Evolution of the Adding Machine: The Story of Burroughs*. London: Burroughs Machines, 1953.

Bush, Vannevar
(1890–1974)
American
Engineer, Inventor

Although largely forgotten today, the analog computer once offered an alternative way to solve mathematics problems. Unlike digital computers with their binary on/off digital circuits, analog computers use natural properties of geometry, fluid flow, or electric current to set up problems in such a way that nature itself can solve them. Vannevar Bush pioneered important advances in analog computing. He then went on explore pathways for future research and the possibility of using computers to link information in new and powerful ways.

Vannevar was born on March 11, 1890, in Everett, Massachusetts, and spent his childhood in Boston. His father was a Universalist minister; both of his grandfathers had been sea captains. Bush graduated from Tufts University in 1913 with a B.S. degree. During World War I, Bush worked for General Electric and the U.S. Navy and then earned his Ph.D. in electrical engineering in a joint program from Harvard University and the Massachusetts Institute of Technology (MIT).

During the latter part of the war, Bush worked on ways to detect the German U-boats that were devastating Allied shipping. He returned in 1919 to MIT as an associate professor of electrical engineering. He became a full professor in 1923, and served from 1932 to 1938 as vice president and dean of engineering.

The 1920s and 1930s saw the transformation of America into a modern technological society. In particular, large electrical grids and power systems were being built around the country, extending the benefits of electricity into rural areas. In order to design and operate intricate power systems, electrical engineers had to deal with complex equations that involved extensive calculation. Bush became involved in the design of devices for automatically solving such problems. In 1927, Bush and his colleagues and students created a device called a product integraph that could solve simple equations through the interaction of specially designed gears, cams, and other mechanical devices.

During the 1930s, Bush and his team steadily improved the device and extended its capabilities so that it could solve a wider range of calculus problems. The 1930 version, which they called a differential analyzer, could solve equations by repeatedly adding and subtracting quantities until variables "dropped out." Unlike the later digital computers, this analog computer received its inputs in the form of tracings of graphs with a stylus. The graphs were then converted into gear motions and the results of operations were obtained by measuring the distance various gears had turned. Numbers were represented in familiar decimal form, not the binary system that would later be used by most digital computers. Using the differential analyzer, engineers could simulate the operation of a power system and test their designs.

The use of mechanical parts made these computers relatively slow. In 1935, however, Bush began work on a new machine, which would become known as the Rockefeller Differential Analyzer (named after the Rockefeller Foundation, which funded part of the development). Although an analog computer, this machine used vacuum tubes and relays for much of its operation, as would the ENIAC and other early digital computers. The machine could solve sixth-order differential equations with up to 18 variables, and the president of MIT declared it to be "one of the great scientific instruments of modern times." The machine did extensive work on ballistics and other problems during World War II, and several copies were made for other laboratories.

The differential analyzer was a triumph of analog computing, but the focus would soon shift to the more versatile and faster digital computer. The inventors of ENIAC, J. PRESPER ECKERT and JOHN MAUCHLY, studied Bush's machine and how it used vacuum tube circuits, but the machine's limitations encouraged them to use digital circuits rather than analog.

Bush, too, changed the direction of his work. During the war he served as the chairman of the National Defense Research Committee (NDRC), and then became head of the Office of Scientific Research and Development, which oversaw wartime research projects including the development of radar and of the atomic bomb.

Bush used his contacts at major universities to bring together top scientists and engineers for projects whose scale, like that of the huge new Pentagon building, represented a dawning era in the relationship between government and science. Physicist Arthur Compton commented that Bush's "understanding of men and science and his boldness, courage and perseverance put in the hands of American soldiers weapons that won battles and saved lives. But his great achievement was that of persuading a government, ignorant of how science works and unfamiliar with the strange ways of scientists, to make effective use of the power of their knowledge."

In 1945, Bush released a report titled *Science: The Endless Frontier.* In it he proposed the formation of a national research foundation, which would be independent of the military and fund research not only in the "hard" sciences of physics and electronics but in areas such as biology and medicine as well. Bush hoped that the organization would be protected from political pressures and have its priorities set by scientists, not Congress. While this proved to be unrealistic, the organization, called the National Science Foundation, would indeed play a vital role in many kinds of research. In the computing field, this would include the development of supercomputers and the Internet.

Bush's impact on the further development of computers was mixed. Because of his attachment to analog computing, Bush opposed funding for the ENIAC digital computer. Fortunately for the future of computing, the army funded the machine anyway. The advantages of digital computing made it dominant by the end of the 1940s, although analog computers survived in specialized applications.

Bush's ability to foresee future technologies would lead him to develop a proposal for a new way to organize information with the aid of computers. Together with John H. Howard (also of MIT), he invented the Rapid Selector, a device that could retrieve specific information from a roll of microfilm by scanning for special binary codes on the edges of the film. His most far-reaching idea, however, was described in an article published in *Atlantic Monthly* magazine in July 1945. It was a device that he called the "Memex"—a machine that would link or associate pieces of information with one another in a way similar to the associations made in the human brain. Bush visualized the Memex as a desktop workstation that would enable its user to explore the world's information resources by following links. This idea became known as hypertext, and in the 1990s it became the way webpages were organized on the Internet by TIM BERNERS-LEE.

Throughout his career, Bush always kept the possible effects of new technologies in mind, and he urged his fellow scientists and engineers to do likewise. As quoted by G. Pascal Zachary, Bush warned that "One most unfortunate product is the type of engineer who does not realize that in order to apply the fruits of science for the benefit of mankind, he must not only grasp the principles of science, but must also know the needs and aspirations, the possibilities and frailties of those whom he would serve."

Bush's numerous awards reflect the diversity of his work. They include medals from the American Institute of Electrical Engineers (1935), American Society of Mechanical Engineers (1943), and the National Institute of Social Sciences (1945). In 1964 Bush was awarded the National Medal of Science.

Further Reading

Bush, Vannevar. "As We May Think." *Atlantic Monthly*, July 1945, p. 101ff. Also available online. URL: http://www.theatlantic.com/unbound/flashbks/computer/bushf.htm. Downloaded on November 26, 2002.

Nyce, James M., and Paul Kahn, editors. *From Memex to Hypertext: Vannevar Bush and the Mind's Machine*. San Diego, Calif.: Academic Press, 1992.

Zachary, G. Pascal. *Endless Frontier: Vannevar Bush, Engineer of the American Century*. New York: Free Press, 1997.

Bushnell, Nolan

(1943–)
American
Inventor, Entrepreneur

Bloop! Bonk! Bloop! Bonk! That was the sound of Pong, a crude videogame introduced in 1972. All it had was two movable "paddles" and a little glowing ball to hit with them. Today, no one would give it a second look, but Pong soon became a multibillion-dollar game machine industry, and later inspired programmers to create similar games for personal computers. Inventor Nolan Bushnell, with energy as restless as his game screens, became one of the first modern Silicon Valley computer pioneers.

Nolan Bushnell was born on February 5, 1943, in Clearfield, Utah, a small town near the Great Salt Lake. In third grade, he became fascinated with electronics when the teacher asked him to create a demonstration of electricity from the odds and ends in the class's "science box." After building the demo, he went home and dug up every switch, flashlight, and other electrical component he could find, and began to tinker.

Nolan earned his ham (amateur) radio license when he was only 10 years old. When he ventured from electronics to more active pursuits, he was less successful. An attempt to build a rocket-powered roller skate succeeded only in nearly burning down the family's garage.

Nolan Bushnell took minicomputer video games out of the MIT labs and built them into consoles that began to replace pinball machines in bars and pizza parlors. His Atari Corporation then invented the cartridge system that connected video games to the home TV set. *(© Roger Ressmeyer/CORBIS)*

Nolan was not a stereotypical nerd, however. In high school, he played basketball, skied, and tinkered with cars. In 1961, Bushnell entered Utah State College in Salt Lake City. He majored in engineering, but he took a broad range of other courses as well, including economics and business classes.

In the next few years, computer scientists and student "hackers" began to experiment with connecting the newly developed, affordable minicomputers to graphics displays. Students at the Massachusetts Institute of Technology (MIT), for example, developed Space War, a game in which players tried to steer blips representing spaceships around a solar system, shooting torpedoes at one another and trying to master the effects of simulated gravity. Bushnell began to wonder whether such a game could be built into a smaller, coin-operated machine for ordinary people to play.

Meanwhile, Bushnell helped pay his college expenses by starting his own business, called the Campus Company, and selling advertising on free calendars. During the summer he worked at an amusement park. At first his job was to guess peoples' weights and ages (a popular novelty at the time). However, he was soon promoted to manager of the park's games. This gave him the chance to study a number of different mechanical games and to figure out what elements made them appeal to people. He concluded that the most popular games were those that tested manual dexterity and skill but were not very complicated.

After graduating from college in 1968 with a B.S. degree in electrical engineering, Bushnell tried to get a job at Disneyland, the ultimate amusement park. When he was turned down, he moved to Santa Clara, California, in the area that would soon become known as Silicon Valley. He worked for two years developing sound equipment for Ampex Corporation, but became restless and decided to pursue his dream of making coin-operated games.

Stuffing a Digital Equipment Corporation (DEC) PDP minicomputer into a box in an arcade would not have worked—the cheapest machines cost about $10,000, and they would not have fit in the box, anyway. Fortunately, by the early 1970s there was a potential alternative: a microprocessor, a single small chip that included all the basic logic and arithmetic functions for a computer. Combined with some additional chips and circuits to process input and display graphics, the microprocessor made table-sized video games possible. Cobbling together the parts on a table in his daughter's bedroom, Bushnell built his first machine, then programmed a simplified version of the MIT Space War game, calling it Computer Space.

Bushnell then made a deal with a small company called Nutting Associates, which had been making mechanical arcade games. They agreed to produce his new game machine, and he became the engineer in charge of the product. Meanwhile, he and a friend from his old job at Ampex had been setting up coin-operated pinball machines in local bars, splitting the proceeds with the bar owners. They made a bit of money by adding the new Computer Space machines to the route, but sales were slow.

Using the money he had earned, Bushnell started a new project. He was convinced that the space game had been too complicated. After all, people in bars generally do not want games that require a lot of concentration. So Bushnell sketched out the game that became Pong—just a moving ball, two movable paddles, and a way to keep score. He and another former Ampex associate, Al Alcorn, put together the new game in only three months.

The game worked, but building a unit cost much more than the $100 that Bushnell thought people would be willing to pay. None of the game companies seemed interested in marketing the game for him. Bushnell then decided to do it on his own, and started a company called Atari, a word that in the Asian game of Go means roughly what "check" does in chess.

Despite the higher price he had to charge, Bushnell's Pong game sold like hotcakes after its introduction in November 1972. Bushnell then worked on a home version of the game that could be hooked up to an ordinary TV set. Sears Roebuck liked that version so much that it bought up the entire supply in 1974.

After millions of people had bought Pong, sales began to drop. Bushnell then came up with a clever idea: Atari designed its machines so that they could play a variety of different games. All the user had to do was buy preprogrammed cartridges and he or she could play everything from a car-racing game to games such as poker or even Scrabble.

People who like to make games also like to play a lot. As it grew, Atari Corporation's headquarters in Los Gatos, California, began to look very different from the usual corporate workplace. Engineers and programmers could wear jeans and T-shirts, and they were even encouraged to brainstorm while relaxing in hot tubs. The "Atari culture" became the forerunner for the strange combination of intense work and laid-back lifestyle that would mark many Silicon Valley companies in the 1980s and 1990s.

By 1976, Atari was a $40 million company, but to reach the next level it had to attract major investors, especially because it now faced competition from other video game makers. Bushnell decided to sell the company to Warner Communications for $28 million. He kept half the money as his share and remained chairman of the company.

Bushnell helped Atari create its first general-purpose personal computer. The Atari PC had better graphics than did competitors from Apple and Radio Shack. Its ability to easily manipulate patterns called "sprites" to move images on the screen attracted many game designers. But he became frustrated because he had to work on projects that Warner wanted, not the new ideas he was constantly thinking up, and he also felt stuck in just one role. Bushnell, however, believed that his strength came from a combination of skills. He would later tell author Robert Slater that "I'm not the best engineer around. I'm not the best marketeer. I'm not the best financial strategist. But I think I'm better than anyone at all three of those at the same time."

So Bushnell left Warner after only two years and launched a new venture. He had observed that modern kids like nothing better than video games and pizza, so why not combine the two? In 1977, he opened Pizza Time Theatre, which entertained families with video games and animated robot characters while they waited for their pizza. By 1982, there were 200 restaurants in the chain.

During the 1980s, Bushnell changed his focus to venture capitalism—finding promising startup companies and giving them money in return for a stake in the ownership. One of the companies he nurtured was Androbot, which made a household robot not unlike the science fiction creations on the 1960s cartoon *The Jetsons*. The robot could detect intruders, do some housecleaning, and make clever conversation—but it turned out to be a novelty rather than a seriously useful product. In the late 1990s, Bushnell entered the popular new arena of Internet gaming with his uWink website.

None of Bushnell's later products was very successful, but that did not seem to diminish his joy in trying new things. He was quoted as saying, "Business is the greatest game of all. Lots of complexity and a minimum of rules. And you can keep score with the money."

Further Reading

Cohen, Scott. *Zap! The Rise and Fall of Atari*. New York: McGraw-Hill, 1984.

Slater, Robert. *Portraits in Silicon*. Cambridge, Mass.: MIT Press, 1987.

C

Case, Steve
(1958–)
American
Entrepreneur

By the 1980s, personal computers (PCs) were starting to appear not only in businesses but in homes. Using a device called a modem, a PC could be connected to a phone line so it could exchange data with other computers. Some enterprising PC owners set up bulletin boards that other users could dial into and read and post messages, as well as downloading the latest game programs and other files. Meanwhile, some college students and engineers were communicating over the Internet, a network that was still unknown to the general public. The ordinary home PC user still was not connected to anything.

By the end of the decade, however, easy-to-use commercial services would change the face of the on-line world and bring it to millions of consumers. An entrepreneur named Steve Case and his company, America Online (AOL), would become the most successful on-line service in history.

Steve Case was born on August 21, 1958, in Honolulu. His father was a lawyer and his mother a teacher. Case became interested in business at a young age, teaming up with his brother Dan, who was about a year older. Together they started a juice stand (using the plentiful free fruit from the trees in their backyard). When the brothers reached their teen years, they named themselves Case Enterprises and branched out into newspaper delivery, selling seeds and greeting cards, and even publishing an advertising circular under the name Aloha Sales Agency.

In high school, Steve continued his interest in publishing, becoming a record reviewer (and incidentally receiving lots of free music for his collection, as well as free concert tickets). He went to the mainland to continue his education, attending Williams College in Williamstown, Massachusetts. There he majored in political science, which he noted was "the closest thing to marketing." He "minored" in rock and roll, singing for two bands—not very well, as he would later admit.

Graduating in 1980 with a B.A. degree, he went to work as a marketer for Procter & Gamble, working on campaigns for a home permanent kit and fabric softener strips. Changing hairstyles seemed to doom the former product. In 1982, Case went to PepsiCo, which not only makes the popular soft drink but owns many other companies. Case was assigned the task of visiting the company's Pizza Hut stores and researching ideas for new flavors of pizza.

Wanting something interesting to do on the road and in endless motel rooms, Case bought a

Kaypro, an early suitcase-sized "portable" computer. Equipping it with a modem, he connected to an early on-line service called the Source, and used it to visit many of the bulletin boards that were springing up around the country. He later told an interviewer that "it was all very painful and time-consuming, but I could glimpse the future. There was something magical about being able to dial out to the world from Wichita."

In 1983, he met up again with his brother Dan, who had become an investment banker. One of Dan's clients was a company called Control Video Corporation, which was trying to set up an on-line service that could deliver Atari video games to PC users. Unfortunately, the company was suffering from poor management, not enough money, and the winding down of the great video game craze of the 1970s.

In 1985, Case tried to help save the company (which was now called Quantum Computer Services) by finding other services besides games that could be delivered on-line. They linked up with the big PC makers Apple, Radio Shack, and IBM to develop on-line software for their respective operating systems. The $5 million coming in from these deals turned out not to be enough to pay the expenses of developing the software. Investors decided that Case had bitten off more than he could chew and called for his firing. However, chief executive officer Jim Kimsey stuck by him. Gradually the company began to sign up enough on-line subscribers to become profitable.

Six years later the company changed its name to America Online (AOL). Like Amazon's JEFFREY P. BEZOS would do a few years later, Case decided to expand the company's user base as fast as possible. AOL spent (and continues to spend) millions of dollars distributing free trial disks by mail and in magazines, as well as advertising on TV and radio. The other part of Case's strategy was to make deals with companies that had software or services that they wanted to offer on-line. Having more things for users to see, do, and buy on-line would in turn make AOL a compelling product for still more users.

Meanwhile, software giant Microsoft had noticed AOL's success. First Microsoft cofounder Paul Allen and then WILLIAM GATES III himself expressed an interest in buying AOL. Case, however, vowed to resist the takeover attempt, despite Gates's threat to develop his own competing on-line service.

By the mid-1990s, the World Wide Web (created in 1990 by TIM BERNERS-LEE) was bringing the Internet into the public eye. Recognizing that people were going to want to access this vast and growing worldwide network, Case bought companies that would provide AOL with the expertise and software needed to connect people to the Internet. AOL became more profitable in 1993 and 1994, but in 1995 the company took a loss despite its growing revenues. This was because AOL had spent even more money on buying other companies and setting up new services than the $394 million it had taken in. With the company reaching 2 million subscribers that year, Wall Street investors remained confident that it would be profitable in the long term. *Inc.* magazine named Case an Entrepreneur of the Year in 1994.

The rise of AOL was accompanied by growth pains, however. Sometimes the company added users faster than it could provide connections and computer capacity. This resulted in slow service and complaints from users. Veteran Internet users tended to look down upon the flood of AOL users who were posting in Internet newsgroups as "clueless newbies" who understood nothing about the sophisticated culture of the Internet. But defenders of the easy-to-use AOL pointed out that it was more accessible and reassuring to consumers who were hesitant to venture directly onto the Internet. Besides, as Case told an interviewer, "Consumers don't want millions of websites. They don't want to click on 'Sports' and have 18,000 sports websites pop up. . . . They will rely on people to package and present the services that are going to be of the highest quality and the most relevant to their particular interest."

In 1995 Microsoft made good on its threat by starting its own on-line service, Microsoft Network (MSN), and including it with its Windows 95 operating system. New PC users now had an icon on their screens that would take them directly to MSN. AOL and the other two leading on-line services, CompuServe (which AOL later bought) and Prodigy, cried foul and asked the government to investigate Microsoft for unfairly using its monopoly on the operating system. Microsoft would eventually face a variety of antitrust challenges in the courts, but meanwhile Case and Gates did not hesitate to make a deal they thought would be mutually advantageous. AOL agreed to include the Microsoft Internet Explorer browser with its introductory software disks, and Microsoft in turn agreed that its next operating system, Windows 98, would include an AOL icon on the desktop.

The 21st century brought a sharp downturn in the on-line economy. Rather than try to cut his losses by reducing services, Case typically took the opposite approach. In January 2001, AOL merged with media giant Time Warner, creating a company, AOL Time Warner, worth more than $180 billion. Time Warner offered access to cable TV (and cable Internet service) as well as a huge amount of "content"—news and other features—that AOL can use to make it a compelling portal, or gateway to the Internet.

Of course, in times of economic uncertainty and rapidly changing technology, AOL's future and Case's own future plans remain to be seen (In January 2003 Case announced that he would resign as chairman of AOL Time Warner, and he admitted that the company's performance after the merger had been disappointing.) Many industry observers believe Case had alienated the "traditional media" people at Time Warner by making extravagant claims for future web initiatives. Case's apparently easygoing manner and laid-back, casual lifestyle has sometimes fooled competitors into thinking that he is not also a hard-driving executive. But Case had proven that he could meet a variety of challenges, keeping AOL the top brand name in online services.

Further Reading

Roberts, Johnnie L. "How It All Fell Apart: Steve Case and Jerry Levin Created AOL Time Warner in a Marriage of Convenience. The Inside Story of What Went Wrong." *Newsweek*, December 9, 2002, p. 52.

Swisher, Kara. *aol.com: How Steve Case Beat Bill Gates, Nailed the Netheads, and Made Millions in the War for the Web.* New York: Times Books, 1999.

Cerf, Vinton

(1943–)
American
Computer Scientist

Something as complex as the Internet has more than one inventor. However, when experts are asked who played the key role in actually getting people around the world to agree on a single way to connect computers, they usually point to Vinton Cerf. Combining technical and human skills, Cerf led the effort to build the "plumbing" that lets a user in Moscow, Russia, send a message to another user in Moscow, Idaho.

Vinton (usually called Vint) Cerf was born on June 23, 1943, in New Haven, Connecticut. He was born prematurely and suffered from a hearing defect. While not deaf, he still had to make a special effort to understand and communicate with other people. Later, he would reflect that this condition may have helped drive him to find ways to use computers to communicate in new ways.

Every time a Web user clicks on a link, the request and the returning data travel over the electronic "pipes" of the Internet. Behind the scenes, Vinton Cerf developed the rules for connecting the electronic plumbing, and persuaded engineers and institutions to work together to implement them. *(Courtesy of MCI)*

Despite his disability, Cerf was a bright student who especially loved math, science, and science fiction. He built "volcanoes" using plaster of paris and chemicals that sparked and fumed. When the Soviets orbited *Sputnik* and America's interest turned to space travel, the teenage Cerf started building model rockets.

Cerf saw his first computer when he was a 15-year-old high school student, but the vacuum-tube-filled monster was strictly off limits to the public. However, two years later, he and a friend persuaded a sympathetic official at the University of California, Los Angeles (UCLA) to let them use the school's Bendix G-15 mainframe. "The bug bit, and I was infected by computers," Cerf later told an interviewer for *Forbes*. However, he kept a wide perspective and was always interested in making new connections between things. As would be recounted in *Current Biography*, one day, while walking by a building at high school, he realized that "the whole universe was hooked together."

Cerf enrolled in Stanford University and earned his B.S. degree in mathematics in 1965. He then went to work as a programmer for IBM, helping design time-sharing systems that allow many users to share the same computer. But Cerf felt the need to explore computer science more deeply, so he enrolled in the graduate program at UCLA. His adviser, Jerry Estrin, was working on the "Snuper Computer"—a system that let a user working on one computer log onto a remote computer. Cerf joined this effort and wrote his thesis on the subject of remote computer monitoring. This experience would be important as Cerf went further into the brand-new field of computer networking.

Cerf's colleagues were impressed by his wide-ranging intelligence, but he found social life to be more difficult. Because of his hearing problems, he could not always understand what other people were saying. He tried to compensate by asking lots of questions. He noted to authors Katle Hafner and Matthew Lyon, however, that "in a group conversation this can backfire embarrassingly if the question you ask is one which has just been asked by someone else. A variation (equally embarrassing) is to enthusiastically suggest something just suggested."

One day Cerf's hearing aid dealer arranged it so that he would meet another hearing-impaired client, Sigrid Thorstenberg. Soon Cerf and Thorstenberg, a designer and illustrator, were chatting about computers, art, and many other subjects. A year later they were married.

In 1968, Cerf joined ARPANET, a computer networking project sponsored by the U.S. Defense Department. Unlike government agencies with more narrowly focused efforts, ARPANET encouraged researchers to brainstorm and develop new ideas and proposals. Cerf and his colleagues came up with some wild ideas, including on-line databases and e-mail.

The actual data transmission for the network would use an idea borrowed from PAUL BARAN, of the RAND Corporation think tank. Data would be broken into individually addressed "packets" that could be sent over many different possible routes from one computer to another. This meant that if a breakdown (or perhaps an enemy attack) brought down one "node," or connected computer, communications could be detoured around it. At the destination computer, a program would reassemble the arriving packets into the complete message or file.

Cerf's first job was to test these ideas for computer networking. He wrote simulation programs that explored how the network would respond as more users were added. Often the simulated network "crashed." But gradually they came up with a workable design, and in late 1969 the first network, the precursor to the Internet, linked computers at UCLA, the Stanford Research Institute (SRI), the University of California at Santa Barbara (UCSB), and the University of Utah. Each of the four computers had an interface message processor, or IMP—essentially a huge, sophisticated modem, the size of a refrigerator.

By 1972, ARPANET was growing rapidly as many other universities came onto the network. Cerf and his colleagues jury-rigged phone lines and connections and put on a demonstration of the new network at the International Conference on Computer Communications in Washington, D.C. The technology was still unreliable, but they managed to make it work, and the importance of networking again grew in the worldwide community of computer scientists and engineers.

In the early days of computer networking, each vendor (such as IBM or Digital Equipment Corporation) developed networking software that ran only on its own computers. If computer users with different kinds of machines were going to be able to communicate with each other, there would have to be an agreed-upon protocol, or set of rules for constructing, routing, and reassembling data packets. There had to be a network that connected all networks—an "Internet." Cerf decided that his next task was to get the different computer manufacturers and their engineers to agree on one protocol. In 1973, Cerf and his colleague Robert Kahn announced such a protocol, called TCP/IP (transmission control protocol/Internet protocol).

In the next few years, Cerf went to numerous meetings, presenting the new system and urging its adoption. He stressed that this universal networking system could run on any computer anywhere. At one meeting, he made this point humorously by taking off his coat, tie, and dress shirt. Underneath was a T-shirt that read "IP on everything."

With the Internet well under way, Cerf decided to enter the commercial sector and work on ways to bring the benefits of digital communications and networking to consumers. In 1982, he took a consulting position with MCI, the long-distance phone company, and helped it develop an e-mail system. But Cerf also remained closely involved with the Internet, which was starting to evolve from a government and university network to a general public network. In 1986, he joined Robert Kahn at the Corporation for National Research Initiatives, a nonprofit group that sought to plan further developments in high-tech research. As of 2003, Cerf is senior vice president of architecture and technology for MCI.

In 1992, Cerf helped found the Internet Society and became its first president. By then, the World Wide Web had been invented (by TIM BERNERS-LEE), though the explosion of the Web and of e-commerce was still a few years away. Already, however, the Internet was raising difficult social issues such as censorship and the need to protect on-line users from fraud. As political leaders began to call for new regulations for the Internet, Cerf suggested a different approach. He wanted software developers to create programs that people could use to control what they (or their children) saw on the Web, protect sensitive personal data from disclosure, and safeguard privacy. Looking at the 21st century, Cerf saw the Internet becoming truly worldwide, and merging with the telephone system to become a seamless web of communication.

Cerf has received many of the highest awards in computer science, including the Association for Computing Machinery (ACM) Software System Award (shared with ROBERT KAHN in 1991) for development of the TCP/IP software that drives the Internet. The two pioneers also shared a National Medal of Technology, presented by President Bill Clinton in 1997.

But one award had a special meaning to Cerf. In 1998, he received the Alexander Graham Bell award for his contribution to improving the lives of the deaf and hard-of-hearing. When he received the award, Cerf noted (as recounted by author Debbie Sklar) that "As an individual who is hearing-impaired, I'm extremely proud of the level-playing-field result the text-based Internet has had on communications among hard-of-hearing and hearing communities alike."

Further Reading

Brody, Herb. "Net Cerfing," *Technology Review* 101 (May–June 1998): 73ff.

Cerf, Vinton. "Cerf's Up." Available on-line. URL: http://www1.worldcom.com/global/resources/cerfs_up. Downloaded on October 31, 2002.

Hafner, Katie, and Matthew Lyon. *Where Wizards Stay Up Late: the Origins of the Internet*. New York: Simon and Schuster, 1996.

Schiek, Elizabeth A., ed. *Current Biography Yearbook, 1998*. "Cerf, Vinton." New York: H. W. Wilson, 1998, pp. 89–92.

Sklar, Debbie L. "The Bell Tolls for Vint Cerf," *America's Network*, August 15, 1998, p. 49.

Church, Alonzo
(1903–1995)
American
Mathematician

Mathematics is perhaps the oldest of the sciences, and computer science is certainly one of the newest. Since the beginning of computer research in the late 1930s, the two fields have had a close relationship and have depended on one another for progress. Engineers need mathematicians not only to provide algorithms, or procedures for solving problems, but to answer a fundamental question: What kinds of problems can be solved by computation? Mathematicians, in turn, need computers if they are to tackle problems involving calculations that could not be finished manually in a lifetime. Logician and mathematician Alonzo Church would do much to put computing on a firm mathematical foundation.

Church was born on June 14, 1903, in Washington, D.C. His father was a municipal judge. The Church family had a long mathematical tradition: Alonzo's great-grandfather, also named Alonzo Church, had been a professor of mathematics at Athens College (later the University of Georgia) and had served as the school's president from 1829 to 1859.

Following in those distinguished footsteps, Church attended Princeton University and received his A.B. degree in mathematics in 1924. Church found that he was very interested in the logical underpinnings of mathematics. Oswald Veblen, a distinguished geometer and expert on mathematical postulates, encouraged young Church's interest in the foundations of mathematics.

Responding to Veblen's encouragement, Church continued into graduate school at Princeton and received his Ph.D. in 1927. He received a National Research Fellowship, and in 1928 he went to Göttingen, Germany, where he discussed his research with David Hilbert and other noted European mathematicians.

At the beginning of the 20th century, Hilbert had made a list of what he considered to be the most important unanswered questions or problems in mathematics. The 10th problem on the list asked whether there was a "mechanical procedure" that, starting from the basic axioms of mathematics, could prove whether any given proposition was true. By "mechanical," Hilbert meant a set of steps that could be carried out routinely by mathematicians. However, by the late 1930s, researchers such as ALAN TURING were beginning to think in terms of actual machines that could carry out such steps.

In 1936, Church used formal logic to prove that no such procedure could be guaranteed to work in every case—that arithmetic was "undecidable." This, of course, did not mean that the answer to a calculation such as $9 \times 7 + 1$ could not be determined precisely. Rather, it meant that there was no way to prove that an arithmetic proposition might not come along that could not be answered.

Church arrived at this result by using recursive functions. A function is an operation that works on a quantity and yields a result—for example, the "square" function returns the results of multiplying a number or quantity by itself. Recursion, a powerful idea used in computer programming today, involves a function calling upon itself repeatedly until it reaches a defined end point.

At about the time Church was writing his proof for Hilbert's 10th problem, Turing was taking a different approach. Instead of using functional analysis like Church, Turing constructed an imaginary machine that could carry out simple operations repeatedly. Turing showed that this machine could carry out any definable calculation, but that some calculations were not definable. Turing and Church arrived at the same conclusion through different routes, but their approaches were complementary. Both received credit for placing the theory of computability on a firm foundation and paving the way for further research into computing machines.

Church also contributed to future computer science by developing the "lambda calculus," first introduced in 1935. This describes the rules for combining mathematical functions, using recursion, and building functions up into complex expressions. Researchers in artificial intelligence (AI) (such as JOHN MCCARTHY) would later use the lambda calculus to design powerful "functional languages" such as LISP.

Church taught at Princeton for many years. In 1961, he received the title of Professor of Mathematics and Philosophy. In 1967, he took the same position at UCLA, where he was active until 1990. He received numerous honorary degrees, and in 1990 an international symposium was held in his honor at the State University of New York at Buffalo.

Through a long career as a professor of mathematics, Church nurtured many students who became great mathematicians, and he edited an influential periodical, *The Journal of Symbolic Logic*. He retired in 1990 and died five years later. He was elected to the American Academy of Arts and Sciences and the U.S. and British Academies of Science.

Further Reading

Barendregt, H. "The Impact of the Lambda Calculus in Logic and Computer Science." *The Bulletin of Symbolic Logic* 3, no. 2 (1997): 181–215.

Church, Alonzo. *Introduction to Mathematical Logic*. Princeton, N.J.: Princeton University Press, 1956.

"The Church-Turing Thesis." Available on-line. URL: http://plato.stanford.edu/entries/church-turing. Updated on August 19, 2002.

Davis, M. *The Undecidable: Basic Papers on Undecidable Propositions, Unsolvable Problems, and Computable Functions*. Hewlett, N.Y.: Raven Press, 1965.

Codd, Edgar F.

(1923–2003)
British
Mathematician, Computer Scientist

The "glue" that makes the Information Age possible is the organization of data into databases, or collections of related files containing records of information. Edgar Codd created the relational database model that is used today by businesses, government agencies, and anyone who needs to organize large quantities of information so that it can be retrieved reliably. Codd also made important contributions to other areas of computing, including multiprogramming and self-reproducing programs.

Codd was born on August 19, 1923, in Portland, England. During World War II, he served as a captain in the Royal Air Force. After the war, Codd completed his B.A. and M.A. degrees in mathematics at Oxford University. He then immigrated to the United States. After a brief stint as a mathematics instructor at the University of Tennessee, Codd went to work at IBM in New York. There he joined JOHN BACKUS in the development of the IBM SSEC (Selective Sequence Electronic Calculator), an early computer.

From 1953 to 1957, Codd worked for Computing Devices of Canada, but he then returned to IBM. Codd joined the team working on STRETCH, an ambitious high-performance computer. Codd's special interest was in multiprogramming, or the ability to run many programs simultaneously. Codd designed an "executive," or control program, that could keep track of the execution of multiple programs. He wrote a number of influential articles that helped bring multiprogramming to the attention of the computer science community.

In 1962, Codd took a leave of absence from IBM to go to the University of Michigan, where he earned a Ph.D. in communications sciences in 1965. By then his research interest had shifted from multiprogramming to the creation of programs that could reproduce themselves by systematically applying simple rules to a grid of uniform "cells." In 1968, Codd published a book called *Cellular Automata* that elaborated on the work he had done in his doctoral dissertation. (The idea of cellular automation would be publicized later by mathematician John Conway, who designed the "Game of Life" to generate and "evolve" visual patterns.)

By the late 1960s, Codd's research took yet another turn. By that time, corporations and government agencies were storing huge quantities of information on giant mainframe computers. Applications ranged from the Social Security Administration rolls to large corporate payrolls, inventory, and transaction records. It was becoming increasingly difficult to manage all this information. One fundamental problem was that there was no standard model or way of organizing and viewing information in a database. Programmers created and updated data files using program code (such as COBOL) often without much regard to how data in one file (such as inventory records) might be related to that in other files (such as customer information and purchase transactions).

Codd decided to focus on the problem of specifying the relationships between data records in a uniform way. His 1970 article "A Relational Model of Data for Large Shared Data Banks" described a database organized as a set of tables. In a table, the columns represent fields such as customer name, customer number, account balance, and so on. Each row contains one record (in this case, the information for one customer). A second table might hold transaction records, with the columns representing customer numbers and item numbers. A relation can be set up between the two tables, since both have a customer number column. This means, for example, that the database user can ask for a report listing the name of each customer with a list of all transactions involving that customer. The customer number serves as the "bridge" by which the two tables can be "joined."

Codd's relational model gave programmers ways to ensure the consistency and integrity of a database. For example, they can make sure that no two customers have the same customer number, which is supposed to be a unique "key" identifying each customer.

Codd's work also led directly to the development of SQL (structured query language), a standard way to specify what information is to be retrieved from a database. Since SQL is not tied to any particular computer language or operating system, someone who masters this language can easily work with many different kinds of database systems.

Codd retired from IBM in 1984. He established two database consulting companies and continued to publish technical papers defending the relational database model against more recent challenges and revisions. Codd died at his Florida home of heart failure on April 18, 2003.

Codd has received a number of important honors, including the Association for Computing Machinery Turing Award (1981) for his work in relational databases, membership in the National Academy of Engineering (1983), and membership of the American Academy of Arts and Sciences (1994).

Further Reading

Codd, Edgar F. *Cellular Automata*. New York: Academic Press, 1968.

———. "Multiprogramming Scheduling: Parts 1 and 2: Introduction and Theory." *Communications of the ACM* 3, no. 6 (1960): 347–350.

———. "Multiprogramming Scheduling: Parts 3 and 4: Scheduling Algorithm and External Constraints." *Communications of the ACM* 3, no. 7 (1960): 413–418.

———. *The Relational Model for Database Management, Version 2*. Reading, Mass.: Addison-Wesley, 1990.

———. "A Relational Model of Data for Large Shared Data Banks." *Communications of the ACM* 13, no. 6 (1970): 377–387.

Corbató, Fernando

(1926–)
American
Computer Scientist

During the 1950s, someone fortunate enough to have access to a computer had to prepare a program on punched cards or tape and hand it over to the machine's operator. The large, expensive mainframe computers could only run one program at a time, so each user had to wait his or her turn. As a result, most users were lucky if they could run their programs once or twice a day.

As computers gradually gained larger amounts of "core," or memory, it was often the case that whatever program was running needed only a fraction of the available memory, with much of the machine's resources going to waste. If someone could come up with a way to load several programs at once and have them share the memory, the computer could be used much more efficiently. Users might even be able to use the computer interactively, typing in commands and getting results almost instantly. This idea, called time-sharing, was the achievement of Fernando Jose Corbató.

Corbató was born on July 1, 1926, in Oakland, California. As a young man, Corbató developed an interest in electronics. In 1941, when he was still in high school, the United States entered World War II. Corbató knew that he would be drafted soon, so he looked for a way that he could serve in a technical specialty. Fortunately, the U.S. Navy had started the Eddy Program, which was designed to train the large number of technicians that would be needed to maintain the new electronic systems that were starting to revolutionize warfare—devices such as radar, sonar, and the loran navigation system. The 17-year-old Corbató therefore enlisted in the navy, went through a one-year electronics training program, and then served as an electronics technician for the rest of the war.

The war had ushered in the nuclear age and made physics into something of a "glamor" field. After he got out of the navy, Corbató enrolled in the California Institute of Technology (Caltech) and earned his bachelor's degree in physics in 1950. He then went to the Massachusetts Institute of Technology (MIT) for his graduate studies, earning his doctorate in physics in 1956.

However, while studying physics at MIT Corbató encountered Philip M. Morse, who would later become director of the MIT Computation Center. In 1951 Morse convinced Corbató to take on a research assistantship in the use of digital computers in science and engineering. The program was funded by the Office of Naval Research, and introduced Corbató to the navy's latest and greatest computer, a machine called Whirlwind. This machine, the "supercomputer" of its time, was in great demand despite its unreliable memory system, which crashed every 20 minutes or so.

Although the Whirlwind was used mainly for classified military research, the navy made it available for civilian projects about three or four hours of each day. Users such as Corbató generally only got one "shot" at the machine per day, so if a program had a bug he would have to try to figure out the problem, make a new program tape, and hope it ran properly the next day. Given how exciting the Whirlwind was to use (it even had a graphics display), this must have been quite frustrating.

After getting his doctorate in physics, Corbató joined the newly formed Computation Center, which was equipped with one of the newest commercial mainframes, an IBM 704, which was shared by MIT, surrounding colleges, and the local IBM staff. Corbató served in a variety of administrative positions, but his attention was increasingly drawn to finding ways that the many users clamoring for access to the computer could share the machine more efficiently.

In 1961, Corbató started a new project, the development of what would become the CTSS, or Compatible Time-Sharing System. By November of that year, Corbató and two fellow researchers, Marjorie Daggett and Bob Daley, had put together a crude time-sharing system on the IBM 709 (which had replaced the 704). The system took a five-kilobyte portion of the machine's memory and inserted into it a tiny operating system which received and processed commands from up to four users at teletype terminals, while the rest of the memory was used by traditional batch programs. This time-sharing system worked only if no one tried to run a program that required all the machine's memory. In other words, it worked only with programs that were "compatible," or well-behaved.

As Corbató described in a conference paper, the new system allowed a user to:

1. Develop programs in languages compatible with the background system,
2. Develop a private file of programs,
3. Start debugging sessions at the state of the previous session, and
4. Set his own pace with little waste of computer time.

In other words, Corbató's time-sharing system introduced most of the features that today's programmers and users expect as a matter of course. The machine efficiently switched back and forth among the running programs, running a few instructions at a time. Because computers are so much faster than human beings, each user experienced little or no delay in the responses to commands. Each user in effect could treat the mainframe as his or her "personal computer." (Modern personal computers were still almost 20 years in the future.)

In 1962, however, MIT acquired a new IBM 7090 computer, which included the ability to execute a "hardware interrupt." This meant that the operating system could serve as a "traffic cop" and enforce the sharing of computing time between programs. By 1963, an improved version of CTSS allowed a single mainframe to be shared by up to 21 simultaneous users. Corbató participated in further development of time-sharing under Project MAC, which eventually adopted a new time-sharing operating system called Multics, an ancestor of today's UNIX operating system.

Meanwhile, Corbató rose steadily in the academic ranks at MIT, becoming an associate professor in 1962, a full professor in 1965, and the associate department head for computer science and engineering from 1974 to 1978 and 1983 to 1993. Corbató retired in 1996. In 1991 he received the prestigious Association for Computing Machinery Turing Award "for his work in organizing the concepts and leading the development of the general-purpose large-scale time sharing system and resource-sharing computer systems CTSS and MULTICS." He is also a fellow of the Institute for Electrical and Electronics Engineers (IEEE) and recipient of the IEEE Computer Pioneer Award (1982) and a number of other fellowships and honors.

Further Reading

Corbató, Fernando, et al. "An Experimental Time-Sharing System," *Proceedings SJCC* 21 (1962): 335–344.

Lee, J. A. N. *Computer Pioneers*. Los Alamitos, Calif.: IEEE Computer Press, 1995.

Crawford, Chris
(ca. 1950–)
American
Programmer

Many early computer and video games relied on simple, compelling action such as chasing, racing, and shooting. However, starting in the 1970s programmers and computer science students began to develop and enjoy a different sort of game. In games such as the original Adventure, the player controlled a character who traveled in an imaginary world and interacted with that world and its inhabitants. When personal computers became widespread in the 1980s, companies such as Infocom offered more elaborate versions of adventure and role-playing games, while other companies specialized in detailed war game simulations.

Unlike film directors, game designers generally worked in anonymity, with players not knowing or caring who was designing their entertainment. However, a few game designers began to stand out because of their creativity and ambition to raise game design from "mere" programming to a true art form. Chris Crawford has been one of the most innovative of these designers.

Crawford is reticent about his origins but says he was born in Texas and was 52 in 2002. Crawford's interest in games began in 1966 when a friend introduced him to Blitzkrieg, a board-and-counters war game. These games tended to have many pieces and special rules, so while they can provide a vivid simulation of war or other activities, they require much dedication to play well.

Crawford studied physics in college, earning his B.S. degree at the University of California, Davis, in 1972 and his M.S. degree at the University of Missouri, Columbia, in 1975. He then taught physics at community college as well as teaching courses in energy policy through University Extension. Meanwhile, the first microcomputer systems had become available. Crawford bought himself a computer kit, built his own machine, and began to experiment with programming computerized versions of his favorite board games. He realized that the computer had the potential of taking care of rules interpretation and other details, leaving players free to think strategically and tactically, more like real generals and leaders. The computer could also offer a challenging artificial opponent.

In 1979, Crawford found an opportunity to turn his longtime hobby of game design into a career. He joined Atari, a company that had become famous for Pong and other video games, and also marketed a line of personal computers that featured a sophisticated color graphics system.

Crawford's first two products for Atari drew upon his background in energy issues. Energy Czar put the player in charge of managing an energy crisis like the one that had occurred only a few years earlier, in the early 1970s, while Scram offered a simulated nuclear power plant that had to be kept under control. (Nuclear power was a hotly debated issue at the time.)

Crawford was then put in charge of training programmers to get up to speed on the Atari computers. During that time he designed two war games, Eastern Front (1941) (set during World War II), and Legionnaire, which featured warfare in the ancient world. Both of these games were popular and would greatly influence later designers. Crawford was perhaps the first designer to successfully translate the concepts of the old cardboard-counter war games into the more flexible and interactive world of the computer while keeping them easy to play and understand. (His 1986 game Patton vs. Rommel continued the evolution of computer war games with more sophisticated game mechanics.) Another popular Crawford title was Excalibur, which explored the Arthurian legends. Crawford also described his approach to game design in a 1984 book, *The Art of Computer Game Design*.

That same year, the video game industry went into a deep recession, taking down Atari and leaving Crawford unemployed. He became a freelance game designer, learned how to program the new Apple Macintosh computer, and in 1985 created what would become his most successful game: Balance of Power. While most game designers until then had focused on war, Crawford's new game featured diplomacy. Set in the cold war milieu of NATO, the Soviet bloc, and nonaligned countries, the nations in Balance of Power formed complex, ever-changing relationships based on their different interests. The game provided a great deal of detail and was absorbing to players who appreciated history and current affairs. It sold a remarkable 250,000 copies. Crawford wrote a book (also titled *Balance of Power*) describing how he designed the game and exposing some of its inner workings.

In 1987, Crawford tried to extend his design skills into a new level of gaming. His game Trust and Betrayal extended the sophisticated relationships found in Balance of Power from the level of nations to that of interpersonal relationships. Each person in the game could take on a number of emotional states, as well as remembering how he or she had been treated earlier by other characters. While innovative and intriguing, Trust and Betrayal seemed to be too far a stretch for most gamers, and did not become a commercial success. (Interestingly, the current game The Sims, with a less sophisticated personality model and elaborate graphics, seems to be quite successful.) In 1989 Crawford also wrote a revised version of Balance of Power that reflected the approaching end of the cold war and a world no longer dominated by the clash between two superpowers and their allies.

At the start of the 1990s, Crawford continued refining earlier game systems and developing new ideas. Guns and Butter put the player in the role of a macroeconomic planner having to create a budget that balances military and social needs, while Balance of the Planet challenges the player to provide resources while saving the environment. Together with Balance of Power, these games provided superb tools for high school and college students in social studies, political science, economics, and other classes to simulate real-world decision making. Meanwhile, Crawford's efforts in the traditional war game field continued with Patton Strikes Back, which attempted (not entirely successfully) to make a war game that was simple enough to appeal to non-war-gamers while remaining a good conflict simulation.

Crawford's game designs have often been controversial because of their emphasis on the "soft" aspects of character interaction and dynamics as opposed to the graphic action that seemed to be demanded by most gamers. In an *Omni* magazine on-line interview, Crawford looked back, noting:

> There were two fundamental "killer problems" for my work. The first was the emphasis on cosmetic demonstration. People who had just bought the latest, greatest video or audio card wanted games that fully utilized these capabilities. This biased game design toward the cosmetic.
>
> The second killer problem was the testosterone-soaked nature of computer games. There is nothing intrinsic in the medium that demands this, but the audience drifted in this direction, and so character interaction got lost in the demand for louder booms and brighter explosions.

In addition to designing games Crawford has been very influential in helping to develop game design as a recognized discipline and to foster the community of game designers. He created, edited, and largely wrote *The Journal of Computer Game Design*, as well as organizing the first seven annual meetings of the Computer Game Developers' Conference. He also has lectured on computer game design at many universities around the world. As he notes in an on-line interview with the website GameGeek Peeks:

> I'm a communicator; I thrive on getting through to people one way or the other. I enjoy public speaking, I enjoy writing, and I enjoy using the computer to express ideas. Computer games people, though, are coming from a completely different angle. Their passion is for the technology; they want to make it jump through hoops in fascinating new ways. So they create games that show off great new hoop-jumping tricks, but they don't have anything to say in their games. My work doesn't jump through any hoops—it's about what I have to say.

Crawford's current project is an interactive storytelling system that he calls the Erasmotron. He has described it as a system based upon characters that can perform actions and have repertoires of responses to the actions of other characters. The "story builder" defines the character attributes and responses. The player then enters that world with a character and experiences the interaction as a unique story.

Crawford's latest book is *Understanding Interactivity*, which he believes sums up the insights gained in more than 20 years of game design. Of course, the theory of interactivity as discussed by Crawford is applicable to many kinds of software other than games. Crawford's seminal work has been noted in many retrospectives on the history of computer games appearing in such magazines as *Computer Gaming World*.

Further Reading

Crawford, Chris. *The Art of Computer Game Design*. Berkeley, Calif.: Osborne-McGraw-Hill, 1984.

———. *The Art of Interactive Design*. San Francisco: No Starch Press, 2002.

———. *Balance of Power: International Politics as the Ultimate Global Game*. Redmond, Wash.: Microsoft Press, 1986.

GameGeek Peeks. "Interview with Chris Crawford." Available on-line. URL: http://mono211.com/gamegeekpeeks/chrisc.html. Downloaded on October 31, 2002.

Omni. "Prime Time Replay: Chat with Chris Crawford." Available on-line. URL: http://www.omnimag.com/archives/chats/em080497.html. Posted on August 4, 1997.

"Welcome to Erasmatazz." Available on-line. URL: http://www.erasmatazz.com. Updated on October 30, 2002.

Cray, Seymour
(1925–1996)
American
Engineer, Inventor

Seymour Cray was an innovative computer designer who pioneered the development of high-performance computers that came to be called supercomputers. Cray was born in Chippewa Falls, Wisconsin, on September 28, 1925, and attended high school there. After serving in World War II as an army electrical technician, Cray went to the University of Minnesota and earned a B.S. degree in electrical engineering and then an M.S. degree in applied mathematics in 1951. (This is a common background for many of the designers who had to combine mathematics and engineering principles to create the first computers.)

Cray then joined Engineering Research Associates (ERA), one of a handful of companies that sought to commercialize the digital computing technology that had been developed during and just after the war. Cray soon became known for his ability to grasp every aspect of computing from logic circuits to the infant discipline of software development. For his first project, he helped design the ERA 1103, one of the first computers specifically designed for scientific applications.

When ERA and its competitor, the Eckert-Mauchly Computer Company, were bought by Remington Rand, Cray became the chief designer of Univac, the first commercially successful computer. He also played a key role in the design of computer circuits using transistors, which were starting to become commercially available. One of Cray's other special projects was the Bogart, a computer designed for cryptography applications for the new National Security Agency (NSA).

In 1957, however, Cray and two colleagues struck out on their own to form Control Data Corporation (CDC). Their CDC 1604 was one of the first computers to move from vacuum tubes to transistors. This powerful machine used 48-bit words and was particularly good for the "number crunching" needed for scientific problems as well as engineering applications such as aircraft design.

Seymour Cray built sleek supercomputers that looked—and performed—like something out of science fiction. Scientists relied on them to understand what was going on in splitting atoms or folding proteins, and engineers used them to design new generations of aircraft. *(Courtesy Cray Corporation)*

Cray's next achievement was the CDC 6600, an even more powerful 60-bit machine released in 1963 and considered by many to be the first "supercomputer." The machine packed together 350,000 transistors and generated so much heat that it required a built-in Freon cooling system. CDC seemed to be well on its way to achieving dominance in the scientific and engineering computing fields comparable to that of IBM in business computing. Indeed in a memo to his staff, IBM chief executive officer THOMAS WATSON JR. reacted with dismay to Cray's achievement:

> I understand that in the laboratory developing the [CDC 6600] system there are only 34 people including the janitor. Of these 14 are engineers and 4 are programmers . . . Contrasting this modest effort with our vast development activities, I fail to understand why we have lost our industry leadership position by letting someone else offer the world's most powerful computer.

Cray believed the very compactness of his team was a key to its success, and suggested that "Mr. Watson [had] answered his own question."

In 1965, IBM began the ASC project in an attempt to regain leadership in supercomputing. By 1969, however, the IBM project had bogged down and was canceled. IBM then announced a forthcoming computer, the Model 90. This machine, however, existed only as a set of schematics, and CDC believed that IBM was unfairly using the announcement to deprive CDC of sales. CDC sued IBM, and the Department of Justice negotiated a consent decree under which IBM paid CDC more than $600 million to compensate it for lost revenue.

By the late 1960s, Cray had persuaded CDC to provide him with production facilities within walking distance of his home in Chippewa Falls. There he designed the CDC 7600, ushering in the next stage of supercomputing. However CDC disagreed with Cray about the commercial feasibility of building the even more powerful computers on Cray's drawing board. In 1972, therefore, Cray formed his own company, Cray Research, Inc.

Cray's reputation as a computer architect was so great by then that investors flocked to buy stock in his company. The design of his revolutionary Cray-1 (released in 1976) resembled the monolith from the science fiction movie *2001: A Space Odyssey*. The Cray-1 was the first supercomputer to use "vector" or parallel processing, where tasks can be assigned to different processors to speed up throughput (the volume of data being processed). Cray then left his position as president of CDC to devote himself full time to an ambitious successor, the Cray-2. This machine would not be completed until 1985 because of problems with the cutting-edge chip technology being used to pack ever-growing numbers of chips together. The Cray-2 would have an unprecedented 2 gigabytes (2 billion bytes) of internal memory and the ability to process 1.2 billion floating point operations per second.

Meanwhile, however, CDC executives had become worried about the delay and the potential loss of revenue and assigned another designer, Steve Chen, to design an upgrade for the original Cray-1.

While costing millions of dollars apiece, the Cray supercomputers made it possible to perform simulations in atomic physics, aerodynamics, and other fields that were far beyond the capabilities of earlier computers. However, Cray Research and its spinoff Cray Computer Corporation ran into financial problems and were bought by Silicon Graphics (SGI) in 1996.

Cray liked to stress the importance for designers of getting simple things right when beginning a new design. When asked what computerized design aids he had used in developing the Cray-1, he replied that he preferred a number three pencil and a Quadrille notepad. And when he was told that Apple had bought a Cray supercomputer to help them design a new Macintosh, Cray replied that he had just bought a Mac to help him design the next Cray.

By the mid-1990s, however, Cray had run into trouble. The company had spent $300 million on developing the Cray-3 without being able to make it commercially feasible. Cray decided to start over with the even more powerful

Cray-4 design (with 64 processors), but the company had to file for bankruptcy in 1995.

The following year, Cray was involved in an auto accident that resulted in his death on October 5, 1996. He left a legacy of powerful ideas and machines that defined for decades the frontiers of computing. Cray received many honors including the Institute of Electrical and Electronics Engineers (IEEE) Computer Society Pioneer Award (1980) and the Association for Computing Machinery/IEEE Eckert-Mauchly Award (1989).

Further Reading

Bell, Gordon. "A Seymour Cray Perspective." Available on-line. URL: http://research.microsoft.com/users/gbell/craytalk. Posted on January 25, 1998.

Breckenridge, Charles W. "A Tribute to Seymour Cray." Available on-line. URL: http://www.cgl.ucsf.edu/home/tef/cray/tribute.html. Posted on November 19, 1996.

Murray, C. J. *The Supermen: The Story of Seymour Cray and the Technical Wizards behind the Supercomputer*. New York: New York: John Wiley, 1997.

Smithsonian Institute. National Museum of American History. "Seymour Cray Interview." Available on-line. URL: http://americanhistory.si.edu/csr/comphist/cray.htm. Posted on May 9, 1995.

D

Davies, Donald Watts

(1924–2000)
British
Computer Scientist

When a user sends an e-mail, the message is broken up into small blocks of data called packets. The packets are routed over the network, then handed off from computer to computer until they reach their ultimate destination, where they are reassembled into the complete message. This versatile communications mechanism, called packet-switching, was just one of the ideas proposed and developed by an equally versatile British computer scientist named Donald Watts Davies.

Davies was born June 7, 1924, in Treorchy in the Rhondda valley in Wales. His father, a clerk at a coal mine, died only a few months after Davies's birth. The family then moved to Portsmouth, England, where Davies attended elementary and high school.

Young Davies developed a keen interest in physics, and earned his B.Sc. degree at Imperial College, London University, in 1943. He added a mathematics degree in 1947. A standout honor student, Davies received the Lubbock Memorial Prize as the leading mathematician at the university. During World War II Davies also worked in the British nuclear research program at Birmingham University.

During his last year at London University, Davies attended a lecture that would shape his future career. He learned about a new project at the National Physical Laboratory (NPL) that would build one of Britain's first stored-program digital computers, the ACE. Davies quickly signed up for the project, which was led by one of the world's preeminent computer scientists, ALAN TURING. Together with colleagues such as Jim Wilkinson and Ted Newman, Davies played an important role in translating Turing's revolutionary computer architecture into a working machine, the ACE, which ran its first program in 1950.

Using the ACE and its successors, Davies developed many interesting computer applications, including one of the first automated language translation programs. (Responding to cold war needs, the program translated technical documents from Russian into English.) Davies also pioneered the use of computers to simulate real-world conditions, developing simulations for road traffic and for the warning and escape systems used in coal mines.

In 1963, Davies became technical manager for the Advanced Computer Technology Project for the British Ministry of Technology (similar to the U.S. Advanced Research Projects Agency, or ARPA). In that role, he developed new computer hardware, including a multiple-processor computer system and a new way to organize and retrieve data in file systems. In 1966, Davies was promoted to director of the NPL's Autonomics Division, which he transformed into a wider-ranging computer science research enterprise.

Davies's interest then turned to the new field of networking, or communications links between computers. He proposed the packet-switching idea as a way to route messages safely and efficiently between computers, and his ideas were well received by the British Post Office (which also ran the telephone system). In 1967, at a conference sponsored by the Association for Computing Machinery (ACM) in Gatlinburg, Tennessee, Davies presented his packet-switching ideas. A similar scheme had been proposed earlier by American communications researcher PAUL BARAN, but it had the rather unwieldy name of "distributed adaptive message block switching" and focused mainly on routing phone messages. Davies, on the other hand, focused specifically on data transmission and provided the simpler term "packet-switching." Davies' presentation convinced LAWRENCE ROBERTS to adopt the system for the proposed ARPANET, which would later become the Internet.

Davies's interests ranged widely over other aspects of computer science. During the 1970s and 1980s, he worked on applications such as voice recognition, image processing, data security, and encryption. In 1984, Davies left the NPL and became a private consultant, helping banks and other institutions devise secure data communications systems.

Donald Davies died on May 28, 2000, while on a trip to Australia. In recognition of his achievements, he was made a Commander of the British Empire (CBE) in 1983 and a fellow of the Royal Society (Britain's premier scientific group) in 1987. (Davies noted that he was particularly pleased at being able to sign the same register that bore the signature of Isaac Newton.) Davies also received the John von Neumann Award in 1983, as well as other honors in the field of computer science.

Further Reading

"Donald W. Davies CBE, FRS." Available on-line. URL: http://www.thocp.net/biographies/davies_donald.htm. Updated on July 30, 2001.

Lee, J. A. N. *Computer Pioneers*. Los Alamitos, Calif.: IEEE Computer Society Press, 1995.

Schofield, Jack. "Donald Davies: Simple Idea That Made the Internet Possible." *Guardian Unlimited*. Available on-line. URL: http://www.guardian.co.uk/Archive/Article/0,4273,4024597,00.html. Posted on June 2, 2000.

Dell, Michael

(1965–)
American
Entrepreneur

In the early days of personal computing, most individuals (as opposed to organizations) bought Personal Computers (PCs) from storefront businesses. There were "name brand" PCs from IBM, Compaq, Apple, and other major companies, and there were "clones" put together in the back rooms of computer shops. Clones offered lower prices and more flexibility in components such as hard drives, but quality was more variable. Getting technical support and repair service could also be a problem if the shop went out of business.

Many industry experts had believed that computers were too complicated to sell by mail, and that users would insist on the assurance that came from dealing personally with a store. A Texas entrepreneur named Michael Dell proved them wrong. His company is now the biggest seller of PCs, which are mainly sold by phone and over the Web.

Michael Dell was born on February 23, 1965, in Houston, Texas. As he would recall later in an interview with *Fortune* magazine, he always knew he wanted to run a business. As a 12-year-old he worked as a dishwasher and water boy at a Chinese restaurant. He used his earnings to build a stamp collection. Soon he became a stamp dealer, selling stamps by mail and running a stamp auction.

By age 16, Dell had saved enough to buy an Apple II, one of the first generation of desktop computers that were revolutionizing information processing. He used the computer to boost his next business, selling newspaper subscriptions. Dell realized that people who had just got married were likely to be moving and setting up new households, and thus likely to want to start newspaper subscriptions. Dell went to the local courthouse and obtained lists of marriage license applicants and their addresses, and used his computer to generate letters with subscription offers. As a high school senior, Dell made enough money selling subscriptions to buy an $18,000 BMW. This ability to sense new market opportunities and harness technology to serve them would be characteristic of Dell's career as an entrepreneur.

Dell's father was an orthodontist and his mother a stockbroker. They harbored professional ambitions for their son, specifically wanting him to be a doctor, and he obliged by enrolling in premedical school at the University of Texas in Austin in 1983. But Dell's heart was not in medicine, and he was soon spending his spare time on a new business enterprise—selling personal computers.

By the mid-1980s, the IBM PC had become the dominant form of personal computer. However, IBM had lost control of the machine's architecture, which meant that companies ranging from Compaq, Hewlett-Packard, and Radio Shack to "no-name" clone shops could make IBM PC–compatible machines and sell them with copies of Microsoft's MS-DOS operating system.

Dell had tinkered with PCs and followed the industry for some time, and he decided that he could assemble and sell PCs on his own. Using remaindered systems and parts, Dell rebuilt and upgraded PCs to his customers' specifications, selling them door to door and by phone and mail. At a time when computers were usually sold either by big corporations or by dealers who often had little technical knowledge, Dell's direct, personalized service proved to be quite successful.

When Dell's parents found out about his "sideline," they reacted angrily. Eventually, they agreed to let Dell finish the current school year and then sell PCs through the summer. If sales fell, Dell would return to school. In the month before the new semester, Dell sold $180,000 worth of PCs, and that ended his higher education.

Having proven the success of his methods, Dell decided that he could achieve much greater volume of sales—and profits—by building PCs from scratch, buying motherboards, chips, hard drives, and other components direct from the manufacturers. His first machine was called the Turbo. By 1986, Dell's annual sales had reached $34 million. The following year the Dell Computer Company was born. In 1988, the company went public, raising $30 million from selling stock. By then, industry observers and magazines such as *Business Week* had begun to write about the fast-growing company. In 1991, annual sales hit $546 million.

Dell's business model relies on several principles. Computers are sold directly to corporate or individual users. A user, for example, can log onto the Dell website and select a featured configuration (perhaps with some changes), or select all the features individually, including the amount of memory, hard drive size, CD or DVD drives, and video and audio cards. Because a machine is not assembled until it is ordered, and Dell uses a "just in time" supply system, cash is not tied up in parts or inventory. Systems are built, tested, and shipped within a few days to a week or so after ordering. User problems are minimized by using consistent quality components and assembly, and by providing prompt technical support, including overnight or even same-day service by technicians. It is this sort of service that has overcome the reluctance of consumers to buy computers by mail.

By 1992, Dell seemed to be at the top of his world. Annual sales reached $2 billion, and Dell had become the youngest chief executive officer (CEO) to head a Fortune 500 company. But then the company "hit the wall." Sales were growing faster than the finely tuned infrastructure could handle, and customers began to complain of delays and bad service. Chasing after growth, Dell had also begun to sell PCs through stores such as Wal-Mart and CompUSA. This was a completely different business model from the company's successful direct consumer sales, and brought new complications that distracted Dell from his core business.

In addition, Dell had not sufficiently realized the importance of the growing laptop computer market, and when the company finally began to develop a line of laptops, they used the less powerful 386 chip instead of the newer 486. The machines sold sluggishly. Finally, Dell decided to scrap his laptop plans and start over.

In an interview with *Fortune* magazine, Dell recalled how things had gotten away from him:

> When you are growing a business you really have little way of determining what the problems are. You had different parts of the company believing they were making their plan, but when you rolled up the results of the company you had a big problem. It was symptomatic of not understanding the relationship between costs and revenues and profits within the different lines of the business.

During 1993, Dell retrenched, bringing in experienced executives to help streamline the company, focusing on profitability rather than raw growth. The company abandoned store sales to focus exclusively on direct sales, and developed a competitive laptop called the Latitude. They also entered the growing market for network servers, ensuring that corporate customers could meet all their computing needs with Dell products. The result was a turnaround: By 1994, Dell had returned to profitability, earning $149 million that year.

Starting in 2001, the computer industry faced a serious economic decline. While Internet companies, or "dot-coms," were hardest hit, PC manufacturers such as Dell also faced a declining demand for PCs. By the second quarter of 2001, Dell dominated the industry, ahead of number-two Compaq and other leading competitors such as Gateway and IBM. While Dell has held onto its commanding market share, its revenues have declined. (Dell personally took a "pay cut" from $225 million in fiscal 2001 to $100 million in 2002.)

Dell has responded to the decline in revenue by pushing for even greater efficiency in the company's manufacturing and other operations. For example, its Parmer Lane, Texas, plant assembled 120 computers an hour in 1995, but by 2001, 700 PCs an hour were coming off the assembly line. Dell has also "outsourced" some operations (such as technical support) to cheaper foreign labor. Since Dell cannot know when demand for PCs will pick up again, he has relentlessly used efficiency improvement and the resulting price cuts to maintain or increase its market share. He seems confident that even the recent merger of Compaq and printer giant Hewlett-Packard (HP) will not keep Dell Computer from surviving—and thriving under today's harsh conditions. (Indeed, in early 2003 Dell announced it would start selling its own name-brand printers.)

Dell has been a successful entrepreneur by any measure. His net worth is somewhere between $3 billion and $4 billion. In 2001, he was chosen CEO of the year by *Chief Executive* magazine. At age 36, he was the youngest executive ever to win the award. Dell also received an Internet Industry Leader award in 2000 from the U.S. Internet Council.

As he told Helen Thorpe of *Texas Monthly* in 1997, Dell has no plans of retiring young: "I don't imagine many jobs could be more fun than this. . . . I think if you left the computer industry after having been in it for a long time, you'd be incredibly bored, and your brain would atrophy."

Further Reading

Jager, Rama Dev, and Rafael Ortiz. *In the Company of Giants: Candid Conversations with the Visionaries of the Digital World.* New York: McGraw-Hill, 1997.

"The Tech Slump Doesn't Scare Michael Dell." *Business Week*, April 16, 2001, p. 48.

Thorpe, Helen. "Michael Dell: From Boy Wonder to Grown-up CEO, He's Changed Personal Computing Forever." *Texas Monthly*, September 1997, pp. 117ff.

Dertouzos, Michael
(1936–2001)
Greek/American
Computer Scientist

Born on November 5, 1936, in Athens, Greece, Michael Dertouzos spent an adventurous boyhood accompanying his father (an admiral) in the Greek navy's destroyers and submarines. Showing a technical bent, the boy became interested in Morse code, shipboard machinery, and mathematics. At the age of 16, Dertouzos became fascinated by an article about Claude Shannon's work in information theory and a project at the Massachusetts Institute of Technology (MIT) that sought to build a mechanical robot mouse. He quickly decided that he wanted to come to America to study at MIT.

The hardships of World War II brought many people in Athens to the brink of starvation. After the war, however, Dertouzos was finally able to go to America to continue his higher education. Dertouzos received a Fulbright scholarship that placed him in the University of Arkansas, where he earned his bachelor's and master's degrees while working on acoustic-mechanical devices for the Baldwin Piano company. He was then able to fulfill his boyhood dream by receiving his Ph.D. in electrical engineering from MIT in 1964. He promptly joined the MIT faculty, where he would remain for his entire career. Starting in 1974, he served as director of MIT's Laboratory for Computer Science (LCS). This lab became a hotbed of new ideas in computing, including computer time-sharing, Ethernet networking, and public-key cryptography.

Dertouzos believed that the key to people getting the most out of computers was not having many people sharing one big machine, but having many machines linked together in networks. As he explained to an interviewer:

> I'd never liked time-sharing. Maybe it was the Greek in me, but it seemed like a socialist kind of sharing, with central control, like forcing everyone to ride a bus as opposed to driving a personal automobile. And I'd never liked the information-utility notion . . . Information isn't natural gas. It doesn't come from one place, or a few places. It comes from all over. It's more of a commodity. . . . So in 1976 or so I was looking for a metaphor of how the machines [in an online information system] would interact with each other. Being Greek, I thought of the Athens flea market, where I used to spend every Sunday, and I envisioned an on-line version: a community with very large numbers of people coming together in a place

Michael Dertouzos helped create the future at MIT's Laboratory for Computer Science. He also wrote vividly about what it will be like to live in a future in which people wear computers and they are "as natural as the air we breathe." *(© Reuters NewMedia, Inc./CORBIS)*

where they could buy, sell and exchange information.

In pursuing his vision, Dertouzos also forged links outside the academic world. Combining theoretical interest with an entrepreneur's eye on market trends, he started a small company called Computek in 1968. It made some of the first "smart terminals" which included their own processors.

In the 1980s, Dertouzos began to explore the relationship between developments in information processing and the emerging information marketplace. However, he noted that the spectacular growth of the information industry took place against a backdrop of the decline of American manufacturing. Dertouzos's 1989 book *Made in America* suggested ways to revitalize American industry.

During the 1990s, Dertouzos brought MIT into closer relationship with the visionary designers who were creating and expanding the World Wide Web. After Dertouzos's death, TIM BERNERS-LEE, inventor of the Web, recalled in a tribute that

> Michael met me first in Switzerland, and later invited me to the LCS. He was totally enthusiastic about the idea of the Web, of the Consortium, and of making the whole thing an international collaboration. Michael had been promoting the vision of the Information Marketplace long before the Web came along. He put in a huge amount of effort to make things happen, but always did it with great warmth as well as strength.

Dertouzos combined technological, teaching, and managerial skills to smooth the path for complex projects. An MIT colleague, FERNANDO J. CORBATÓ, remarked that

> Michael had a broad understanding of technology and a teacher's knack for explaining ideas. One direction in which this shone was his skill in interfacing with government sponsors of research. He was skillful in evoking the best research ideas from within the laboratory; he could educate without being condescending,

Dertouzos was dissatisfied with operating systems such as Microsoft Windows and with popular application programs. He believed that their designers made it unnecessarily difficult for users to perform tasks, and spent more time on adding fancy features than on improving the basic usability of their products. In 1999, Dertouzos and the MIT LCS announced a new project called Oxygen. Working in collaboration with the MIT Artificial Intelligence Laboratory, Oxygen was intended to design new interfaces and systems to make computers "as natural a part of our environment as the air we breathe."

As a futurist, Dertouzos tried to paint vivid pictures of possible future uses of computers in order to engage the general public in thinking about the potential of emerging technologies. His 1995 book *What Will Be* paints a vivid portrait of a pervasively digital environment in the near future. His imaginative future is based on actual MIT research, such as the design of a "body net," a kind of wearable computer and sensor system that would allow people to not only keep in touch with information but to communicate detailed information with other people similarly equipped. This digital world will also include "smart rooms" and a variety of robot assistants, particularly in the area of health care.

However, *What Will Be* and his 2001 publication *The Unfinished Revolution* are not unalloyed celebrations of technological wizardry. Dertouzos has pointed out that there is a disconnect between technological visionaries who lack understanding of the daily realities of most people's lives, and humanists who do not understand the intricate interconnectedness (and thus social impact) of new technologies. "We made a big mistake 300 years ago when we separated technology and humanism," Dertouzos told an interviewer for *Scientific American*. "It's time to put the two back together." Dertouzos's lifelong hobbies of sailing and woodworking also expressed his interest in finding trends and shaping them toward new goals.

Dertouzos received an Institute of Electrical and Electronics Engineers Fellowship and was awarded membership in the National Academy of Engineering. He died on August 27, 2001, after a long bout with heart disease. He was buried in Athens near the finish line for the Olympic marathon.

Further Reading

Dertouzos, Michael. *What Will Be: How the New World of Information Will Change Our Lives*. San Francisco, Calif.: HarperSanFrancisco, 1997.

———. *The Unfinished Revolution: Human-Centered Computers and What They Can Do For Us*. New York: HarperBusiness, 2001.

Leutwyler, Kristin. "What Will Really Be." *Scientific American*, July 1997, pp. 28ff.

"Remembering Technology's Humanist." *Technology Review* (MIT). Available on-line. URL: http://www.techreview.com/articles/dertouzos090601.asp. Posted on September 6, 2001.

Schwartz, John. "Michael L. Dertouzos, 64, Computer Visionary, Dies," *New York Times*, August 30, 2001, p. B9.

Diffie, Bailey Whitfield

(1944–)
American
Mathematician, Computer Scientist

Bailey Whitfield Diffie was born on June 5, 1944, in the borough of Queens, New York City. The following day was D-day, the beginning of the end of World War II. During that war, cryptography—the making and breaking of codes—had come of age. During the cold war years that would follow, cryptography would become a vast, secret arena, the province of contending governments. It would remain so until Whitfield Diffie (who did not use his original first name) played a key role in developing public key cryptography. This invention would eventually put the privacy-guarding powers of cryptography into the hands of millions of ordinary people, much to the consternation of espionage and law enforcement agencies.

Diffie's parents were both intellectuals. His father taught Iberian (Spanish) history and culture at City College in New York, and his mother had written papers on French history. Liberal in outlook, both parents seemed comfortable with the fact that Whit was not following the usual course of childhood development. Although he seemed to be very bright, he showed no desire to learn to read, preferring to listen endlessly as stories were read to him. Finally, when he was 10, he suddenly decided to read a book called *The Space Cat*, and having mastered reading, went on to devour the Oz books.

One day that same year, Diffie's fifth-grade teacher introduced the class to cryptography, explaining how to solve a substitution cipher—a simple system in which each letter of the alphabet in a message is replaced by a different letter. Diffie was fascinated by the mysterious process by which messages could be hidden from prying eyes. But as he later recalled to author Steven Levy, "I never became a very good puzzle solver, and I never worked on solving codes very much then or later." Rather, he was drawn to the challenge of creating codes (cryptosystems) that would protect privacy and resist cracking.

Although he was an indifferent high school student who barely qualified for graduation, he scored so high on standardized tests that he won admission to the University of California, Berkeley in 1962. He studied mathematics there for two years, but then transferred to the Massachusetts Institute of Technology (MIT) and obtained his B.S. degree in mathematics in 1965.

In the early 1960s, MIT was a hotbed of innovation in computers. The earliest hackers (adventurous but largely benign explorers of the computing frontier) were taking advantage of time-sharing and the new, relatively inexpensive minicomputers such as the PDP-1. Diffie was reluctant at first to become involved with computer programming. As he told Levy, "I thought of computers as very low class. I thought of myself as a pure mathematician and was interested in partial differential equations and topology and things like that."

However, by 1965 the Vietnam War was well under way. Diffie, whose childhood interest in military matters had been transformed into a peace-oriented hippie lifestyle, took a job at Mitre Corporation, a defense contractor, as a way to escape the draft. At Mitre, Diffie plunged into computer programming, helping create Mathlab, a program that allowed mathematicians to not merely calculate with a computer, but also to manipulate mathematical symbols to solve equations. (The program would eventually evolve into Macsyma, a software package used widely in the mathematical community.) Flexible work arrangements allowed Diffie to also spend time in the MIT artificial intelligence lab, where freewheeling hackers under the direction of artificial intelligence pioneer MARVIN MINSKY were creating more intelligent software.

By the early 1970s, Diffie had moved to the West Coast. He worked at the Stanford Artificial Intelligence Laboratory (SAIL) where he met LAWRENCE ROBERTS, who was head of information processing research for ARPA, the Defense Department's research agency. His main project was the creation of the ARPANET, the computer network that later evolved into the Internet.

Roberts was interested in providing security for the new network, and (along with artificial intelligence pioneer JOHN MCCARTHY) helped revive Diffie's dormant interest in cryptography. Diffie began to read books such as David Kahn's *The Codebreakers*, a popular and richly detailed history of cryptography, as well as some of the more technical works that were publicly available.

Diffie learned that modern cryptography operated on the fly, with a computer program called a black box generating a "key stream," a series of digits that was logically combined with the stream of plain text to generate the encrypted message, or ciphertext.

By 1974, Diffie had learned that IBM was developing a more secure cipher system, the DES (data encryption standard), under government supervision. What this really meant, Diffie discovered, was the supervision of the National Security Agency (NSA), the largest and most secretive of government spy agencies. Diffie became frustrated with the way the NSA doled out or withheld information on cryptography, making independent research in the field very difficult. He therefore traveled widely, seeking both relevant information and informed people.

Diffie found one such person in a colleague, Martin Hellman, a Stanford professor who had also been struggling on his own to develop a better cryptosystem. They decided to pool their ideas and efforts. Finally, one afternoon in May 1975 while Diffie was house-sitting for John McCarthy, the pieces came together. As Diffie later told Steven Levy, "The thing I remember distinctly is that I was sitting in the living room when I thought of it the first time and then I went downstairs to get a Coke and I almost lost it," he says. "I mean, there was this moment when I was thinking about something. What was it? And then I got it back and didn't forget it."

Diffie's brainstorm became known as public key cryptography. It combined two important ideas that had already been discovered by other researchers. The first idea was the "trap-door function"—a mathematical operation that is easily performed "forward" but that is very hard to work "backward." Diffie realized, however, that a trap-door function could be devised that could be worked backward easily—if the person had the appropriate key.

The second idea was that of key exchange. In classical cryptography, a single key is used for both encryption and decryption. In such a case, it is absolutely vital to keep the key secret from any third party, so arrangements have to be made in advance to transmit and protect the key.

Diffie, however, was able to work out the theory for a system that generates pairs of mathematically interrelated keys: a private key and a public key. Now each participant publishes his or her public key, but keeps the corresponding private key secret. If a user wants to send an encrypted message to someone, the user obtains that person's public key from the electronic equivalent of a phone directory. The resulting message can be decrypted only by the intended recipient, who uses the corresponding secret, private key.

The public key system can also be used in reverse, where it becomes a form of "digital signature" for verifying the authenticity of a message. Here a person creates a message encrypted with his or her private key. Since such a message can be decrypted only by using the corresponding public key, any other person can use that key (together with a trusted third-party key service) to verify that the message really came from its purported author.

Diffie and Hellman's 1976 paper in the *IEEE Transactions on Information Theory* began boldly, with the statement, "We stand today on the brink of a revolution in cryptography." This paper soon came to the attention of three researchers who would create a practical implementation called RSA (for Rivest, Shamir, and Adelman).

Through the 1980s, Diffie, resisting urgent invitations from the NSA, served as manager of Secure Systems Research for the phone company Northern Telecom, designing systems for managing security keys for packet-switched data communications systems (such as the Internet).

In 1991 Diffie was appointed Distinguished Engineer for Sun Microsystems, a position that has left him free to deal with cryptography-related public policy issues. The best known of these issues has been the Clipper Chip, a proposal that all new computers be fitted with a hardware encryption device that would protect users' data but include a "back door" that would allow the government to decrypt data, presumably after obtaining a court order. Along with many civil libertarians and privacy activists, Diffie does not believe users should have to trust largely unaccountable government agencies for the preservation of their privacy. Their opposition was strong enough to scuttle the Clipper Chip proposal by the end of the 1990s. Another proposal, using public key cryptography but having a third party "key escrow" agency hold the keys for possible criminal investigation, also has fared poorly.

In testimony before Congress, Diffie summed up his work by saying, "My feeling was that cryptography was vitally important for personal privacy, and my goal was to make it better known. I am pleased to say that if I have succeeded in nothing else, I have achieved that goal."

Diffie has received a number of awards for both technical excellence and contributions to civil liberties. These include the Institute of Electrical and Electronics Engineers (IEEE) Information Theory Society Best Paper Award (1979), the IEEE Donald Fink Award (1981), the Electronic Frontier Foundation Pioneer Award (1994), and the National Computer Systems Security Award (1996), given by the National Institute of Standards and Technology and the NSA.

Further Reading

Diffie, Whitfield. "Interview with Whitfield Diffie on the Development of Public Key Cryptography." Conducted by Franco Furger; edited by Arnd Weber, 1992. Available on-line. URL: http://www.itas.fzk.de/mahp/weber/diffie.htm. Updated on January 16, 2002.

Diffie, Whitfield, and Susan Landau. *Privacy on the Line: The Politics of Wiretapping and Encryption*. Cambridge, Mass.: MIT Press, 1998.

Kahn, David. *The Codebreakers: The Story of Secret Writing*. Rev. ed. New York: Scribner, 1996.

Levy, Steven. *Crypto: How the Code Rebels Beat the Government: Saving Privacy in the Digital Age*. New York: Viking Penguin, 2001.

Dijkstra, Edsger
(1930–2002)
Dutch/American
Computer Scientist

By the late 1960s, operating systems and applications programs had become complex affairs consisting of hundreds of thousands or even millions of lines of code. The simple stand-alone programs that followed a basic "input, process, output" sequence had given way to systems in which many procedures or modules depended on one another in complex ways.

The discipline of computer science was just beginning to be formalized and its practitioners were starting to apply theory systematically to the design of programs. The first generation of high-level computer languages, such as FORTRAN and COBOL, had given programmers a crucial ability to represent data and operations symbolically. Beginning with Algol, the next generation of languages would offer cleaner program structure and improved definition and organization of data and program modules. In the 1970s, Edsger Dijkstra would become one of the leading advocates of "structured programming" and help spur the creation of new langauges and programming practices.

Edsger Dijkstra's ideas about structured programming helped develop the field of software engineering, enabling programmers to organize and manage increasingly complex software projects. *(Photo courtesy of the Department of Computer Sciences, UT Austin)*

Dijkstra was born in Rotterdam, the Netherlands, in 1930 into a scientific family. His mother was a mathematician and his father was a chemist. He received intensive and diverse intellectual training, studying ancient Greek, Latin, several modern languages, biology, mathematics, and chemistry. While majoring in physics at the University of Leiden, in 1951 he attended a summer school at Cambridge University in England that kindled what soon became a major interest in programming.

Returning to the Netherlands the following year, Dijkstra obtained a position at the Mathematical Center in Amsterdam, while continuing to pursue a physics degree. There he worked with Bram J. Loopstra and Carel S. Scholten on a computer called ARMAC. His main task was to write the detailed, precise operational specifications that would guide the engineers who were designing the machine.

Dijkstra's work soon took him further into the design of the machine's control software—what would become known as an operating system. He developed software that could handle interrupts—that is, signals from peripheral hardware such as card readers or disk drives that needed attention from the central processor. By then he had abandoned physics in favor of the emerging discipline that would become computer science, and wrote his doctoral dissertation on interrupt handling. He received his Ph.D. in 1959 from the University of Amsterdam.

Meanwhile, Dijkstra had discovered a fundamental algorithm (or mathematical "recipe") for finding the shortest path between two points.

He applied the algorithm to the practical problem of designing electrical circuits that used as little wire as possible, and generalized it into a procedure for traversing treelike data structures.

In 1962, Dijkstra became a professor of mathematics at the Technical University in Eindhoven. Dijkstra began to explore the problem of communication and resource-sharing within computers. He developed the idea of a control mechanism called a semaphore. Like the railroad signaling device that allows only one train at a time to pass through a single section of track, the programming semaphore provides *mutual exclusion*, ensuring that two processes do not try to access the same memory or other resource at the same time.

Another problem Dijkstra tackled involved the sequencing of several processes that are accessing the same resources. He found ways to avoid a deadlock situation, in which one process had part of what it needed to finish a task but was stuck because the process holding the other needed resource was in turn waiting for the first process to finish. His algorithms for allowing multiple processes (or processors) to take turns gaining access to memory or other resources would become fundamental for the design of new computing architectures.

In 1973, Dijkstra immigrated to the United States, where he became a research fellow at Burroughs, one of the major manufacturers of mainframe computers. During this time, he helped launch the "structured programming" movement. His 1968 statement "GO TO Considered Harmful" had criticized the use of that unconditional "jump" instruction because it made programs hard to read and verify. The decade's newer structured languages such as Pascal and C affirmed Dijkstra's belief in avoiding or discouraging such haphazard program flow. Dijkstra's 1976 book *A Discipline of Programming* provided an elegant exposition of structured programming that is still studied today.

Beginning in the 1980s, Dijkstra was a professor of mathematics at the University of Texas at Austin. He held the Schlumberger Centennial Chair in Computer Science. His interests gradually shifted from computer science to mathematical methodology, which finds computer applications in such areas as automatic generating of programs and proving program correctness. In 1972, Dijkstra won the Association for Computing Machinery's Turing Award, one of the highest honors in the field. Dijkstra died on August 6, 2002.

Further Reading

Dijkstra, Edsger Wybe. *A Discipline of Programming*. Upper Saddle River, N.J.: Prentice Hall, 1976.

Shasha, Dennis, and Cathy Lazere. *Out of the Minds: The Lives and Discoveries of 15 Great Computer Scientists*. New York: Springer-Verlag, 1995.

Drexler, K. Eric

(1955–)
American
Engineer, Futurist

Since the 1960s, computer processors have become ever smaller yet more powerful. Today, the equivalent of millions of transistors can be fabricated onto a single silicon chip. But if K. Eric Drexler's provocative vision is correct, technology will soon enter a far smaller realm where molecules and even single atoms become both building blocks and the machines to assemble them.

Drexler was born on April 25, 1955, in Oakland, California, and grew up in the small town of Monmouth, Oregon. A true child of the space age, he became fascinated with astronomy as well as space exploration and colonization. While he was still a teenager, his talent brought him to the Massachusetts Institute of Technology (MIT), where he studied aerospace engineering and computer science. Drexler soon became an active member in the movement to encourage space exploration and colonization, participating in NASA seminars with such visionaries as

Shown behind a model of a molecular "nanomachine," K. Eric Drexler believes that engineers will soon be building tiny devices directly from atoms. Besides leading to the ultimate computers, nanomachines may someday "grow" houses or swim in the bloodstream to eliminate cancers or repair diseased organs. *(© Ed Kashi/CORBIS)*

Gerard K. O'Neill, who was designing and promoting space colonies. In 1975, when he was still only 19, Drexler published his first professional scientific paper, dealing with the possible mining of resources from asteroids.

After getting his bachelor's degree, Drexler continued into the MIT graduate program, where he wrote a master's thesis on space propulsion using solar sails. He also worked at MIT's Space Systems Laboratory.

While part of Drexler's mind peered out into the vastness of space, another part became intrigued by the world of the very small. Drexler became interested in genetic engineering, an exciting field that had begun to emerge in the late 1970s. Around the same time, he also read a talk that the endlessly imaginative physicist Richard Feynman had given in 1959. In it, Feynman suggested that simple machines could be built on a molecular scale and used, among other things, to store vast amounts of information in pinhead-sized spaces.

Starting in the mid-1980s, Drexler began to extend and expand Feynman's work, working out more of the physics and engineering needed to build at the atomic level. He coined the term *nanotechnology* (*nano* is short for *nanometer*, one-billionth of a meter). In addition to writing scientific papers, Drexler introduced nanotechnology to the general public with his 1986 book, *Engines of Creation*. In 1987, he founded an organization called the Foresight Institute, to provide a clearinghouse for nanotech researchers as well as to publicize the field.

In 1991 Drexler received the first MIT Ph.D. ever given in the field of molecular nanotechnology. He also founded the Institute for Molecular Manufacturing as a nonprofit nanotech research organization. That same year, Drexler, Chris Peterson, and Gayle Pergamit published *Unbounding the Future: The Nanotechnology Revolution*.

Although Drexler's predictions that nanotechnology would become widespread in only a decade or two seemed to be overoptimistic, there was steady progress through the 1990s in a variety of fields relating to nanotech capabilities. For example, IBM scientists have been able to spell out the company's name in individual atoms, while genetic and biochemical engineers are gaining an increasingly fine ability to manipulate tiny pieces of DNA and protein molecules.

Nanotechnology clearly has implications for both computer design and robotics. As the limits of existing integrated circuit technology approach, further progress may require that computer components consist of individual molecules or even atoms (the latter involves what has been called quantum computing.) Robots, too, could be made small enough to, for example, travel through the human bloodstream, repairing an aneurysm or cleaning out a blocked vessel.

However, science fiction writers as well as some critics have suggested a possible horrific side to nanotech. One of the goals of nanotech is to create not merely tiny machines but tiny machines that can reproduce themselves or even evolve into new forms of "nano-life." Such machines might be landed on Mars, for example, and spread across the planet, releasing oxygen from the soil and otherwise preparing the red planet for eventual human habitation. Here on Earth, nanomachines might perform the more prosaic but nonetheless valuable task of breaking down oil spills or other forms of pollution into their harmless constituents. But suppose the nanomachines got out of control, possibly mutating and changing their behavior so they "disassemble" everything in sight, including humans and their civilization? While this fear may seem farfetched, Drexler has been careful to educate people about the perils as well as the promises of what might be the ultimate technology of the 21st century.

Drexler's work has gained him honors as well as attention. For example, the Association of American Publishers named Drexler's *Nanosystems: Molecular Machinery, Manufacturing, and Computation* the 1992 Outstanding Computer Science Book. And in 1993, Drexler received the Young Innovator Award given by the Kilby Awards Foundation, named for JACK KILBY, inventor of the integrated circuit.

Further Reading

Drexler, K. Eric. *Engines of Creation: The Coming Era of Nanotechnology*. New York: Anchor Books, 1986. Also available on-line. URL: http://www.foresight.org/EOC/index.html.

———. *Nanosystems: Molecular Machinery, Manufacturing, and Computation*. New York: John Wiley, 1992.

Drexler, K. Eric, Chris Peterson, and Gayle Pergamit. *Unbounding the Future: The Nanotechnology Revolution*. New York: William Morrow, 1991.

Foresight Institute. Available on-line. URL: www.foresight.org. Downloaded on October 31, 2002.

Terra, Richard. "A Visions Profile: Eric Drexler and Molecular Nanotechnology." Available on-line. URL: http://home.earthlink.net/~rpterra/nt/DrexlerProfile.html. Updated on April 2, 1998.

Dreyfus, Hubert
(1929–)
American
Philosopher

As the possibilities for computers going beyond "number crunching" to sophisticated information processing became clear starting in the 1950s, the quest to achieve artificial intelligence (AI) was eagerly embraced by a number of innovative researchers. For example, ALLEN NEWELL, HERBERT A. SIMON, and Cliff Shaw at the RAND Corporation attempted to write programs that could "understand" and intelligently manipulate symbols rather than just literal numbers or characters. Similarly, the Massachusetts Institute of Technology's (MIT's) MARVIN MINSKY was attempting to build a robot that could not only perceive its environment, but in some sense understand and manipulate it.

Into this milieu came Hubert Dreyfus, who had earned his Ph.D. in philosophy at Harvard University in 1964. Dreyfus had specialized in the philosophy of perception (how people derive meaning from their environment) and phenomenology (the understanding of processes). When Dreyfus began to teach a survey course on these areas of philosophy, some of his students asked him what he thought of the artificial intelligence researchers who were taking an experimental and engineering approach to the same topics the philosophers were discussing abstractly.

Philosophy had attempted to explain the process of perception and understanding. One tradition, the rationalism represented by such thinkers as Descartes, Kant, and Husserl, took the approach of formalism and attempted to elucidate rules governing the process. They argued that in effect the human mind was a machine, albeit a wonderfully complex and versatile one.

The opposing tradition, represented by the phenomenologists Wittgenstein, Heidegger, and Merleau-Ponty, took a holistic approach in which physical states, emotions, and experience were inextricably intertwined in creating the world that people perceive and relate to.

If computers, which in the 1950s had only the most rudimentary "senses" and no emotions, could perceive and understand in the way humans did, then the rules-based approach of the rationalist philosophers would be vindicated. But when Dreyfus had examined the AI efforts, he published a paper in 1965 titled "Alchemy and Artificial Intelligence." His comparison of AI to alchemy was provocative in its suggestion that like the alchemists, the modern AI researchers had met with some initial success in manipulating their materials (such as by teaching computers to perform such intellectual tasks as playing checkers and even proving mathematical theorems). However Dreyfus concluded that the kind of flexible, intuitive, and ultimately robust intelligence that characterizes the human mind could not be matched by any programmed system. Like the alchemists who ultimately failed to create the "philosopher's stone" that could transform ordinary metals into gold, the AI researchers would, Dreyfus felt, fail to find the key to true machine intelligence.

Each time AI researchers demonstrated the performance of some complex task, Dreyfus examined the performance and concluded that it lacked the essential characteristics of human intelligence—the "philosopher's stone" continued to elude them. Dreyfus expanded his paper into the 1972 book *What Computers Can't Do*. But critics complained that Dreyfus was moving the goalposts after each play, on the assumption that if a computer did it, it must not be true intelligence.

Two decades later, Dreyfus reaffirmed his conclusions in *What Computers Still Can't Do*, while acknowledging that the AI field had become considerably more sophisticated in creating systems of emergent behavior, such as neural networks.

Currently a professor in the Graduate School of Philosophy at the University of California, Berkeley, Dreyfus continues his work in pure philosophy (including a commentary on phenomenologist philosopher Martin Heidegger's *Being and Time*). He also continues his critique of the computer, taking aim at its use in education. In his book *On the Internet*, Dreyfus warns that computer users are being drawn away from the richness of experience advocated by phenomenologists and existentialists in favor of a Platonic world of abstract forms. He suggests that online relationships lack real commitment because they lack real risk.

Further Reading

Dreyfus, Hubert. *What Computers Can't Do: A Critique of Artificial Reason*. New York: Harper and Row, 1972.

———. *What Computers Still Can't Do*. Cambridge, Mass.: MIT Press, 1992.

Dreyfus, Hubert, and Stuart Dreyfus. *Mind over Machine: The Power of Human Intuitive Expertise in the Era of the Computer*. Rev. Ed. New York: Free Press, 1988.

———. *On the Internet*. New York: Routledge, 2001.

Dyson, Esther

(1951–)
American
Futurist

Events in the computer industry move at a breakneck pace. At the start of the 1980s, the personal computer (PC) was pretty much a hobbyist's toy; by the end of the decade it seemed to be on every desk. When the 1990s began, *Internet* was an unfamiliar word to most people, and the World Wide Web did not exist. Ten years later, however, having Internet access (and often, one's own website) seems almost as commonplace as having a TV or a phone.

The inventors, managers, entrepreneurs, and investors who make things happen in the computer industry generally focus on one particular application—web commerce, for example, or wireless networking. It takes a special sort of person to make sense out of the broader trends, the

Esther Dyson is the daughter of a cutting-edge physicist, and for more than two decades she has been on the leading edge of the computer industry. Her newsletter *Release 1.0* has predicted technologies such as the personal digital assistant (PDA) years before they come to market. *(Courtesy of Edventure Holdings)*

emerging technologies that will shape the industry one, five, or 10 years down the road. Such a person must be a journalist, analyst, visionary, and an "entrepreneur of ideas" who can package insight and make it accessible to others. Esther Dyson is all of these things. Virtually since the birth of the PC industry, her newsletter *Release 1.0* and its successors have told the industry what would happen next, and why it is important.

Dyson was born on July 14, 1951, in Zurich, Switzerland, but grew up in Princeton, New Jersey, home of the Institute for Advanced Studies, where many of the world's greatest physicists work. Her father, Freeman Dyson, was one such physicist, whose imagination and writings have ranged from nuclear space propulsion to the dynamics of possible alien civilizations. Dyson's mother, Verena Huber-Dyson, was an accomplished Swiss mathematician. She and Freeman were divorced when Esther was five. Two of their neighbors in Princeton were Nobel Prize winners.

Considering the availability of such role models, it is probably not surprising that Dyson was an avid student. Dyson recalled to an interviewer that she and her brother George often played among "the derelict remains of one of the first computers." When she was 14, she began to study Russian.

At the age of 16, Dyson entered Radcliffe College, the prestigious women's college that is associated with Harvard University. At Radcliffe, Dyson studied economics but also gained journalistic experience writing for the *Harvard Crimson*. After getting her degree in economics in 1972, Dyson embarked on a journalism career, working as a reporter for *Forbes* for three years. More than just reporting the news, Dyson showed early talent as an analyst. For example, she wrote an article that predicted that the U.S. computer chip industry would soon be facing a major challenge from Japan, which would indeed happen about a decade later.

In 1977, Dyson shifted roles from reporter-analyst to investment analyst, advising investment firms about trends, particularly those involving new technology. In 1979, however, she returned to journalism, writing for the *Rosen Electronics Newsletter*, an influential industry publication. When Benjamin M. Rosen, the newsletter's founder, decided to move into the venture capital field, Dyson bought the newsletter from him and renamed it *Release 1.0*. (The name refers to the nomenclature used for the first commercial version of a software product.) She also took over responsibility for the PC

Forum, an annual conference attended by key figures in the new personal computer industry.

In the mid-1980s, Dyson dropped *Release 1.0* for a time, having decided to start a more ambitious newspaper-sized publication called *Computer Industry News*. However, she found she was spending most of her time managing the newspaper rather than exploring and writing about the future. The newspaper never really caught on, and after it failed, Dyson revived *Release 1.0*.

Release 1.0 quickly gained a reputation for spotting and highlighting industry trends, often years before the actual products arrived. For example, in a 1990 issue she featured the concept of the PDA (personal digital assistant), a handheld computer that, in the form of machines such as the Palm Pilot, would be carried by millions of businesspeople and travelers in the late 1990s.

In the late 1980s, Dyson, who had studied Russian language and culture, became very interested in the developments in Russia and eastern Europe as the Soviet Union began to collapse. She would later recall to interviewer Elizabeth Corcoran that "watching a market emerge from the swirling chaos is one of the awe-inspiring experiences." She started a newsletter called *Rel-East* to report on the technology sector in Russia and eastern Europe, as well as founding the East-West High-Tech Forum. She also founded EDventure Holdings, a venture-capital fund that focuses much of its attention on new technology companies in that area.

Moving beyond technology, Dyson has also been involved in organizations dealing with global and cultural issues. These include the Santa Fe Institute, an innovative multidisciplinary scientific research effort; the Institute for East-West Studies; and the Eurasia Foundation.

During the 1990s, Dyson, along with many others in the industry, forged new roles involving the Internet, the Web, and e-commerce. She chaired the Internet Corporation for Assigned Names and Numbers (ICANN), the organization responsible for developing and defining the addresses and protocols used to route traffic on the Internet. She has also served on the board of the Electronic Frontier Foundation, an organization involved with civil liberties and privacy issues in cyberspace. In particular, she sees education and tools for parents, not legislation such as the Communications Decency Act of 1995, to be the answer to on-line problems such as easy access to pornographic or violent content.

Dyson's life is extremely hectic and she logs hundreds of thousands of frequent-flyer miles each year traveling to conferences. She likes to start each day with an hour's swimming. Dyson has not married or otherwise put down roots. When asked, as many women are, whether her gender has made her career more difficult, she suggested that being a woman might actually be an advantage. She told interviewer Claudia Dreifus, "In the computer world, I find, being a woman, you are not so pressed to conform. There's a broader range of character traits that are acceptable."

In 1996, Dyson was awarded the Hungarian von Neumann Medal for "distinction in the dissemination of computer culture." She has also been named by *Fortune* magazine as one of the 50 most powerful women in American business.

Further Reading

Dyson, Esther. "Esther Dyson on Internet Privacy." *CNET News Online*. Available on-line. URL: http://news.com.com/2009-1017-893537.html. Posted on April 27, 2002.

———. *Release 2.1: A Design for Living in the Digital Age*. New York: Broadway Books, 1998.

Kirkpatrick, David. "Esther Dyson: Living the Networked Life." *Fortune*, May 27, 2002, pp. 168ff.

Edventure Holdings, Inc. "Release 1.0." Available on-line. URL: http://www.edventure.com/release1. Updated on October 2002.

E

Eckert, J. Presper

(1919–1995)
American
Inventor, Engineer

Today we think of computing as containing well-defined roles such as engineer, programmer, and user. But the people who created the first computers had to invent not only the machines, but also the ways of working with them. The computing pioneers had to make many fundamental decisions about design and architecture under the pressure of World War II and the nuclear-fueled cold war that soon followed. Since there were no premade parts or even technicians trained in digital circuitry, they and their assistants had to do most of the assembly and wiring themselves. As a result, they created what amounted to hand-crafted machines, each one demonstrating something new and bearing a cautionary tale. Together with JOHN MAUCHLY, J. Presper Eckert played a key role in the design of what is generally considered to be the first general-purpose electronic digital computer, then went on to pioneer the commercial computer industry.

Eckert was born on April 9, 1919, in Philadelphia. An only child, he grew up in a prosperous family that traveled widely and had many connections with such Hollywood celebrities as Douglas Fairbanks and Charles Chaplin. Eckert was a star student in his private high school and also did well at the University of Pennsylvania, where he graduated in 1941 with a degree in electrical engineering and a strong mathematics background.

Continuing at the university as a graduate student and researcher, Eckert met an older researcher named John Mauchly. They found they shared a deep interest in the possibilities of electronic computing, a technology that was being spurred by the need for calculations for such wartime applications as fire control systems, radar, and the top secret atom bomb program. After earning his master's degree in electrical engineering, in 1942 Eckert joined Mauchly in submitting a proposal to the Ballistic Research Laboratory of the Army Ordnance Department for a computer that could be used to calculate urgently-needed firing tables for guns, bombs, and missiles. The army granted the contract, and the two men organized a team that grew to 50 people, with Eckert, who was only 24 years old, serving as project manager.

Begun in April 1943, their computer, the ENIAC (Electronic Numerical Integrator and Computer) was finished in 1946. While it was too late to aid the war effort, the room-size machine filled with 18,000 vacuum tubes demonstrated the practicability of electronic computing. Its computation rate of 5,000 additions per second far exceeded the mechanical calculators of the time. While impressive, the ENIAC did not feature radical new forms of architecture. Rather, it was built as an electronic analog to the mechanical calculator, with tubes and circuits replacing cogs or relays.

A noted mathematician named JOHN VON NEUMANN helped Eckert and Mauchly to begin to develop a new machine, EDVAC, for the University of Pennsylvania. This next generation of computers had more flexible binary logic circuits and new forms of memory storage, such as a mercury delay line, which stored data as pulses generated by vibrating crystals and circulating through liquid mercury. (Fortunately for today's environmental concerns, this form of memory has long since vanished from computing.)

While this effort was still underway, Eckert and Mauchly formed their own business, the Eckert-Mauchly Computer Corporation, and began to develop the BINAC (BINary Automatic Computer), which was intended to be a relatively compact and lower-cost version of ENIAC. This machine demonstrated a key principle of modern computers—the storage of program instructions along with data. The ability to store, manipulate, and edit instructions greatly increased the flexibility and ease of use of computing machines. Among other things, it meant that a program could execute defined sections (subroutines) and then return to the main flow of processing.

By the late 1940s, Eckert and Mauchly began to develop Univac I, the first commercial implementation of the new computing technology. Financial difficulties threatened to sink their company in 1950, and it was acquired by Remington Rand. Working as a division within that company, the Eckert-Mauchly team completed Univac I in time for the computer to make a remarkably accurate forecast of the 1952 presidential election results. Eckert went on to manage the development of later generations of UNIVAC, including the LARC, a machine designed specifically for scientific applications.

Eckert continued with the Sperry-Rand Corporation (later called Univac and then Unisys Corporation) and became a vice president and senior technical adviser. However, his focus always remained on technical possibilities rather than business considerations, and corporate colleagues often considered him impractical—a situation that would be faced by a later generation who wanted to build personal computers.

Eckert received an honorary doctorate from the University of Pennsylvania in 1964. In 1969 he was awarded the National Medal of Science, the nation's highest award for achievement in science and engineering. He retired in 1989 and died on June 3, 1995.

Further Reading

Eckstein, P. "Presper Eckert." *IEEE Annals of the History of Computing* 18, no. 1 (spring 1996): 25–44.

McCartney, Scott. *Eniac: The Triumphs and Tragedies of the World's First Computer.* New York: Berkley Books, 1999.

Smithsonian Institution. National Museum of American History. "Presper Eckert Interview." Available on-line. URL: http://americanhistory.si.edu/csr/comphist/eckert.htm. Posted on February 2, 1988.

Eckert, Wallace J.

(1902–1971)
American
Astronomer, Inventor

By the early 20th century, scientists had begun to probe both the tiny world of the atom and the vast reaches of the universe with ever more powerful and sophisticated instruments, devising more sophisticated theories that required lengthy, complicated calculations with many variables. Yet the scientist's mathematical toolkit still consisted of little more than paper and pencil, aided perhaps by a slide rule or a simple mechanical calculator.

Wallace John Eckert (who is no relation to ENIAC pioneer J. PRESPER ECKERT), played a key role in bringing automated computing to science, as well as in inducing an office tabulator company called International Business Machines (IBM) to build its first computer.

Eckert was born in Pittsburgh, Pennsylvania, on June 19, 1902, and grew up on a farm in a small town called Albion. After receiving an A.B. degree at Oberlin College and master's degrees from both Amherst College and the University of Chicago, he entered the doctoral program in astronomy at Columbia University, completing his Ph.D. at Yale in 1931.

Eckert did his graduate work under Ernest W. Brown, an astronomer who had developed a theory for understanding the orbit of the Moon based on elaborate differential equations expressing the changes in position over time. Although Kepler and Newton had provided the basic theory for planetary orbits, the path of the Moon was in reality very complex, being influenced not only by the gravity of the Earth but also that of the Sun, as well as the effects of tidal forces. After developing his equations, Brown had to create hundreds of pages of tables that could be used to look up approximations of values that could then be plugged into the equation to obtain the lunar position for a given date.

Having become familiar with the elaborate calculations required by Brown's lunar theory, Eckert soon became convinced that mechanical help would be increasingly necessary if science was to continue to make progress. Soon after he became an assistant instructor at Columbia University in 1926, he became the first faculty member to introduce his students to the use of mechanical calculators.

Meanwhile, another Columbia professor, Benjamin D. Wood, had established the Columbia University Statistical Bureau. Its researchers had access to punch card tabulators donated by IBM. These tabulators, the descendants of the 19th century machine invented by HERMAN HOLLERITH that had automated the U.S. census, were increasingly being used by government and business to automate the collation and summarization of data.

Eckert believed that this technology could also be applied to scientific computing. In 1933, Eckert suggested to IBM's chief executive officer, THOMAS J. WATSON SR. that IBM provide equipment to set up a more elaborate computing facility at Columbia. Watson was enthusiastic about the project, and the Thomas J. Watson Astronomical Computing Bureau, often simply called the Watson Lab, was born. Eckert became the facility's first director, coordinating projects with the American Astronomical Society as well as IBM.

Eckert's key insight was that separate pieces of equipment such as card readers, tabulators, and calculating punches could be linked into a kind of data processing pipeline, together with a "mechanical programmer," a switching system controlled by mechanical disks that could be set to direct the machinery to carry out multistep calculations automatically. While not a full-fledged computer in the modern sense, Eckert's calculating assembly demonstrated the value of automated calculation for science. His 1940 book, *Punched Card Methods in Scientific Computation*, introduced many scientists to the new technology.

Starting in 1940, Eckert served as director of the U.S. Nautical Almanac Office in Washington, D.C., and also worked at the Naval Observatory, where he introduced automated computing methods as wartime pressure began to make the preparation of navigation aids an urgent priority. One product of his work was the American Air Almanac, a navigational guide that would be used by pilots for many years to come.

Following the war Eckert returned to Columbia as directory of the Watson Laboratory as well as working for IBM. During the war J. Presper Eckert and JOHN MAUCHLY had created ENIAC, the first large-scale electronic digital computer, a machine far in advance of anything at IBM. IBM asked Wallace Eckert to help the company catch up; and he designed the Selective Sequence Electronic Calculator (SSEC). This 140-foot-long, U-shaped machine was a hybrid, using 12,000 vacuum tubes but also including 21,000 mechanical relays. As the name implied, the SSEC could choose different sequences of operations depending on the results of data tests. IBM eventually moved the SSEC to the lobby of its New York headquarters where passersby could marvel at the blinking, whirring "Giant Brain." In 1954, Eckert also designed the NORC (Naval Ordnance Research Calculator), a machine that set new standards in scientific computing power and which remained in service until 1968.

Eckert used the SSEC to make more accurate calculations of the orbits of the Moon as well as the planets Jupiter, Saturn, Uranus, Neptune, and Pluto. His lunar data was so accurate that in 1969, when Apollo astronauts blasted off for the Moon, their path was guided by the data Eckert had calculated many years earlier.

Eckert retired from Columbia in 1967. During more than 20 years of his directorship at Columbia, more than 1,000 astronomers, physicists, statisticians, and other researchers received training there in scientific computation. He received many honors from the astronomical and computing communities, including an IBM fellowship (1967), the IBM Award (1969), and the James Craig Watson Medal of the National Academy of Science. Eckert died on August 24, 1971.

Further Reading

da Cruz, Frank. "The IBM Selective Sequence Electronic Calculator." Available on-line. URL: http://www.columbia.edu/acis/history/ssec.html. Posted on January, 2001.

———. "Professor Wallace J. Eckert." Available on-line. URL: http://www.columbia.edu/acis/history/eckert.html. Posted on September 1, 2002.

Eckert, Wallace J. *Punched Card Methods in Scientific Computation*. Charles Babbage Institute Reprint Series, vol. 5. Cambridge, Mass.: MIT Press, 1984.

Ellison, Larry (Lawrence John Ellison)

(1944–)

American

Entrepreneur

If the word processor is the software that drives the modern office, the information-processing engine under the hood of modern business is the database program. Every employer, commodity, and transaction involved with a business is represented by records in a complex system of interconnected databases. Lawrence (Larry) Ellison and his Oracle Corporation parlayed database software into a leading market position, competing head to head against giants Microsoft and IBM.

Ellison was born in 1944 in New York City, but his mother, an unmarried teenager, gave the boy to her aunt and uncle, Lillian and Louis Ellison, to raise. Larry Ellison would later describe his adoptive father, a Russian immigrant, as tough and hard to get along with. The family lived on the south side of Chicago in what Ellison has called a "Jewish ghetto." In high school, Ellison was bright but seldom took much interest in his studies. However, when something attracted Ellison's attention, he could be quite formidable. His friend Rick Rosenfeld would later tell *Vanity Fair* interviewer Byran Burrough that "[Ellison] was very intense, very opinionated. Whatever he was talking about, he was loud about it. Larry just had an answer to everything. [Today], he's the same guy I knew in high school."

After high school Ellison attended the University of Illinois, where he did well in science classes. When he learned that his adoptive mother had died suddenly, Ellison dropped out of school. He later attempted to resume his college education, but made little headway. Ellison's adoptive father further discouraged him, saying he would never amount to anything.

However, the toughness from his upbringing and a certain stubborn streak kept prodding Ellison to prove his adoptive father wrong. Back in high school, Ellison had been exposed to some introductory programming classes, and the 1960s was a time of high demand for programmers as businesses began to automate more of the data processing with mainframe computers. Ellison went to San Francisco and became a programmer for a bank. He then moved south of the city to the area called Silicon Valley, which was just starting to become prominent in the rise of integrated circuit technology and microelectronics. By the end of the 1960s Ellison's skills had advanced to the point where he was making a very nice living and he bought a home, a sports car, and a sailboat.

In the early 1970s, Ellison received his first intensive exposure to databases. Ellison worked for Ampex Corporation to design large database systems, including one for the Central Intelligence Agency (CIA), which had the code name "Oracle."

In 1977, Ellison and a partner, Bob Miner, used that name to start their own company, Oracle Corporation. Their first customer was the CIA. Ellison and Miner ran into a financial crunch because they did not realize how long it would take to get money from the government, but they persevered and soon received a contract from the U.S. Air Force.

By then programmers were feeling the need to go beyond simple "flat file" databases to relational databases (invented by EDGAR F. CODD) that could connect data fields from different files together. IBM in particular was developing SQL (structured query language), a way that users could easily extract and process information from many databases with a single command. For example, a user could ask for the records for all salespersons with more than $1 million in annual sales and sort them by region.

Ellison and Miner took the SQL specifications and created their own implementation of the relational database system, developing a prototype in 1979, two years before the huge but slower-moving IBM. Ellison recruited a sales force that shared his hard-driving attitude, and their product, also called Oracle, seized the lead in the database market. During the 1980s, Oracle became the database of choice for many large corporations, with its market position so dominant that it seemed secure for years to come. Meanwhile, Ellison had gained a reputation for acquisitiveness (building a $40 million mansion) and ruthlessness, leading to his friend Steve Jobs (chief executive officer [CEO] of Apple Computer) dubbing him the "outrageous CEO poster child."

However, near the end of 1990 the bubble burst. Oracle had a disaster that would become familiar to investors more than a decade later: The company's accounting was too aggressive. In particular, the company had inflated its earnings reports by recording revenues before the money had actually been received. Auditors demanded that Oracle restate its earnings, which resulted in a $28.7 million loss. With investors losing confidence, the value of Oracle stock plunged by about 80 percent. The crash also revealed other problems: In its rush for market dominance, Oracle had neglected many of the basics of customer service, support, and the fixing of software bugs. The company was now threatened with the loss of many of its biggest corporate customers.

Ellison did not give up, however. He decided that he would change the way both he and his company did business. He hired new management, quality control, and customer support staff. He also decided that he would meet with many of his customers and listen to their concerns. The new methods turned the company around, and it recorded earnings of $600 million in 1996.

Ellison's personal life shared much of the roller-coaster nature of his business career. Fond of active outdoor sports, Ellison had broken his neck in 1990 while bodysurfing in Hawaii, and the following year his elbow was smashed in a bicycle accident. The required recuperation gave Ellison time for, as he told interviewer Burroughs, "confronting all sorts of things in my life and trying to order them and understand them. I had to understand my role in life. Who am I? Who is my family?" This quest led him to, among other things, finding and reconnecting to his birth mother, Florence Spellman.

During the later 1990s, Ellison took his business in new directions. One was the promotion of the "network computer," or NC, a stripped-down, diskless, inexpensive personal computer (PC) that relied on the network for software and file storage. Besides saving money for businesses, Ellison thought that the NC, at a cost of less than $500, might enable schools to provide computing power at each student's desk. Ellison even considered mounting a corporate takeover of Apple so that he could use their technology to produce network computers, but STEVE JOBS's return as CEO of Apple apparently forestalled this plan.

As it happened, the falling prices of regular PCs at the end of the 1990s made the network computer something of a moot point. One advantage touted for the new technology is that users could run applications via a Web-based language called Java without having to pay Microsoft for a copy of Windows for each PC. This never materialized on a large scale, but in 2000–1, Ellison took a high-profile role in testimony in the legal actions against Microsoft that accused the software giant of unfairly stifling competition through what was seen as its monopolistic control of PC operating systems.

In recent years, Ellison has aggressively pursued the integration of his database software with the growing use of the World Wide Web in business, with sales and services increasingly being handled on-line. He has also created integrated on-line systems for sectors such as the automobile industry, signing agreements with Ford, General Motors, and Daimler Chrysler.

Ellison continues to live a high-energy, high-risk life, flying his own jet and sailing world-class sailboats. Ellison remains confident about the future of his company, having declared that "IBM is the past, Microsoft the present, Oracle is the future."

Further Reading

Schick, Elizabeth A., ed. "Ellison, Lawrence J." *Current Biography 1998*. New York: H. W. Wilson, 1998, pp. 21ff.

Stone, Florence M. *The Oracle of Oracle: The Story of Volatile CEO Larry Ellison and the Strategies Behind His Company's Phenomenal Success*. New York: AMACOM, 2002.

Wilson, Mike. *The Difference Between God and Larry Ellison: Inside Oracle Corporation*. New York: Morrow, 1998.

Engelbart, Douglas

(1925–)
American
Inventor, Engineer

Before the 1970s, using a computer meant typing line after line of cryptic commands. If one got the incantation exactly right, there was a reasonable chance of getting the expected results. Of course, it was hard to type the correct command, command options, and the often long, complicated file names. The result was often an equally cryptic error message.

Starting in the 1970s, an engineer named Douglas Engelbart began to change all that. He made it possible for a user to simply point to icons for programs and run them, or to pick up a file in one folder and move it to another. The user could do this simply by moving a little box that became known as a "mouse."

Engelbart grew up on a small farm near Portland, Oregon. His father owned a radio store in the city, and the boy acquired a keen interest in electronics. In high school he read about new inventions including something that put moving pictures on a screen—television. After graduation, he went to Oregon State University to major in electrical engineering, but his studies were interrupted by World War II. During the war, Engelbart served in the Philippines as a radar technician. Radar also interested him because like television, it showed meaningful information on a screen. During that time, he read a seminal article entitled "As We May Think," written by VANNEVAR BUSH. Bush suggested that information could be stored on microfilms that could be electronically scanned and linked together to allow the reader to move from one document to a related document. This system, later called "hypertext," would be the concept behind the development of the World Wide Web by TIM BERNERS-LEE.

Bush's idea of a mechanical microfilm-based information library was not very practical. Having been exposed to the visual display of information in television and radar, Engelbart began to think about what might be done with a television-like cathode ray tube (CRT) display and the new electronic digital computers that were just starting to come into service.

Engelbart received a B.S. degree in electrical engineering from Oregon State University in 1948. He then worked for the National Advisory Committee for Aeronautics (the predecessor of NASA) at the Ames Laboratory. Continuing to be inspired by Bush's vision, Engelbart conceived of a computer display that would allow the user to visually navigate through information. Engelbart then returned to college, receiving his doctorate in electrical engineering in 1955 at the University of California, Berkeley, taught there for a few years, and went to the Stanford Research Institute (SRI), a hotbed of futuristic ideas.

By 1962 Engelbart had developed a new approach to the use of computers. By then, the machines had proven their worth as very fast calculators. There had also been some interesting work in artificial intelligence (AI), the notion that computers might in some sense "think." Engelbart suggested a third possibility, midway between dumb calculators and smart "electronic brains." He wrote a seminal paper titled "Augmenting Human Intellect: A Conceptual Framework." In this paper, Engelbart emphasized the computer as a tool that would enable people to better visualize and organize complex information to meet the increasing challenges of the modern world. In other words, the computer would not replace the human brain, but amplify its power. But for a computer to expand the scope of the human intellect, it would have to be much easier and more natural to use.

In 1963, Engelbart left SRI and formed his own research lab, the Augmentation Research Center. Five years later, he and 17 colleagues from the Augmentation Research Center demonstrated a new kind of computer user interface to an audience of about 1,000 computer professionals at a conference in San Francisco. In this 90-minute demonstration, Engelbart presented what would become the future of personal computing. It included a small box whose movement on a flat surface correlated with the display of a pointer on the screen, and whose button could be used to select menus and objects. The demonstration included another exciting new technology: networking. Two users 40 miles apart could access the same screen and work with a document at the same time.

This breakthrough in user interfaces was not the result of a single inspiration, but represented many hours of work by Engelbart and his colleagues, who had tried many other devices including joysticks, light pens, and even a pedal installed inside a desk, intended to be pushed by the user's knee. Engelbart found that nothing seemed to work as well as his simple device, which was patented in 1970 as an "X-Y Position Indicator." But the device clearly needed a simpler name, and apparently someone at the lab noticed that the small box with its cord trailing like a tail looked a bit like a familiar rodent. Soon everyone was calling it a "mouse."

By the mid-1970s, a group at the Xerox Palo Alto Research Center (PARC) was hard at work refining this interface, creating an operating system in which open programs and documents were displayed in windows on the screen, while others could be selected from menus or by clicking on icons. Xerox marketed the new system in the form of a computer workstation called the Alto, but the machine was too expensive to be considered a true personal computer. It would take Apple Computer and later Microsoft to make it truly ubiquitous in the 1980s and 1990s (see JOBS, STEVE.)

Besides user interface design, Engelbart also took a keen interest in the development of the ARPANET (ancestor of the Internet) and adapted NLS, a hypertext system he had previously designed, to help coordinate network development. (However, the dominant form of hypertext on the Internet would be Tim Berners-Lee's World Wide Web.) In 1989 Engelbart founded the Bootstrap Alliance, an organization dedicated to improving the collaboration within organizations, and thus their performance. During the 1990s this nurturing of new businesses and other organizations would become his primary focus.

Engelbart has been honored as a true pioneer of modern computing. In 1997, he received the Lemelson-MIT Award (and an accompanying $500,000) as well as the Association for Computing Machinery Turing Award. In 2000 he received the National Medal of Technology.

Further Reading

Bardini, Thierry. *Bootstrapping: Douglas Engelbart, Coevolution, and the Origins of Personal Computing*. Stanford, Calif.: Stanford University Press, 2000.

Bootstrap Alliance. "Douglas Carl Engelbart." Available on-line. URL: http://www.bootstrap.org/engelbart/index.jsp. Downloaded on November 1, 2002.

Estridge, Philip Donald

(1937–1985)
American
Entrepreneur

Most people associate the personal computer (PC) with famous names such as BILL GATES and STEVE JOBS. Far fewer have heard of Philip Estridge. Chances are good, though, that on many desks is a direct descendant of the machine that Estridge and his team designed in 1981: the IBM PC.

Estridge was born on June 23, 1937, and grew up in Jacksonville, Florida. His father was a professional photographer. In 1959, Estridge graduated from the University of Florida with a bachelor's degree in electrical engineering. He joined IBM in 1959 as a junior engineer at the Kingston, New York, facility, working on a variety of government projects, including the SAGE computerized early-warning defense system and, in 1963, programming for the NASA Goddard Space Flight Center. In 1969, Estridge moved into the area of system design within IBM, serving from 1975 to 1979 as programming manager for the Series/1 minicomputer.

By the late 1970s, however, the even smaller microcomputer had started to appear in the computer market in the form of machines such as the Radio Shack TRS-80, Commodore PET, and particularly, the sleek, all-in-one Apple II. Equipped with useful software such as the VisiCalc spreadsheet, invented by DANIEL BRICKLIN, machines that many IBM executives had thought of as hobbyist toys were starting to appear in offices.

Estridge was able to convince IBM's upper management that desktop computers represented a genuine market that IBM should enter, both for its profit potential and to avoid the threat to the company's supremacy in office machines. Further, Estridge's proposed concept won out over two competing company teams, and he was placed in charge of IBM's new Entry Systems

Unit. He was told to design a desktop computer system and bring it to market in only a year. In an industry where systems typically took at least several years from concept to finished product, such a schedule was unprecedented. Equally unprecedented was IBM's willingness to give Estridge almost complete independence from the usual corporate bureaucracy, allowing him to make key decisions quickly and without interference.

In designing and managing the PC project, Estridge had to cope with the IBM corporate culture, which was far different from that of Silicon Valley and the businesses run by young entrepreneurs such as Microsoft's Bill Gates and Apple's Steve Jobs. IBM had vast resources in money, manufacturing plants, and personnel, but it was not geared to design an entirely new machine in just a year. Estridge realized that he would have to use microprocessors, memory, disk drives, and other "off the shelf" components. He chose the Intel 8088 processor because it belonged to a family of chips closely compatible with already existing software written for the CP/M operating system.

As quoted by *PC Week* in 1985, Bill Gates recalled that Estridge "had a keen sense of what IBM was super-good at and what it should use outside people for." Gates also noted that Estridge was willing to stretch the technical capabilities of the design. At a time when most personal computers were eight-bit (that is, they moved eight binary digits of data at a time), Estridge insisted on a more powerful 16-bit design.

There was no time to develop operating system software from scratch, so he looked for an existing system that could be quickly adapted to the new machine. A new version of CP/M (Control Program for Microcomputers) seemed the logical choice, but after negotiations with CP/M author GARY KILDALL seemed to stall, Estridge hedged his bets by offering Microsoft's PC-DOS and the University of California, Santa Cruz's p-System as well. Microsoft's product was the cheapest and it quickly became the overwhelming choice among PC users.

Another of Estridge's key decisions was to make the IBM PC an "open platform." This meant that, like the Apple II, the PC would not be restricted to using hardware provided by the manufacturer. The motherboard was designed with standard connecting slots into which could be plugged a growing variety of expansion boards by third-party manufacturers, including additional memory (the original machine came with only 64 kilobytes), serial and parallel ports, sound and video display cards, and other useful features. This openness to third-party products represented a radical departure from IBM's practice of trying to discourage third-party manufacture of compatible hardware.

As promised, the IBM PC appeared in April 1981. The company had projected sales of 250,000 machines in the next five years, but by 1984 1 million PCs were already in use. Although Apple would answer with its innovative Macintosh that year, the IBM PC and its successors, the PC-XT and PC-AT, would quickly dominate the business computing market. Estridge's Entry Systems Unit had grown from its initial force of 12 employees to nearly 10,000, gaining status as a full corporate division.

By the mid-1980s, however, IBM was starting to be challenged by competition from companies such as Compaq. The same open design and specifications that gave the IBM PC an initial boost in the marketplace also allowed competitors to legally "clone," or reverse-engineer, functionally compatible machines, which were often more powerful or less expensive than those from IBM.

No one will know how Estridge might have faced these new challenges. His life was tragically cut short on August 2, 1985, when Delta Airlines flight 191, carrying him, his wife, Mary Ann, and many IBM employees, crashed on landing at Dallas–Fort Worth Airport, the victim of wind shear from a thunderstorm. Estridge was only 47 years old. By then, though, his IBM PC was changing the way millions of people worked, learned, and played, and he has been widely honored as an innovator by publications ranging from *Computerworld* and *InfoWorld* to *Time* and *Business Week*.

Further Reading

Bradley, D. J. "The Creation of the IBM PC." *Byte*, September, 1990, pp. 414–420.

Bunnell, David. "Boca Diary: April–May 1982." *PC Magazine*. Available on-line. URL: http://www.pcmag.com/article2/0,4149,874,00.asp. Posted on April 1, 1982.

Freiburger, Paul, and Michael Swaine. *Fire in the Valley: The Making of the Personal Computer*. 2d ed. Berkeley, Calif.: Osborne/McGraw-Hill, 1999.

Eubanks, Gordon

(1946–)
American
Programmer, Entrepreneur

In the early days of personal computing there was little distinction between professional programmers and technically savvy users. Since there was little commercial software, if one wanted one's more or less hand-assembled machine to do something useful, one wrote programs in BASIC or (if more ambitious) assembly language. Gordon Eubanks was one of a number of pioneers who created the better programming tools and utilities that helped turn a hobby into one of the world's most important industries.

Eubanks was born on November 7, 1946. Unlike many computer entrepreneurs, he had no particular childhood interest in electronics as such, but he does recall dreaming that he would one day own his own computer—something that in the 1950s was rather like owning one's own jumbo jet. The next best thing was getting to work with someone else's computer, and so Eubanks studied engineering at Oklahoma State University, receiving his bachelor's degree in 1968.

Eubanks spent the 1970s in the navy as a submarine officer. Later, in a reference to a popular thriller, he jokingly referred to his service as "*Hunt for Red October* stuff." He believes that the high-pressure situations faced by cold war submariners gave him good preparation for the pressure he would face in creating business opportunities in Silicon Valley. Meanwhile, he took advantage of the opportunities the navy had to offer, earning a master's degree in computer science from the Naval Postgraduate School in Monterey, California. His thesis adviser was GARY KILDALL, the talented programmer who had created CP/M, the first widespread operating system for microcomputers.

By the late 1970s, the personal computer (PC) was just starting to arrive on the consumer market in the form of machines such as the Tandy TRS-80 and Apple II. The more serious enthusiasts, however, tended to buy more powerful machines, many of which used a standard motherboard called the S-100 bus. These machines had the capacity to run significant applications, but the programming tools were rather primitive. Working under Kildall's direction at Digital Research, Eubanks developed a new BASIC version called EBASIC, which later became CBASIC. While most early microcomputer BASICs had only primitive flow control statements, such as simple IF–THEN statements, FOR loops and the GOTO that had been condemned as sloppy by EDSGER DIJKSTRA. CBASIC added an ELSE part to the IF statement and provided a more general WHILE loop. The language also allowed program code to be better organized into functions and subroutines, breaking the processing into more logical and manageable portions.

Meanwhile, important events were about to reshape the nascent personal computer industry. By 1980, IBM had undertaken a large, secret project to build its own personal computer under the direction of PHILIP DONALD ESTRIDGE. Gary Kildall's CP/M was the logical choice for the new machine's operating system. Earlier, when Kildall had asked Eubanks what should be done about the future of CP/M, Eubanks had urged him to make it the cornerstone for a business plan. However, Kildall seemed to be interested only in pursuing his technical ideas, not in developing business relationships.

As recounted in an interview by Clive Akass, Eubanks then received a phone call from Tim Patterson, developer of a new disk operating system (DOS). Patterson warned Eubanks to start developing a version of CBASIC for DOS. When Eubanks asked why, Patterson replied "I can't tell you, but a big Seattle company has licensed it, and licensed it to a hardware company that's bigger than anyone you can think of." Eubanks asked him to be more specific: "You are telling me that IBM licensed it from Microsoft." Tim replied: "I didn't say that but you should definitely support it." At this point, Eubanks knew that Kildall had missed one of the biggest opportunities in the history of computing.

Once the IBM PC came on the market in 1981, CP/M would fade into obscurity and DOS would be king. In 1982, however, Eubanks and a partner bought Symantec, a small company that developed system utilities. Symantec struggled for a time—one payday the company gave out stock instead of a paycheck. By the later 1980s and 1990s, however, Symantec grew steadily, finding its niche in selling the kind of unglamorous but necessary software that PC users increasingly demanded. Symantec also bought Norton, one of the best-known names in system utilities—programs that perform such tasks as diagnosing system problems, reorganizing hard drives, or detecting and eradicating computer viruses. (Today Symantec is one of the best-known names in antivirus software.)

In 1992, Eubanks and several other Symantec executives were accused of stealing trade secrets from software developer Borland International. However, in 1996 the charges were dropped. Eubanks also entered the legal news in 1999 when he became one of the witnesses to testify on behalf of BILL GATES and Microsoft in the federal antitrust proceedings against the software giant. Eubanks argued that with the software market changing so rapidly, competition is forcing even the biggest companies such as Microsoft to continually innovate, and thus monopoly was not really an issue.

In 1999, Eubanks left Symantec and became chief executive officer of a company called Oblix, which specializes in software for securing and controlling access to websites. With the increased concern about security and the need to reliably verify a person's identity on-line, this sector of the market is likely to do well in coming years. In 2000, *PC Week* listed Oblix among key companies for "21st Century Infrastructure." Whatever coming years may bring, Gordon Eubanks is widely recognized as an industry pioneer.

Further Reading

Akass, Clive. "Legend of the Fall." *Personal Computer World*, September 1996. Available on-line. URL: http://images.vnunet.com/v6_image/pcw/pcw_images/history/eubanks.PDF.

Jaeger, Rama Dev, and Rafael Ortiz. *In the Company of Giants: Candid Conversations with the Visionaries of the Digital World*. New York: McGraw-Hill, 1997.

F

⊠ Fanning, Shawn

(1981–)
American
Inventor, Entrepreneur, Activist

Late in 1999, thousands of mainly young Internet users found a new way to get music from their favorite bands—for free! All they had to do was download some software called Napster, which let them search for other users who had put tracks from music CDs on their personal computer's (PC's) hard drive. In turn, they could share their own music collection by "ripping" (copying) the tracks from CD to their hard drive, where other Napster users could find and download them. To the young music fans who had grown up with mouse in hand and Walkman on head, Napster was just one of the many cool technologies they had seen in the ever-changing world of cyberspace. To the companies that made those music CDs, however, Napster was like a big siphon poised to suck away their profits. After all, why would people pay $15.00 for a CD if they could get the same music for nothing?

The source of all this controversy was a college student named Shawn Fanning. Fanning was born in 1981 in Brockton, Massachusetts. His mother, an unmarried teenager, lived with her own parents, making for a rambunctious two-generation family. Eventually she married an ex-marine turned bakery truck driver, and young Fanning soon had four new siblings.

The growing family was often short of money, but Fanning's Uncle John took an interest in his education, in particular supporting his interest in computers by buying him his own machine and then an Internet connection. Uncle John had seen computers as a road to a good future for the boy, who quickly took to the world of the Internet and programming and made it his own. When his parents had a falling out, Fanning was put for a time in a foster home, but he simply immersed himself further in the world of computers and on-line chat. During summer vacations in high school, Fanning worked at his uncle's company, NetGames, absorbing advanced programming techniques from the Carnegie Mellon University computer science students who also worked at the firm.

After graduation, Fanning had planned to enter the Carnegie Mellon computer science program himself, but he was not accepted. Instead, he went to Northeastern University in Boston, where he was given advanced placement. His college roommate introduced him to MP3 files, the increasingly popular way to store and play digitized music on PCs. Fanning decided to learn Windows programming so he could write programs to organize and play MP3 files.

When he was barely old enough to vote, Shawn Fanning came up with a way for computer users to multiply their music libraries by swapping MP3 files over the Internet. But recording companies saw Fanning's Napster as a threat to their business, and legal action eventually forced the company out of business. *(CORBIS Sygma)*

Fanning was disappointed by the hit-and-miss process of finding MP3 files on the Internet—often a file could not be found or downloaded properly. He got the idea of programming a file-sharing service where users could provide many access points for a given MP3 file, avoiding the glitches and bottlenecks caused by relying on a single site. He spent every waking hour on the program for several weeks. In June 1999 he distributed the software to 30 friends as a test. He called the program Napster, which had been his high school nickname and chat room "handle"—originally referring to his curly hair.

Although Fanning had asked his friends not to tell anyone else about Napster, after a few days the program began to appear on various websites, where it was downloaded by about 15,000 people. Before he knew it, Fanning found that his program had been featured in the "download spotlight" at the popular site Download.com, and after that it mushroomed into one of the most popular Internet downloads ever.

Fanning's Uncle John saw that Napster had potential as a business, and he set up the Napster Corporation for his nephew. Soon a venture capitalist named Eileen Richardson was on board, and the company opened an office in San Mateo, California. Fanning joined the new generation of Internet pioneers (such as JERRY YANG and David Filo of Yahoo!) in leaving college to work on his business full time.

However, by December 1999 Napster had begun to attract a different kind of attention. The Recording Industry Association of America (RIAA) sued the tiny firm because much of the music being traded over Napster was from copyrighted commercial CDs. Like other publishers, the recording companies paid royalties to the creators of the music and, in turn, had the sole right to copy and sell it for a profit. Copying a book is expensive and tedious, but copying an online music track is simply a matter of telling a program to make an exact copy of all the bits in a file.

The recording companies were generally not very up-to-date with regard to the possibilities of the Internet—most had not even begun to think about how they might sell music on-line. They did understand, however, that Napster enabled anyone to get for free the same music they were trying to sell. If enough people did that, it would put them out of business.

Napster's defense was both moral and legal. The moral defense was based on the argument that the recording companies were vastly overcharging for music anyway, and that little of the revenue went to the musicians themselves. Perhaps Napster might spur record companies to be more innovative and to find a way to lower prices and offer extra value to consumers. Further, the recording companies often ignored what fans considered to be the most interesting and innovative music groups. With Napster, listeners might be exposed to a much greater variety of musicians. If someone liked a track or two they found on Napster, it was argued, he or she might well buy the CD for convenience if it were reasonably priced.

Napster's legal defense was based on the argument that Napster itself was not violating anyone's copyright. Napster did not store music files on its own server—the actual music was on the hard disks of thousands of users. Napster only facilitated the exchange of music, and after all, the music could be public domain or freely donated by musicians—it did not have to be copyrighted.

The courts, however, proved unsympathetic to Napster's defense. On July 26, 2000, federal judge Marilyn Hall Patel ordered Napster to remove all links to copyrighted music from its database in only two days. The result was a rush by hundreds of thousands of Napster users to download as much music as they could get in the time remaining. Then, as the deadline approached, the U.S. Court of Appeals for the Ninth Circuit temporarily blocked the judge's ruling. In October, Napster's lawyers and those of the RIAA made their formal arguments before a panel of judges of the appeals court.

On February 12, 2001, the appeals court ruled that Napster must prevent its users from accessing copyrighted material. As a result, the company was forced to rewrite its software to block much of the music that had made Napster so attractive to users. Although the company claimed that it would still be able to create a viable music service by paying recording companies fees for legal access to the music and then charging its users a small subscription fee, the number of Napster users plummeted, as did the company's revenue. In May 2002, both Napster's chief executive officer, Konrad Hillers, and Fanning, as chief technology officer, left the company. On June 3, the company filed for bankruptcy.

The company's future viability is much in doubt, as it owes potentially hundreds of millions of dollars in damages for copyright violations and there has been little sign that major recording labels are interested in making deals with Napster. In September 2002, a court ruled out Napster's proposed acquisition by the giant German publisher Bertelsmann for $8 million, citing a conflict of interest.

But the problems for the recording industry, too, are far from over. Music is still being shared on-line under such services as Gnutella and BearShare. Unlike Napster, these services have no central site or database at all, so it is difficult for record companies to find someone to sue.

At any rate, Shawn Fanning still has plenty of time to launch a new career. As of 2002, he's barely of legal age. He has appeared on the cover of magazines such as *Fortune* and *Business Week*, and received a 2000 Technical Excellence "Person of the Year" award from *PC Magazine*. The technology of file-sharing that underlies Napster might well find new applications in the ever-changing Internet future.

Further Reading

Evangelista, Benny. "Napster Files For Bankruptcy." *San Francisco Chronicle*, June 3, 2002, on pp. B1, B6.

"Fanning, Shawn." *Current Biography Yearbook*. New York: H. W. Wilson, 2000, pp. 185–189.

Menn, Joseph. *All the Rave: The Rise and Fall of Shawn Fanning's Napster*. New York: Crown Publishers, 2003.

Wood, Chris, and Michael Snider. "The Heirs of Napster: If Music-Sharing Site Goes Down, Others are Ready to Move In." *Maclean's*, February 26, 2001, p. 52.

Feigenbaum, Edward

(1936–)
American
Computer Scientist, Inventor

When the first large-scale electronic computers were developed in the late 1940s and early 1950s, they were given the label "giant brains." However, while computers could calculate faster than any human being, getting a computer program to actually reason through the steps of solving a problem turned out to be far from easy. Edward Feigenbaum, a pioneer artificial intelligence (AI) researcher, came up with a way to take the knowledge acquired by human experts such as doctors and chemists and turn it into rules that computers could use to make diagnoses or solve problems.

Feigenbaum was born on January 20, 1936, in Weehawken, New Jersey. His father, a Polish immigrant, died before Feigenbaum's first birthday. His stepfather, an accountant and bakery manager, was fascinated by science and regularly brought young Feigenabum to New York City to visit the Hayden Planetarium and the vast American Museum of Natural History. The electromechanical calculator his stepfather used to keep accounts at the bakery particularly fascinated the boy. His interest in science gradually turned to a perhaps more practical interest in electrical engineering.

While at the Carnegie Institute of Technology (now Carnegie Mellon University), Feigenbaum was encouraged by one of his professors to venture beyond the curriculum to the emerging field of computation. He became interested in JOHN VON NEUMANN's work in game theory and decision making and met HERBERT A. SIMON, who was conducting pioneering research into how organizations made decisions. This in turn brought Feigenbaum into the early ferment of artificial intelligence research in the mid-1950s. Simon and ALAN NEWELL had just developed Logic Theorist, a program that simulated the process by which mathematicians proved theorems through the application of heuristics, or strategies for breaking problems down into simpler components from which a chain of assertions could be assembled, leading to a proof.

Feigenbaum quickly learned to program IBM mainframes and then began writing AI programs. For his doctoral thesis, he explored the relation of artificial problem solving to the operation of the human mind. He wrote a computer program that could simulate the human process of perceiving, memorizing, and organizing data for retrieval. Feigenbaum's program, the Elementary Perceiver and Memorizer (EPAM), was a seminal contribution to AI. Its "discrimination net," which attempted to distinguish between different stimuli by retaining key bits of information, would eventually evolve into the neural network. Together with Julian Feldman, Feigenbaum edited the 1962 book *Computers and Thought*, which summarized both the remarkable progress and perplexing difficulties encountered during the field's first decade.

Early computers could do arithmetic like lightning, but they could not think like mathematicians. That changed when Edward Feigenbaum and his colleagues created programs that could draw new conclusions by reasoning from axioms. Later, Feigenbaum would create the expert system, which reasons from a "knowledge base" encoded from the experience of human researchers. *(Photo by William F. Miller)*

During the 1960s, Feigenbaum worked to develop systems that could perform induction (that is, derive general principles based on the accumulation of data about specific cases).

Working on a project to develop a mass spectrometer for a Mars probe, Feigenbaum and his fellow researchers became frustrated by the computer's lack of knowledge about basic rules of chemistry. Feigenbaum then decided that such rules might be encoded in a "knowledge base" in such a way that the program could apply it to the data being gathered from chemical samples. The result in 1965 was Dendral, the first of what would become a host of successful and productive programs known as expert systems. A further advance came in 1970 with Meta-Dendral, a program that could not only apply existing rules to determine the structure of a compound, it could also compare known structures with the existing database of rules and infer new rules, thus improving its own performance.

During the 1980s Feigenbaum coedited the four-volume *Handbook of Artificial Intelligence*. He also introduced expert systems to a lay audience in two books, *The Fifth Generation* (coauthored with Pamela McCorduck) and *The Rise of the Expert Company* (coauthored with McCorduck and H. Penny Nii).

Feigenbaum combined scientific creativity with entrepreneurship in founding a company called IntelliGenetics and serving as a director of Teknowledge and IntelliCorp. These companies pioneered the commercialization of expert systems. Feigenbaum and his colleagues created the discipline of "knowledge engineering"—capturing and encoding professional knowledge in medicine, chemistry, engineering, and other fields so that it can be used by an expert system. In what he calls the "knowledge principle," Feigenbaum asserts that the quality of knowledge in a system is more important than the algorithms used for reasoning. Thus Feigenbaum has tried to develop knowledge bases that might be maintained and shared as easily as conventional databases.

Remaining active in the 1990s, Feigenbaum was second president of the American Association for Artificial Intelligence and (from 1994 to 1997) chief scientist of the U.S. Air Force. He contributed his expertise by serving on the board of many influential research organizations, including the Computer Science Advisory Board of the National Science Foundation and the National Research Council's Computer Science and Technology Board. The World Congress of Expert Systems created the Feigenbaum Medal in his honor and made him its first recipient in 1991. In 1995, Feigenbaum received the prestigious Association for Computing Machinery's Turing Award.

Further Reading

Feigenbaum, Edward, Julian Feldman, and Paul Armer, eds. *Computers and Thought*. Cambridge, Mass.: MIT Press, 1995.

Feigenbaum, Edward, Pamela McCorduck, and H. Penny Nii. *The Rise of the Expert Company: How Visionary Companies Are Using Artificial Intelligence to Achieve Higher Productivity and Profits*. New York: Vintage Books, 1989.

Shasha, Dennis, and Cathy Lazere. *Out of Their Minds: The Lives and Discoveries of 15 Great Computer Scientists*. New York: Copernicus, 1995.

Felsenstein, Lee

(1945–)

American

Inventor, Engineer

During the 1960s the counterculture espoused imagination, unconventional lifestyles, and a romantic rebellion against the Establishment, or mainstream society. There seemed to be little in common between the hippies and activists and the young computer hackers at MIT and elsewhere who were stretching the limits of computer systems and their application—except perhaps the fact that both tended to think and live unconventionally.

However, there were some individuals who were able to combine attitudes of the counterculture with its social activism and the emerging computer culture with its intense devotion to technology. One such person is Lee Felsenstein, who found that computers and on-line communications could be turned into powerful tools for community activism.

Felsenstein was born in 1945 in Philadelphia. He attended the University of California, Berkeley, and received his B.S. degree in electrical engineering in 1972. Dropping out for a while in 1967, he worked as an electronic design engineer for Ampex, building minicomputer interfaces. By 1972, Felsenstein was working as chief engineer at a nonprofit community collective called Resource One in San Francisco. He came up with the idea of setting up teletype computer terminals connected to a minicomputer at Resource One to enable ordinary members of the public to access information and post messages. As described on Felsenstein's historical web page, a notice on the terminals explained that: "COMMUNITY MEMORY is a kind of electronic bulletin board, an information flea market. You can put your notices into the Community Memory, and you can look through the memory for the notice you want."

There were already computer networks in use at some large universities and companies. Indeed, ARPANET, ancestor of the Internet, was growing. But Community Memory was the first on-line service that could be used by ordinary people with no special computer knowledge. It was used, for example, by musicians seeking to buy or sell guitars or to find a gig.

By the mid-1970s, experimenters were starting to build computer systems using the newly available microprocessor. Felsenstein quickly saw both the technical and social possibilities of these devices, and from 1975 to 1986 he was moderator of a unique organization, the Homebrew Computer Club. This club, founded by Fred Moore and Gordon French, met at first in French's garage. At a typical meeting, experimenters might demonstrate their latest systems or give a talk about an engineering problem. (It was at one such meeting that an enthusiastic crowd of electronics hobbyists first got to see STEVE JOBS and STEVE WOZNIAK demonstrate a prototype of what would soon become the Apple II, perhaps the most successful of early commercial microcomputers.) All in all, more than 20 separate computer-related companies can trace their genesis back to contacts made at Homebrew meetings.

In 1975, Felsenstein applied his own design ideas to a computer called the Sol-20 (named for Les Solomon, one of its original designers). The Sol-20 demonstrated the use of a complete microcomputer system, combining processor, keyboard, and a video interface circuit that Felsenstein designed called the VDM-1. This interface allowed output to be displayed on an ordinary TV set. By allowing a separate chip to do some of the work of preparing video images, the VDM-1 took considerable burden off the main processor. This principle is still used in today's video cards. The Sol-20 was a capable machine, but it would be overshadowed the following year by the Apple II.

Another Felsenstein design was the Pennywhistle 103, a computer modem kit that let the owners of early microcomputers get online to the bulletin board systems (BBS) that were starting to spring up in the late 1970s.

Felsenstein also designed the first portable computer, the Osborne-1, in 1981. This was not a laptop or notebook computer in the modern sense, but more like a suitcase. (An example is on display at the Smithsonian Museum of American History, along with a Sol-1.) Adam Osborne rewarded Felsenstein by making him a founder and chief engineer of Osborne Computer, Inc. Unfortunately, the company went bankrupt in 1983.

In 1984, Felsenstein went back to the social side of the computing field. He was asked by Stewart Brand (founder of *The Whole Earth*

Catalog) to organize the first annual Hacker's Conference. At this time, the word *hacker* still meant an unconventional but highly talented programmer who sought to make computers do exciting new things, not a malicious cyber-thief or vandal.

The late 1980s saw sweeping changes in the Soviet Union under the movement called glasnost, or "openness." In 1989, Felsenstein went to the Soviet Union (soon to be Russia), to found a company called Glav-PC and provide consulting services to technology companies seeking to do business in the newly opened Russian market. During the 1990s, Felsenstein served as senior researcher with Interval Research, a company started by Microsoft executive Paul Allen to help develop new computing technologies. By 2003, Felsenstein had invented a pedal-powered wireless computer to connect some of the world's poorest, most isolated people to the Internet.

Lee Felsenstein was honored in 1994 with the Pioneer Award from the Electronic Frontier Foundation and inducted into the Computer Museum of America Hall of Fame in 1998. His career has demonstrated advances both in hardware technology and the "soft" technology of social organization and activism.

Further Reading

"Community Memory: Discussion List on the History of Cyberspace." Available on-line. URL: http://memex.org/community-memory.html. Updated on April 4, 1997.

Fagan, Kevin. "Pedal-Powered e-mail in the Jungle." *San Francisco Chronicle*, January 17, 2003, pp. A1, A15.

Felsenstein, Lee. "How Community Memory Came to Be, Part 1 and 2." Available on-line. URL: http://madhaus.utcs.utoronto.ca/local/internaut/comm.html, http://madhaus.utcs.utoronto.ca/local/internaut/comm2.html. Updated on January 18, 1994.

Freiberger, Paul, and Michael Swaine. *Fire in the Valley: The Making of the Personal Computer*. 2nd ed. New York: McGraw-Hill, 1999.

Levy, Steven. *Hackers: Heroes of the Computer Revolution*. Updated ed. New York: Penguin, 2001.

Forrester, Jay W.

(1918–)
American
Inventor, Engineer

By the late 1940s, the success of ENIAC, the first large-scale electronic digital computer, had spurred great interest in the possible application of computing to science and business. However, the technology as pioneered by J. PRESPER ECKERT and JOHN MAUCHLY faced obstacles that threatened to limit its usefulness. The vacuum tubes used for processing in early computers were unreliable and the memory used to store data was slow and expensive. These and many other challenges would be met by Jay W. Forrester, whose pioneering efforts would do much to establish the computer industry.

Forrester was only a generation away from a different sort of pioneer. He was born on July 14, 1918, on a remote cattle ranch outside Climax, Nebraska, which his parents had homesteaded. As a boy, Forrester attended a one-room schoolhouse, but he soon learned to supplement his education. Young Forrester had a boundless interest in mechanical and especially electrical matters. In his senior year in high school, he used scavenged auto parts to build a wind-driven 12-volt electrical generator that brought power to the family ranch for the first time.

Following the expected career path for a bright boy in rural Nebraska, Forrester planned to attend the agricultural college at the University of Nebraska. Just as he was about to enroll there, however, he decided that he wanted to pursue his real interest. Instead of agriculture school, he enrolled in the electrical engineering program. He graduated in 1939 as the top student in the program and promptly signed up for graduate studies at the Massachusetts Institute of Technology (MIT).

At MIT, Forrester first worked as a student assistant in the MIT High-Voltage Laboratory, but then switched to the Servomechanisms (Servo) Laboratory, where he learned about the complicated electrically-actuated controls that would be part of a coming revolution in automation.

Like most students of the time, Forrester found his studies diverted by wartime needs. In December 1944, he went to work on an MIT project to build devices for aeronautical research for the U.S. Navy. The main device, called an Aircraft Stability and Control Analyzer (ASCA), was essentially an analog computer. (Unlike today's more familiar digital computers, an analog computer uses the interaction of mechanical or electronic forces to solve problems.)

The problem with the ASCA was that it was supposed to simulate the real-time behavior of an airplane's rudder, elevator, and other controls as a pilot moved them in a plane traveling hundreds of miles an hour. This proved to be difficult to simulate with an analog computer (which is more "tuned" than "programmed") and further, the existing electromechanical mechanisms simply were not fast enough to respond in real time.

A colleague suggested that Forrester learn about a new technology called digital computing, and eventually he was put in touch with the laboratory where J. Presper Eckert, JOHN VON NEUMANN, and other researchers were building an electronic digital computer—the machine that would be known as ENIAC. Forrester was immediately impressed by both the speed of electronic computing (which used no mechanical parts) and the relative ease of use and versatility of programmable machines. He persuaded Gordon Brown, his supervisor and thesis adviser, to shift the Servo Lab's efforts from analog to digital computing, and Forrester became head of a new division, the Digital Computer Laboratory.

ENIAC and its successor EDVAC were basically sequential machines—they performed one task at a time. However, simulating many real-world activities (such as flying a plane) requires calculating and updating many different variables at virtually the same time. Forrester and his assistant Robert R. Everett realized that existing sequential machines, even electronic ones, would be too slow to keep up with the real world. Instead, they began to design a parallel machine, in which different units could perform operations at the same time. Because of its expected speed, the machine would be called Whirlwind.

However, a fundamental problem with the first generation of large-scale computers was the vacuum tube that was its key component. The typical vacuum tube had a life of about 500 hours. The more tubes one put into a machine, the more frequently a tube would blow just when it was needed in a calculation. As a result, the Whirlwind would grind to a halt several times a day.

Forrester addressed this problem in two ways. First, he developed an improved electrostatic storage tube for use in the computer memory, extending the average tube life to more than a month. He further improved reliability by designing a circuit that could sense the changes in electronic characteristics that indicated that a tube would fail soon. These tubes could then be replaced during regular maintenance times rather than disrupting calculations.

As a result of these improvements, by March 1951 the Whirlwind was running reliably for 35 hours a week. Using 16-bit data words, it could perform more than 20,000 multiplications a second—roughly comparable in performance to an early personal computer.

However, Forrester believed that even the improved storage tube was not the real answer to the computer memory problem. For one thing, tubes were relatively slow and consumed a great deal of power. One existing alternative was a cathode ray tube (CRT) similar to that found in television sets. CRTs in the form of the Williams tube were already in use for computer memory, but they were not very reliable. Forrester designed an improved version, the MIT Storage Tube, which stored 1,024 kilobytes' worth of data that could be randomly accessed (that is, any piece of data could be directly fetched). The only problem was that the tubes cost $1,000 each, and only lasted about a month. As recounted by Robert Slater, Forrester later recalled, it "was not feasible computer storage. There was simply nothing that was suitable, and I had a project and a reputation that rested on our solving the problem. So it was very much a case of necessity being the mother of invention."

The CRT was a two-dimensional storage system, "painting" data on the tube much like the tiny pixels that make up a TV picture. Forrester visualized using a three-dimensional storage system instead. If the data could be stored on tiny points within a cubelike lattice, much more data could be fit into the same physical space. But how could the data actually be stored?

Reading through a technical magazine in 1949, Forrester had come upon an advertisement for a material called Deltamax that was used to make cores for magnetic amplifiers. He decided that storing computer data as spots of magnetism rather than electric charge could provide a reliable, fast, relatively compact memory system. Deltamax proved unsuitable, but after experimenting with various materials he hit upon the use of ferrite (iron) cores—tiny doughnut-shaped discs that would be connected to a three-dimensional crisscross grid of wires. To store a binary "1" at a given memory location, electric pulses are sent down the appropriate two wires (something like going to a particular street intersection).

By 1953, the Whirlwind had replaced the storage tubes with two banks of 1024 byte (1 kilobyte) ferrite core memory. (Although it would take a few years for the new commercial computer industry to adopt core memory, for several decades thereafter computer people would use the word *core* as a synonym for computer memory.) The Whirlwind was also fitted with a magnetic drum memory and a tape drive for mass storage.

Forrester then went to work on another military computer project. In 1949, the Soviet Union had exploded its first atomic bomb, and suddenly the task of tracking enemy bombers and directing their interception by fighters became a pressing military problem. Given the speed of modern aircraft and missiles and the complexity of radar systems, the U.S. Air Force decided that an effective air defense system would have to be integrated through the use of high-speed computers. The result was SAGE (Semi-Automated Ground Environment), which had a number of regional air defense centers, each with its own Whirlwind computer, with the machines networked together.

In 1951, Forrester was appointed the project's director. SAGE was a massive and secret undertaking, a sort of computer version of the Manhattan Project which had built the first atomic bomb. Each SAGE installation had a huge Whirlwind II computer weighing 250 tons and using more than 50,000 vacuum tubes for processing. The computer drew a full megawatt of power and needed a massive cooling system to prevent overheating. By 1963, SAGE, with somewhat upgraded computers, had 23 air defense control centers, three combat centers, and a programming center. The system operated until 1983, when it was finally replaced with solid-state technology.

Fortunately, the ability of SAGE to defend against an atomic attack was never tested. The project, however, paid considerable dividends for the development of computer technology and programming methodology. SAGE was not only the biggest vacuum tube computer ever built, it was the first large-scale real-time computer. It pioneered the use of networking and was thus an ancestor of ARPANET and later the Internet. The development of large, complex software programs required new approaches to organization and management, leading to the birth of a new field, software engineering. Further, SAGE operators worked at consoles that displayed information about targets visually on a screen, controlled by a light pen. SAGE thus also contributed to the development of today's personal computer user interface.

By 1956, however, Forrester's career had made a sharp turn from computer engineering to the use of computers to study social interactions, including the operation of business management. He was appointed professor of management at the new MIT Sloan School of Management and then became head of its System Dynamics program. Forrester went on to write many influential papers and books on system dynamics and management.

By the 1970s, Forrester had extended his work on system dynamics into many areas, including economics and the influence of human activity on the environment. In 1970, he developed a computer model for the Club of Rome which predicted that if current trends continued, humanity would face a dismal future marked by overpopulation, pollution, starvation, and poverty. (Forrester's student Dana Meadows summarized these predictions in a book called *The Limits to Growth.*) Critics such as economist Julian Simon, however, challenged the model as skewed and overly pessimistic.

Forrester retired from the Sloan School in 1989, but remained active in applying system dynamics to areas such as economic behavior and education. Forrester has been inducted into the Inventors Hall of Fame and has received numerous honorary degrees and other awards.

Further Reading

"Jay W. Forrester." Available on-line. URL: http://sysdyn.mit.edu/people/jay-forrester.html. Downloaded on November 2, 2002.

Redmond, Kent C., and Thomas M. Smith. *From Whirlwind to MITRE: The R&D Story of the SAGE Air Defense Computer.* Boston, Mass.: MIT Press, 2000.

"SAGE: Cold-War Forerunner to the Information Age." Available on-line. URL: http://www.eskimo.com/~wow-ray/sage28.html. Downloaded on November 2, 2002.

Slater, Robert. *Portraits in Silicon.* Cambridge, Mass.: MIT Press, 1987.

G

⊠ Gates, Bill (William Gates III)
(1955–)
American
Entrepreneur, Programmer

Bill Gates built Microsoft, the dominant company in the computer software field, and in doing so, became the world's wealthiest individual, with a net worth measured in the tens of billions.

Born on October 28, 1955, to a successful professional couple in Seattle, Gates was a teenager when the first microprocessors became available to electronics hobbyists. Gates showed both technical and business talent as early as age 15, when he developed a computerized traffic-control system. He sold his invention for $20,000, then dropped out of high school to work as a programmer for TRW. By age 20, Gates had returned to his schooling and become a freshman at Harvard, but then he saw a cover article in *Popular Electronics*. The story introduced the Altair, the first commercially available microcomputer kit, produced by a tiny company called MITS. This machine consisted of a case with an Intel 8080 processor, a small amount of memory, and supporting circuitry. The "display" consisted of a panel of blinking lights. Users had to come up with their own keyboard and other peripherals.

Although the Altair looked more like a science fair project than a business machine, Gates believed that microcomputing would soon become a significant industry. He recognized that to be useful, the new machines would need software, and whoever could supply what would become "industry standard" software would achieve considerable financial success.

Gates and his friend Paul Allen began by creating an interpreter for the BASIC language that could run on an Altair with only 4 kilobytes of memory. Since the Altair itself did not have enough memory for the editor and other tools needed to develop the program, Gates and Allen used a minicomputer at the Harvard Computation Center to simulate the operation of the Altair's processor. Surprisingly, the program worked when they first tried it out on an actual Altair.

The BASIC interpreter made it possible for people to write useful applications without having to use assembly language. This first product was quite successful, although to Gates's annoyance it was illicitly copied by Altair users at the Homebrew Computer Club and distributed for free. Although the teenage Gates had been a sort of proto-hacker, freely exploring whatever computer systems were available, the businessperson Gates accused the copiers of stealing intellectual property for which programmers were entitled to be paid.

Bill Gates is the multibillionaire CEO of Microsoft Corporation, the leader in operating systems and software for personal computers. His energy and market savvy put him in the same league as Ford and Rockefeller. However, in 2002 a federal court ordered the company to stop using its market position unfairly. *(Courtesy of Microsoft Corporation)*

MITS and the Altair soon disappeared, supplanted by the late 1970s by full-fledged microcomputers from Apple, Radio Shack, Commodore, and other companies. But Gates had shown a shrewd business insight in his early dealings. When he developed the Altair version of BASIC, he insisted on license terms that let him distribute what became known as Microsoft BASIC to other computer companies as well. Thus, as Apple and the other companies developed new microcomputers, they turned to Gates and Allen's new company (originally named Micro-Soft) for this essential software. Even at this early stage, whenever Gates was asked what his business goal was, he replied (as recalled by industry pundits such as Robert X. Cringely) "A computer on every desk and in every home, running Microsoft software." By the end of the 1970s, Microsoft had become the dominant company in computer languages for microcomputers, offering such "grown-up" languages as FORTRAN and even COBOL.

However, the big breakthrough came in 1980, when IBM decided to market its own microcomputer, the IBM PC. IBM had decided that it needed to enter the burgeoning personal computer (PC) market quickly, so contrary to its usual process of developing its own hardware and software, "Big Blue," as IBM is known, decided to buy a third-party operating system. At the time, the PC operating system market was dominated by GARY KILDALL's Digital Research, makers of CP/M (Control Program/Microcomputer), an operating system that ran on the majority of the larger microcomputer systems. However, when negotiations between IBM representatives and Kildall broke down due to miscommunication, IBM decided to hedge its bets by looking for other operating systems it could offer its users.

Gates agreed to supply IBM with a new operating system. Buying an operating system from a small Seattle company, Microsoft polished it a bit and sold it as MS-DOS 1.0. The IBM PC was announced in 1981, and by 1983 it had become the standard desktop PC for business and many home users. Although PC users could buy the more expensive CP/M or Pascal operating systems, most stuck with the relatively inexpensive MS-DOS, so virtually every IBM PC sold resulted in a royalty payment flowing into the coffers of Microsoft. Sales of MS-DOS further exploded as many other companies rushed to create clones of IBM's hardware, each of which also needed a copy of the Microsoft product.

Now dominant in both languages and operating systems, Gates and Microsoft looked for new software worlds to conquer. In the early 1980s, Microsoft was only one of many thriving competitors in the office software market. Word processing was dominated by such names as WordStar and WordPerfect, Lotus 1-2-3 ruled the spreadsheet roost, and dBase II dominated databases. But Gates and Microsoft used the steady revenues from MS-DOS to undertake the creation of Windows, a much larger operating system, or OS, which offered a graphical user interface. The first versions of Windows were clumsy and sold poorly, but by 1990 Windows 3.0 had become the new dominant OS and Microsoft's annual revenues exceeded $1 billion. Gates relentlessly leveraged both the company's technical knowledge of its own OS and its near monopoly in the OS sector to gain a dominant market share for Microsoft's word processing, spreadsheet, and database programs.

By the end of the decade, however, Gates and Microsoft faced formidable challenges. The most immediate was the growth of the Internet and in particular the World Wide Web, invented by TIM BERNERS-LEE, and the Netscape Web browser, developed by MARC ANDREESSEN. The use of JAMES GOSLING's Java language with Web browsers offered a new way to develop and deliver software that was independent of the particular machine upon which the browser was running. Gates feared this might give competitors a way around Microsoft's OS dominance. That dominance itself was being challenged by Linux, a version of Unix created by Finnish programmer LINUS TORVALDS. Linux was essentially free, and talented programmers were busily creating friendlier user interfaces for the operating system.

Gates, who had focused so long on the single desktop PC and its software, admitted that he had been caught off guard by the Internet boom, but said that he would focus on the Internet from now on. In his 1995 book *The Road Ahead*, he proclaimed a vision in which the Internet and network connectivity would be integral to modern computing, and Microsoft programmers were soon hard at work improving the company's Internet Explorer Web browser. Gradually, Explorer displaced Netscape, and other Microsoft software acquired Web-related features.

However, Netscape and other competitors believed that Microsoft had an unfair advantage. They accused Microsoft of using its dominance to virtually force PC makers to include Windows with new PCs. Then, by including web browsers and other software "free" as part of Windows, Microsoft made it very difficult for competitors who were trying to sell similar software. Antitrust lawyers for the U.S. Department of Justice and a number of states agreed with the competitors and began legal action in the late 1990s. In 2000, a federal judge agreed with the government and later an appeals court essentially ratified a proposed settlement that would not break up Microsoft but would restrain a number of its unfair business practices.

Gates's personality often seemed to be in the center of the ongoing controversy about Microsoft's behavior. Positively, he has been characterized as having incredible energy, drive, and focus in revolutionizing the development and marketing of software. But that same personality is viewed by critics as showing arrogance and an inability to understand or acknowledge the effects of its actions. Gates often appears awkward and even petulant in his appearance in public forums. On the other hand, Gates and his wife, Melinda, have through their foundation quietly given more to charity than anyone else in the world, awarding grants for everything from providing computer access to public libraries to developing vaccines to fight AIDS. Gates's achievements and resources guarantee that he will be a major factor in the computer industry and the broader American economy for years to come.

Further Reading

Cringely, Robert X. "The New Bill Gates: A Revisionist Look at the Richest Man on Earth." Available on-line. URL: http://www.pbs.org/cringely/pulpit/pulpit20001123.html. Posted on November 23, 2000.

Erdstrom, Jennifer, and Martin Eller. *Barbarians Led by Bill Gates: Microsoft from the Inside, How the World's Richest Corporation Wields Its Power*. New York: Holt, 1998.

Gates, Bill, Nathan Myhrvold, and Peter M. Rinearson. *The Road Ahead*. Rev. ed. New York: Penguin, 1996.

Lowe, Janet C. *Bill Gates Speaks: Insight from the World's Greatest Entrepreneur*. New York: Wiley, 1998.

Gelernter, David Hillel

(1955–)
American
Computer Scientist, Writer

It was July 24, 1993, a bit after eight in the morning. Computer science professor David Gelernter was sitting at his desk in his Yale University office. He reached for a package, thinking that it held one of the many student dissertations that professors were often called upon to review. He started to open it and, as he would later recall: "There was a hiss of smoke, followed by a terrific flash." The force of the explosion was so great that it embedded shrapnel in steel file cabinets.

Gelernter was badly hurt by the blast, and staggered into the campus health center. He was rushed by ambulance to the emergency room, where doctors fought to stop him from bleeding to death. He recovered with the aid of a series of surgical operations, but he was left with lasting injuries and ongoing pain. His right hand was maimed, and he would lose the hearing in one ear and part of the sight of one eye. As recounted in his 1997 book *Drawing Life*, surgery and physical therapy gradually restored some use to the remainder of his right hand such that he could type and even draw—which, given the importance of art to his life, was very significant.

The package had come from a reclusive former mathematician named Theodore Kaczynski, who would become known as the Unabomber. By the time he was arrested in 1996, Kaczynski's bombs, targeted mainly at people in computer and other high-tech fields, had killed three people and injured 23.

Gelernter was born in 1955 and grew up in Long Island, New York. Although his father had been an early computer pioneer, Gelernter at first was more interested in the humanities. As a child he showed considerable talent for art, winning a youth drawing competition, though he decided not to go to art school. Later, as a Yale undergraduate, he studied classical Hebrew literature as well as art and music. However, feeling the need for better career prospects, he turned to computer science. Gelernter received his B.A. degree in computer science at Yale in 1976, and earned his Ph.D. at the State University of New York (where his father taught) at Stony Brook in 1982. Ever since, he has been on the Yale faculty.

In the late 1970s, Gelernter became interested in parallel processing, the parceling out of programs so that a number of processors can work on various parts of the problem while exchanging data and avoiding conflicts. Together with Nicholas Carriero, Gelernter designed a "coordination language" called Linda. A Linda program uses data and variables organized into structures called "tuples," and a mechanism to "glue" or bind the separately running program components together. By harnessing the available computing power more efficiently, Linda allowed researchers to tackle problems that would otherwise require an expensive supercomputer.

Gelernter spent at least as much energy thinking about how the arrival of powerful computing systems would change human life. In his book *Mirror Worlds* (1992), he describes a new use for the massive computing power that parallel processing can bring. He envisions a world in which communities and organizations create virtual reality counterparts on-line.

For example, a student considering attending Yale might visit its mirror world and attend lectures, walk around the campus, explore its history, and otherwise experience it much as a real-world visitor would. This virtual world would be dynamic, not static like a book or even a website. Rather than being some sort of simulation, it would be created constantly from the real-time data being captured by digital video cameras and other devices.

To create and maintain this world, Gelernter's pioneering techniques for coordinating many computer processors would have to be used to their utmost in order to keep up with the vast the streams of data constantly being carried from the real campus to its virtual counterpart. He was eager to accept the challenge. In "The Metamorphosis of Information Management," published in the August 1989 issue of *Scientific American*, Gelernter declared that "What iron, steel and reinforced concrete were in the late 19th and early 20th centuries, software is now: the preeminent medium for building new and visionary structures." By 1990, Gelernter was working on new ways to search, filter, and organize information, outlining techniques that in the later 1990s would become known as "data mining."

Gelernter also made it clear that the pervasive connection between people and computers was not without its problems and challenges. His interest in the arts and humanities led him to urge that humans bring as much of their feelings and experience as possible into the process of designing technology. Unfortunately, in the mind of Theodore Kaczynski, the Unabomber, Gelernter was a just a high profile "techno-nerd." He became the 23rd target of Kaczynski's years-long terror campaign.

Following the bomb attack, Gelernter's hands were bandaged and he was unable to hold a book to read. Instead, he drew upon a rich store of poetry that he had learned by heart, as well as listening to the Beethoven string quartets that he believes are "probably among the spiritually deepest of all human utterances."

However, Gelernter also became angry at the way many people in the media were treating him. In particular, he did not want to be viewed as a victim. As he recounts in his book, *Drawing Life: Surviving the Unabomber*:

> When a person has been hurt and knocked down, he wants to stand up and get on with his life and as best he can, without denying what's happened to him. But the idea that you would enjoy wallowing in victimhood—and this is—it's difficult to overstate how relentless this obsession is in the press. It's not a word that comes up once or twice but something you hear again and again.

Gelernter was particularly upset when some accounts in the press suggested that his views on the need for a human dimension to technology were somehow similar to those found in the Unabomber's antitechnology manifesto.

In his writings during the years following the attack, Gelernter emphasized a culturally conservative critique of modern intellectuals. He argued that they had taken the virtue of tolerance and elevated it to an absolute refusal to make moral judgments. As a result, according to Gelertner, society has released the floodgates that restrain violence and destructiveness.

Like cyber-critic CLIFFORD STOLL, Gerlernter became increasingly skeptical about the ways in which computers were actually being used. But unlike the Unabomber's "solution" of blowing up the machines and their adherents, Gelertner suggests in his later works that technology can and should be humanized. In *The Muse in the Machine* (1994) he provides a counterpoint to the bitter sharpness of *Drawing Life*. He suggests that the quest for an abstract artificial intelligence would fail until researchers began to create "artificial emotions" that would allow machines to experience senses and feelings more like those out of which human consciousness arises. To do so, he suggests that science become reacquainted with the world of artists, poets, and musicians.

Consciousness, Gelernter said, was like a continuum or spectrum. At one end is the "high focus" of logical reasoning, typified by abstract mathematics and the logic encoded in most computer programs. At the other end, that of "low focus," attention broadens to encompass feelings, symbolism, and analogy. He believes that "When people suddenly place right next to each other two things that they never used to think of together, that seem completely unrelated, all of a sudden the creative insight emerges." For artificial intelligence to be real intelligence, it would have to somehow accommodate both ends of the spectrum.

Gelernter believes that current computer systems fall far short of helping humans manage their lives better. He admits that the Internet is useful for buying things and finding particular pieces of information, but says that it in its present form it is crude and "prehistoric" because of the lack of thoughtful structuring of knowledge. He has also suggested replacing the idea of a computer user's electronic desktop, full of poorly organized files, with a sort of database called a lifestream. It would organize all work and data in relation to the passage of time, allowing people to find something according to what was happening in their lives around that time. He has developed a software program called Scopeware that provides such a facility for PC and palm computer users.

Further Reading

Gelernter, David Hillel. *Drawing Life: Surviving the Unabomber*. New York: Free Press, 1997.

———. *Machine Beauty: Elegance and the Heart of Technology*. New York: Basic Books, 1999.

———. *Mirror Worlds: Or the Day Software Puts the Universe in a Shoebox: How It Will Happen and What It Will Mean*. New York: Oxford University Press, 1992.

———. *The Muse in the Machine: Computerizing the Poetry of Human Thought*. New York: Free Press, 1994.

Gibson, William
(1948–)
American/Canadian
Writer

Many people think that the branch of literature called science fiction (SF) is mainly about predicting the future, but SF writers generally believe their work is more about the *idea* of the future than the attempt to foresee it in detail. SF authors did more or less predict such achievements as space travel and even the atomic bomb (described by H. G. Wells in 1914), but the literary crystal ball was far less successful when it came to computers. The "Golden Age" science fiction of the 1930s usually had its space pilots navigating with the aid of slide rules and steering their ships manually. Even after the digital computer was invented, science fiction computers tended to resemble bigger ENIACs or Univacs that might control the world. The idea of personal computers used by millions of ordinary people in their daily lives was scarcely anticipated.

Meanwhile, science fiction writers had created a rich tapestry of stories about space travel, alien encounters, future cities, and robots. But by the 1960s and 1970s, SF writers such as Harlan Ellison, John Brunner, and Philip K. Dick had begun to experiment with new forms. Epic quests and the "space opera" battles between good and evil were replaced by stories that gave greater emphasis to the social and psychological aspects of future changes.

In the 1980s, a writer named William Ford Gibson would touch off a new literary movement. Ironically, a writer who insists on using a typewriter rather than a word processor would become known for depicting a dark, surreal future in which computer technology both amplifies and transforms the human mind.

Gibson was born on March 17, 1948, in Conway, South Carolina. His father, a contractor at the Oak Ridge, Tennessee, facility where the atomic bomb was built died when William was only six years old. The family moved to Whytheville, near the Appalachian foothills in Virginia where Gibson's mother helped set up the small town's first library and served as its volunteer librarian. This naturally encouraged Gibson's love of books. As a young teenager, he became an enthusiastic science fiction fan and started contributing articles and drawings to fanzines. However, when Gibson went to an Arizona boarding school for his later high school years, his interest in science fiction diminished.

When he was 19, Gibson's mother died. The Vietnam War was at its height, and like many thousands of other American young men he moved to Canada to escape the draft, making only occasional trips back to the United States. Gibson earned a B.A. degree in literature at the University of British Columbia in 1977. He and his wife (a language instructor) and their children have lived in Vancouver, British Columbia, since 1972.

When their first child was born, Gibson's wife had better career prospects than did her husband, so Gibson became a stay-at-home dad. He began to use his spare time to write short stories. His first notable success was "Burning Chrome," published in 1982 by *Omni* magazine. In this story, Gibson coined the term *cyberspace* to refer to the reality that computer-linked users experienced.

With his first novel, *Neuromancer* (1984) Gibson launched something that could be called either a new kind of science fiction or a new genre of literature entirely. The novel begins with this line: "The sky above the port was the color of television, tuned to a dead channel." In its desolation, this image echoes that used by the poet T. S. Eliot in his long poem *The Love Song of J. Alfred Prufrock,* in which the reader is invited to go "When the evening is spread out against the sky / Like a patient etherised upon a table." Gibson further develops the contrast between the flashy, fast-paced technology used by elite cyber-"cowboys" who "jack" their minds into the computer network, and the emotionally dead world that they inhabit. It is a world of fragments and flashes of connection. It is also a world where the artificial and the natural have merged or interpenetrated each other, where the sky is television and people live inside the machine.

In the novels that followed, *Count Zero* (1986) and *Mona Lisa Overdrive* (1988), Gibson further develops his near-future world. The world is culturally diverse (many striking scenes are set in high-tech Tokyo), but dominated by global corporations that have secrets they will readily kill to protect. The physical environment is decayed and polluted. Discovery, pursuit, and escape drive the plot at a rapid pace, and there are echoes of the noir film and the hard-boiled detective made famous by Raymond Chandler.

The word *cyberpunk* had been coined as early as 1980 to refer to stories with amoral, gritty, high-tech characters. (It also drew from the punk rock movement of the 1970s.) But Gibson's work led to a vigorous literary movement, as other writers such as Bruce Sterling, Rudy Rucker, and John Shirley began to explore the impact of computer technology on society.

Gibson himself was not particularly comfortable with the cyberpunk label. He continued to explore divergent aspects of the cyberworld and to evoke different atmospheres. His *The Difference Engine* (1990) is set in an alternate history in which the mechanical computer invented by CHARLES BABBAGE was actually built (powered by steam), and high technology was combined with a Victorian worldview. A later, loosely connected group of novels includes *Virtual Light* (1993), which portrays the survivors of a millennial cataclysm in a divided California; *Idoru* (1996) describes a new kind of media star—an artificially created celebrity; and *All Tomorrow's Parties* (2000) continues the chase to a post-quake San Francisco Bay Bridge inhabited by a sprawling community of outcasts.

Gibson's work has been adapted to other media. His story *Johnny Mnemonic* became a motion picture in 1995, and numerous video games of the late 1990s portray post-apocalyptic cyberpunk-style settings.

By then, cyberpunk as a distinctive movement had faded, with its imagery becoming part of the stock-in-trade of science fiction. However, the idea of cyberspace had taken on a life of its own in the computer field. Although Gibson has claimed to know nothing about computers (and little about science in general), his depiction of the social effects of computer networking were seen as quite relevant to life in Internet chat rooms and on-line games. Everyone from computer scientists to teenage hackers began to think in terms of cyberspace. Gibson's work influenced the creation of virtual reality software by such pioneers as JARON LANIER, as well as the efforts of privacy advocates and activists such as the "cypherpunks" who want ordinary people to have access to encryption to protect their privacy from governments and corporations.

Gibson has won the highest honors in science fiction, including both the Nebula and Hugo Awards for *Neuromancer* as best novel, 1984.

Further Reading

Gibson, William. *Burning Chrome*. New York: Arbor Books, 1986.

———. *Neuromancer*. New York: Ace Books, 1984.

———. *Virtual Light*. New York: Bantam Books, 1993.

"Gibson, William." In *Contemporary Novelists*. 7th ed. Chicago: St. James Press, 2001.

"History of Cyberpunk." Available on-line. URL: http://project.cyberpunk.ru/idb/history.html. Updated on November 2, 2002.

Olsen, Lance. *William Gibson*. San Bernardino, Calif.: Borgo Press, 1992.

Goldstine, Adele

(1920–1964)
American
Mathematician, Programmer

During the 1940s, the abstruse world of theoretical mathematics and the switches and vacuum tubes of electrical engineers came together in the development of the modern computer. While it seemed logical to hire mathematicians to program the first computers, this did not mean that it was easy for them to adjust to a new technology and in the process create a new discipline. Adele Goldstine was one of the pioneers who built the bridge between mathematics and technology.

Goldstine was born on December 21, 1920, and grew up in New York City. She went to Hunter College High School and then attended Hunter College, earning a B.A. degree in mathematics. She then went to the University of Michigan for graduate study, earning her master's degree in mathematics. While there she met her future husband, HERMAN HEINE GOLDSTINE, also a mathematician. They were married in 1941.

As the United States entered World War II, Herman Goldstine, a first lieutenant in the army, was assigned to the Aberdeen Proving Grounds in Maryland. His task was to prepare the ballistic tables needed for aiming guns and bombs. He quickly realized that people using the mechanical calculators of the time could never keep up with the demand for these tables, which had to be prepared for every combination of gun, shell, and fuse. Indeed, a single complete calculation needed 20 hours of work by a skilled operator. (The differential analyzer, a mechanical analog computer invented by VANNEVAR BUSH, was considerably faster, but not fast enough.) Herman Goldstine then learned about the work of J. PRESPER ECKERT and JOHN MAUCHLY, who were proposing that a new kind of calculator, called ENIAC, be built using electronic (vacuum tube) technology.

Meanwhile, Adele Goldstine had joined the war effort. At first she worked as one of the mathematicians, or "computers," assigned to manually calculate tables at the Moore School in Philadelphia (at the time, *computer* referred to a person, not a machine.) By the time the group numbered 75 it had exhausted the local pool of mathematically trained people, so she was sent on a tour of universities throughout New England in search of more recruits. Eventually this effort was supplanted by the WACs (Women's Army Corps) which provided additional trained personnel.

Adele Goldstine, however, was about to get a challenging new job. When ENIAC was finally ready in 1946, the war was over but the appetite for calculation was as great as ever. People were clamoring for computers for everything from designing sleek new jet planes to understanding nuclear reactions in proposed power plants. ENIAC was a huge machine, filling a 30-by-50-foot room with rows of cabinets along each wall. When it was running, the ENIAC's 18,000 vacuum tubes drew 160,000 watts of power, dimming lights in the surrounding neighborhood. ENIAC was about 1,000 times faster than any mechanical calculator, but it was hard to access that power. The reason is that ENIAC was not like modern computers in one important respect. There were no instructions that could be typed into a terminal or compiled into programs to direct the machine. In fact, ENIAC could not store its own instructions at all. To set up ENIAC to tackle a problem, the machine had to be literally rewired by setting hundreds of switches and plugging in cables. The machine had "drawers" called digit trays full of numbers where the constant values needed for the program were input.

Scientists who wanted to take advantage of ENIAC's capabilities and the growing number of people learning to program the machine (including women such as JEAN BARTIK and FRANCES ELIZABETH HOLBERTON) needed to have a clear description of this new and unfamiliar technology. Adele Goldstine was given the job of documenting the operation of ENIAC. As she wrote in her diary (quoted in *Notable Women Scientists*):

> At first I thought I would never be able to understand the workings of the machine since this involved a knowledge of electronics that I did not have at all. But gradually as I lived with the job and the engineers helped to explain matters to me, I got the subject under control. Then I began to understand the machine and had such masses of facts in my head I couldn't bring myself to start writing.

Goldstine eventually wrote a complete set of manuals for the machine, ranging from a technical overview to detailed information for the technicians who would have to maintain it.

Later, Goldstine and mathematician JOHN VON NEUMANN worked together to give ENIAC the ability to "understand" and store instructions similar to those found in modern computers. With 50 instructions permanently wired into the machine, ENIAC could now be programmed from punch cards instead of having to be set up by hand. Goldstine also developed flow diagrams—charts that became the most common tool for designing programs and describing their operation. She also wrote many of the major programs for ENIAC to enable it to perform calculations for the atomic laboratory at Los Alamos, New Mexico.

Adele Goldstine's career was tragically cut short when she developed cancer and died in November 1964. In recent years, her pioneering work has been honored in retrospective, at meetings and in historical publications.

Further Reading

"Goldstine, Adele." In *Notable Women Scientists*, Proffitt, Pamela, editor. Detroit: Gale Group, 2000.

Goldstine, Adele. "Report on the ENIAC." Ordnance Department, U.S. Army, 1946. Available online. URL: http://ftp.arl.army.mil/~mike/comphist/46eniac-report. Downloaded on December 1, 2002.

McCartney, Scott. *ENIAC: The Triumphs and Tragedies of the World's First Computer*. New York: Berkeley Books, 1999.

Goldstine, Herman Heine

(1913–)
American
Mathematician, Writer

Like most inventions, the modern digital electronic computer resulted from a combination of circumstances—in this case, the wartime needs of the military, the availability of suitable technology, and an insight into how that technology might be used. The other key ingredient was the ability to link innovators with the military and government authorities who might be willing to finance their research. Herman Goldstine played an important role in the birth of the computer through both technical contributions and leadership.

Goldstine was born on September 13, 1913, in Chicago. He received his B.S. degree in mathematics at the University of Chicago and continued there, earning his M.S. degree in 1934 and Ph.D. in 1936. He then served as a research assistant and instructor. In 1939, he moved to the University of Michigan, becoming an assistant professor.

Shortly after the United States entered World War II, Goldstine joined the U.S. Army. In August 1942, he was transferred to the Ballistic Research Laboratory as a first lieutenant. His assignment was to speed up the process of calculating trajectories for guns and bombs, a tedious and exacting process that had to be repeated for each variation in gun, elevation, shell type, wind speed, humidity, and so on.

At the time, the calculations were performed by teams of skilled mechanical calculator operators, often women with mathematics degrees, who were known as "computers." (Given the prevailing attitude of the times, it was rare for women to work as full-fledged mathematicians.) Manually calculating just one ballistics table took 20 person-hours. The differential analyzer, a mechanical analog computer invented by VANNEVAR BUSH, could reduce this time to 15 minutes, but with thousands of tables needed this was not fast enough.

A month after arriving at his new job, however, Lieutenant Goldstine learned that two researchers at the University of Pennsylvania's Moore School of Electrical Engineering, J. PRESPER ECKERT and JOHN MAUCHLY, were looking into the possibility of building a new kind of calculator. Instead of using mechanical or even electrical relays, they proposed using vacuum tubes. This meant that the switching between the binary values of 1 and 0 would be at the speed of electrons, allowing for calculations to be at least 1,000 times faster.

Goldstine went to the Moore School and got up to speed on Eckert and Mauchly's ideas. He was in a key position to get them the funding they needed, thanks to his work at the Ballistic Research Laboratory. Ordinarily the army officials, or "brass," tended to be skeptical about apparently outlandish proposals from civilians. But by late 1942, the war was reaching a crucial point. Facing challenges such as the German U-boats in the Atlantic and locked in a "wizard's war" fought over radar systems, military leaders were starting to realize how vital technological superiority would be. Being able to "crunch numbers" thousands of times faster just might be the difference that would make the hundreds of thousands of guns and bombs rolling off America's assembly lines effective enough to turn the tide.

Thus, when Goldstine brought Eckert and Mauchly to his superiors, they quickly agreed to fund would be known as Project PX. They would build ENIAC (Electronic Numerical Integrator and Computer). Goldstine then used his mathematical knowledge to help with designing the machine. By late 1945, ENIAC was performing calculations for the secret nuclear weapons project. The machine was officially unveiled on February 18, 1946.

Much of the operation of ENIAC was handled by a group of talented (and now largely forgotten) women, including JEAN BARTIK, FRANCES ELIZABETH HOLBERTON, and Goldstine's wife, ADELE GOLDSTINE, who wrote the system documentation and reports and supervised the training of programmers.

Herman Goldstine left the army in 1945, and became a researcher with the Institute for Advanced Study at Princeton, New Jersey, from 1946 to 1957. Together with JOHN VON NEUMANN, Goldstine worked on the development of the institute's own computer. Their research included improving computer architecture by giving a machine the ability to store its program instructions along with data. This "stored program concept" made computers more flexible and easier to program. For example, a program could repeat an operation (loop) without having to continually reload cards or tape. In the debate between von Neumann and Mauchly over who had originated the stored program concept, Goldstine strongly defended von Neumann's claim.

In 1958, Goldstine moved to IBM, as a research planner and then as director of mathematical sciences at the Thomas J. Watson Research Center (1958–65), director of scientific development of the data processing division (1965–67) and consultant to the director of research (1967–69). Goldstine was appointed an IBM fellow in 1967, serving until his retirement in 1973.

Goldstine's interests ranged beyond the technical to the historical and philosophical. His 1972 book *The Computer from Pascal to von Neumann* recounts the development of the key ideas of computing, especially those of John von Neumann, which led to the modern computer's ability to store programs as well as data. The discussion of ideas is spiced with personal stories from the early days of computing.

Goldstine was elected to the American Philosophical Society and served as its director from 1984 to 1997. He has also received the Harry Goode Award from the American Federation of Information Processing Societies (1979), the Institute of Electrical and Electronics Engineers Pioneer Award (1982), and the National Medal of Science (1985).

Further Reading

Goldstine, Herman. *The Computer from Pascal to von Neumann*. Princeton, N.J.: Princeton University Press, 1972.

McCartney, Scott. *ENIAC: The Triumphs and Tragedies of the World's First Computer*. New York: Berkeley Books, 1999.

Gosling, James

(1955–)

Canadian

Computer Scientist

Users interact with today's webpages in many ways, ranging from simple forms to elaborate online games and simulations. TIM BERNERS-LEE established the basic structure for webpages with his hyptertext markup language (HTML). Developers could add animation and interactive features using facilities such as CGI (common gateway interface), but writing larger, more elaborate programs to run on webpages was difficult. This changed when Canadian computer scientist James Gosling turned a failed software project into Java, one of the most successful programming languages of all time.

Gosling was born in Canada in 1955. He earned a bachelor's degree at the University of

Calgary in 1977 and in 1983 received his doctorate at Carnegie Mellon University, an institution noted for advanced work in computer science.

During the 1980s, Gosling worked on some important extensions to the UNIX operating system. These included a version of UNIX for computers with multiple processors and "Andrew," a programming system and toolkit that gives UNIX users a graphical interface similar to that in Windows and Macintosh systems. In 1984, Gosling joined Sun Microsystems, a developer of powerful graphic workstation computers.

In 1991, Gosling and other researchers began to develop a programming language code-named Green. This language was object-oriented, encapsulating data and functions together into components that could communicate with each other. Sun then combined Gosling's project with another effort called First Person. The new focus was to develop a programming system that would allow applications to be written for "smart" consumer products such as TV remote controls. However, the consumer electronics industry proved to be less than enthusiastic about the proposed products.

Meanwhile, another Sun researcher, David Ungar, was involved in a project called Self. It was an effort to develop a programming system in which programs could be run on a variety of different computer systems by providing a "virtual machine" for each system. The virtual machine on each computer executed the program's instructions and tied them into the particular codes needed by that platform. This meant that a developer could write one program and have it run on many different machines—unlike Microsoft Windows, for example, which runs only on Intel-compatible PCs.

Gosling decided to create a language that would be suitable for such multiplatform programming. His language, code-named Oak, was also object-oriented, and took much of the structure and syntax of BJARNE STROUSTRUP's popular C++ language. Gosling streamlined the somewhat complex C++ language by removing what he felt were seldom used, misunderstood, and puzzling features.

In 1995, the public was given an overview of the new language, now called Java. Java was vigorously promoted by another key Sun computer scientist, BILL JOY. Sun soon made a Java Development Kit available for free to software developers to encourage them to begin to write Java applications. The timing could not have been better: by the mid-1990s the graphical web browser first developed by MARC ANDREESSEN

James Gosling invented Java, a language ideal for brewing up applications to run on Web browsers and servers. For a time, Java's ability to run on any sort of hardware or operating system seemed to threaten the dominance of Microsoft Windows on the desktop. *(Courtesy of Sun Microsystems)*

was making the Web attractive and accessible to millions of PC users. Web developers were thus eager to find a way to quickly write small programs that users could download and run from websites.

Java offered many advantages for this purpose. Many programmers already knew C++, so they could quickly pick up Java. Java programs could be developed using the free development system from Sun, and once linked to a webpage, anyone using Netscape, Microsoft Internet Explorer, or other browsers could access the page, click on a link or button, and the program would download and then be run by the "Java virtual machine" in the Web browser. (Java could also be used to write stand-alone programs that did not need a Web browser to run.)

The splash made by Java alarmed BILL GATES, chief executive officer of Microsoft. Since Java programs could run on any desktop, laptop, or handheld computer that had a virtual machine, it did not matter what operating system the machine used. If Java became popular enough, software developers might start writing major business software such as word processors, databases, and spreadsheets using Java. A computer maker could then sell computers with a free operating system such as LINUS TORVALDS's Linux, and users might have everything they need without having to ever buy a copy of Microsoft Windows.

Gates responded by creating new versions of programming languages and tools that made it easier for Web developers to write software for Windows, taking advantage of the huge number of machines already running that operating system. Although Java was embraced by IBM and many other companies, Microsoft tried to discourage its use. So far, while Java has not replaced Windows, it has maintained its strength as the preferred language for Web development.

Gosling continues as a vice president and researcher at Sun, holding a fellowship. Recently, he has been working on specifications for a version of Java for real-time programming as well as on creating Java systems tailored to the small hand-held computers called personal digital assistants (PDAs).

Like many other computer scientists, Gosling is also interested in the growing importance of "distributed computing"—having many computers and computerlike devices in homes and workplaces communicating with one another and sharing data. But this new world of computing brings risks as well as powerful capabilities. As Gosling said at a 2002 conference: "When consumer electronics companies start to attempt networking, it's frightening. You read their specifications and say, 'That was a bad idea 20 years ago—and it still is.' You can't just add a coat of 'secure' paint when you're done." Gosling's security concerns extend to all the new kinds of Web services that have been hyped in recent years, such as "pushing" content onto users' screens.

Microsoft's latest salvo in the battle to gain control of Web development is called .Net (or dot-Net), a computer language framework designed to let programs and components work seamlessly over the network. Although .Net has the advantage of being a single structure backed by the resources of the world's most powerful software company, Gosling believes that the J2EE (Java 2 Enterprise Edition) offers a more flexible range of technologies for users with varying needs. Besides, as he noted at a developer's conference, "the J2EE market is a community, while .Net is the product of a corporation."

When asked by an interviewer what it is like being a researcher at Sun, Gosling replied: "The goal of any research lab is to do things that are kind of weird and outlandish and risky. In most IT [information technology] organizations, the big goal is to succeed, which means, 'Don't take risks.' But in a research lab, if you aren't failing often enough, you aren't taking enough risks."

Gosling has been the recipient of numerous awards, including the *PC Magazine* 1997 "Person of the Year" award and the *InfoWorld* Innovators Hall of Fame (2002).

Further Reading

Gosling, James. "Java: an Overview." Available on-line. URL: http://java.sun.com/people/jag/OriginalJavaWhitepaper.pdf. Posted on February 1995.

"James Gosling Home Page." Available on-line. URL: http://java.sun.com/people/jag. Downloaded on November 2, 2002.

Sliwa, Carol. "Gosling: 5 Years on, Diversity of Java Has Been Surprise," *Computerworld*, June 26, 2000, p. 86.

Grove, Andrew S. (András Gróf)
(1936–)
Hungarian/American
Engineer, Entrepreneur

Andrew Grove's family in Hungary fled the Nazis. He then got himself smuggled across the border to escape the communists. Once in the United States, Grove zeroed in on the new semiconductor industry and eventually built Intel, the world's largest maker of computer chips, weathering technical challenges and competition from Japan. *(Photo provided by Intel Corporation)*

Diamonds are lumps of carbon, and computer chips are (mostly) lumps of sand, except that nature shaped the former and human ingenuity the latter. The modern computer industry would not have been possible without researchers such as JACK KILBY and ROBERT NOYCE, who discovered how to pack thousands of functional circuits onto tiny silicon chips. But equally, without entrepreneurs like Andrew Grove, microprocessors would not have become an integral part of not only personal computers (PCs) but calculators, cell phones, microwave ovens, and dozens of other appliances used daily.

Grove was born András Gróf on September 2, 1936, in Budapest to a Jewish family. His father, George, owned a dairy and his mother, Maria, was a bookkeeper. In 1940, young Grove contracted scarlet fever and almost died from the deadly disease. He was left with damaged hearing, although a series of operations in his adulthood restored it.

Grove's family was disrupted by the Nazi conquest of Hungary early in World War II. His father was conscripted into a work brigade that amounted to slave labor. Andrew and his mother then faced the Nazi roundup in which many Hungarian Jews were sent to their death in concentration camps. The Groves were taken in by a Christian family, who risked their own lives to shelter them. Grove later recalled to *Time* interviewer Joshua Cooper Ramo that his mother told him that "I had to forget my real name and that . . . if they said 'Write your name,' I couldn't write [my real name] down."

Although the family survived and was reunited after the war, Hungary had come under Soviet control. Food and heating fuel were hard to come by. Because the Groves had owned a business before the war, communists persecuted them as capitalists as well as Jews. Nevertheless, Grove was a good student in high school, particularly excelling in chemistry.

In 1956, while Grove was studying at the university in Budapest, another crisis arrived. A number of Hungarian activists rebelled against the Soviet-supported communist government. The Soviets responded by invading the country, sending tanks into the streets of Budapest. Like many other university students, Grove feared that he would be arrested by the occupying forces. He and a friend made a dangerous border crossing into Austria, paying a smuggler to take them over little-used routes. Grove then came to the United States aboard a refugee ship, arriving with only about $20 in his pocket. Changing his name to Andrew Grove, he lived with his uncle in New York and studied chemical engineering at the City College of New York, where he received his bachelor's degree with honors in 1960.

Grove then earned his Ph.D. in chemical engineering at the University of California at Berkeley (in only three years) and became a researcher at Fairchild Semiconductor in 1963 and then Assistant Director of Development in 1967. Meeting Robert Noyce and GORDON MOORE, Grove soon became familiar with the work on an exciting new technology—the integrated circuit, which places many tiny components such as transistors to form extremely dense, compact circuits. Grove eventually compiled his accumulated knowledge into a textbook, *Physics and Technology of Semiconductor Devices*, which became a standard in the field.

The research team Grove joined had been trying to use silicon to create semiconductor chips, but the electronic characteristics varied too much. Eventually, Grove and the other researchers found out that the sodium used in the curing process was the source of the problem. Both Noyce and Moore were impressed by Grove's scientific skills and also by the way he had stepped into managing the research effort. Thus, when they decided to leave Fairchild and start their own semiconductor company, they invited Grove to join them.

Grove had been intended to become director of engineering, but the small company, called Intel, needed a director of operations so badly that he soon moved to that post. He established a management style that featured what he called "constructive confrontation"—characterizing it as a vigorous, objective discussion where opposing views could be aired without fear of reprisal. However, Grove, who had not yet had his hearing surgically fixed, sometimes turned off his hearing aid when he no longer wanted any input. Critics characterized the confrontations as more harsh than constructive.

Grove always strove to nail down the key factors that contributed to efficiency and production, trying to make the way operations were run as precise as the placement of transistors on the silicon chips that were revolutionizing computing. In doing so, he sometimes angered employees, as in 1971, when he reminded employees that they were expected to work a full day on December 24.

Scrooge or not, Grove earned the respect of both colleagues and competitors as an ace problem solver who finished projects on time and under budget. He also took the initiative in responding to changing market conditions. In the late 1970s, desktop computers were starting to come into use. It was unclear whether Intel (maker of the 8008, 8080, and subsequent processors) or Motorola (with its 68000 processor) would dominate the market for processors for these new machines.

To win the competition, Grove emphasized the training and deployment of a large sales force, and by the time the IBM PC debuted in 1982, it and its imitators would all be powered by Intel chips, which would also be used in many appliances and other devices. Later, when other companies attempted to "clone" Intel's chips, Grove did not hesitate to respond with lawsuits. Although competitors such as Advanced Micro Devices (AMD) and Cyrix eventually won the right to make competing chips, Intel has held on to the lion's share of the market through a combination of pushing the technological bar higher and maintaining brand consciousness through the "Intel Inside" ad campaign.

As late as the mid-1980s, however, most of Intel's revenue came not from microprocessors but from memory chips, called DRAMs. However, Japanese companies had begun to erode Intel's share of the memory chip market, often "dumping" products below their cost. Instead of trying to compete with the superefficient Japanese manufacturing technology, Grove took a more radical step. He decided to get Intel out of the memory market, even though it meant downsizing the company until the growing microprocessor market made up for the lost revenues. In 1987, Grove had weathered the storm and become Intel's chief executive officer (CEO). He summarized his experience of intense stress and the rapidly changing market with the slogan "Only the paranoid survive."

During the 1990s, Intel introduced the popular Pentium line and then had to deal with mathematical flaws in the first version of the chip that resulted in Intel having to pay $475 million to replace the defective chips. The Pentium was steadily improved, with the Pentium IV reaching a clock speed of more than 2.5 gigahertz, or billions of cycles per second, in 2002. Meanwhile, Grove had to fight a personal battle against prostate cancer that had been diagnosed in 1994. He eventually opted for an experimental radiation treatment that put the disease into remission. Although his overall health remained good, Grove relinquished his CEO title at Intel in 1998, remaining chairman of the board.

Through several books and numerous articles, Grove has had considerable influence on the management of modern electronics manufacturing. He has received many industry awards, including the Institute of Electrical and Electronic Engineers Engineering Leadership Recognition Award (1987), and the American Electronics Association Medal of Achievement (1993). In 1997 he was named CEO of the Year by *CEO Magazine* and was *Time* magazine's Man of the Year.

Further Reading

Burgelman, Robert, and Andrew S. Grove. *Strategy Is Destiny*. New York: Simon and Schuster, 2001.

Grove, Andrew. *High Output Management*. 2nd ed. New York: Vintage Books, 1995.

———. *One-on-One with Andy Grove*. New York: Putnam, 1987.

———. *Only the Paranoid Survive*. New York: Doubleday, 1996.

Intel Corporation. "Andy Grove." Available on-line. URL: http://www.intel.com/pressroom/kits/bios/grove.htm. Downloaded on November 2, 2002.

Ramo, Joshua Cooper. "Man of the Year: Andrew S. Grove—A Survivor's Tale." *Time*, December 29, 1997, pp. 54ff.

H

Hamming, Richard Wesley

(1915–1998)
American
Mathematician, Computer Scientist

A flood of data bits pours through the Internet every second of every day. But whether they make up an e-mail message, a photo image, or a software program, those bits are subject to numerous possible mishaps on their way to a personal computer. If interference between messages or from some nearby piece of electronic equipment "flips" a bit from zero to one, or one to zero, the meaning of the data will change. The results could be minor, such as an e-mail containing the sequence of letters "thk" instead of "the," or a picture having a tiny dot of the wrong color. On the other hand, a wrong bit could translate to a garbled instruction, making a downloaded program "crash."

Richard Hamming found a way not just to detect data errors, but to fix them so that the data received exactly matches the data sent. As a result, users of the world's computer networks may have to cope with network "traffic jams" slowing down their downloads, but they can be confident that the data they receive is reliable.

Hamming was born on February 11, 1915, in Chicago. His father was a native of the Netherlands who had fought in the Boer War. Left for dead on the battlefield, he somehow recovered and later immigrated to the United States, where he worked for some time as a cowboy in Texas before ending up with a more sedentary job as a credit manager.

As a student at Chicago's Crane Technical High School, Hamming became interested in mathematics after, as he later recalled, he "realized that he was a better mathematician than the teacher." Also interested in engineering, Hamming had begun to pursue that career at Crane Junior College when that institution closed because of the financial problems of the Great Depression. His only scholarship offer came from the University of Chicago, which had a mathematics department but no engineering department, so Hamming switched to mathematics. He received his B.S. degree in 1937, then earned an M.A. degree at the University of Nebraska in 1939 and a Ph.D. from the University of Illinois in 1942.

Hamming taught for the next two years but in April 1945, as recounted by J. A. N. Lee, he received a letter from a friend who wrote "I'm in Los Alamos, and there is something interesting going on down here. Come down and work." Intrigued and wanting to contribute to the war effort, Hamming took the train to Los Alamos, followed by his wife, Wanda, a month later. While Wanda worked as a human "computer" at a desk calculator, Richard Hamming was introduced to a room full of IBM machines that were sort of a cross between calculators and modern computers. These electromechanical relay calculators clicked and clacked furiously as they ran, putting Hamming in mind of "the mad scientist's laboratory."

Hamming's job was to fix and restart these mechanical computers whenever they broke down. While he had only the vaguest notion that the calculations involved the design of an atomic bomb of incredible power, he did see how the use of computers meant "that science was going to be changed" by the ability to perform calculations automatically that were too lengthy for humans to compute.

After the United States dropped atomic bombs on Japan and the war ended, most of the researchers left Los Alamos for other jobs. Hamming, however, remained for six months because he had become intrigued by the process that had so accurately predicted the performance of the bombs. He traced out the chains of calculations and observed "feedback loops" that had detected and corrected errors, refining the numbers to make them more and more accurate. He also observed how the scientists had learned from their work and from one another as they became a team. Hamming's study revealed how a stupendous engineering problem had been solved not by trained engineers but by imaginative scientists. Hamming realized that becoming a mathematician rather than an engineer had been fortunate, because, as he recalled, to J. A. N. Lee "As an engineer, I would have been the guy going down manholes instead of having the excitement of frontier research work."

Hamming's Los Alamos experience would shape his career in other ways as well. It established his interest in the theory of automatic error correction and also his interest in the teaching of the discipline that would eventually become known as computer science.

In 1946, Hamming went to work at Bell Telephone Laboratories, one of the nation's premier private research organizations. There he joined three other mathematicians: communications theorist CLAUDE E. SHANNON, Donald P. Ling, and Brockway McMillan. Together they called themselves the Young Turks.

As he expressed in his interview with Lee, Hamming believed that he and other young mathematicians who had come of age during the war represented a generation with a new approach to scientific research:

> During the war we all had to learn things we didn't want to learn to get the war won, so [we] were all cross-fertilized. We were impatient with conventions, and often had responsible jobs very early.

Given their attitude, it was not surprising that the Young Turks "did unconventional things in unconventional ways and still got valuable results. Thus management had to tolerate us and let us alone a lot of the time." Hamming, for example, had been assigned to work on elasticity theory, but he was soon paying much more attention to the laboratory's computers themselves than to the work he was supposed to do with them.

One of Hamming's first discoveries came from his work in training researchers who had used analog computers to use the new digital computers. His work and research resulted both in a textbook and a patent for the "Hamming window," a way to examine part of a signal or spectrum while minimizing interference from other regions. Hamming found cutting-edge research to be challenging and satisfying, telling Lee that "if you don't work on important problems, it's not likely that you'll do important work." He also found it exhilarating: "The emotion at the point of technical breakthrough is better than wine, women, and song put together." However, Hamming's brashness and intensity sometimes irritated colleagues, who accused him of not listening enough to what others had to say.

By the late 1940s Hamming had focused his research on finding a way for computers to find and fix errors automatically. Claude Shannon's work had shown that errors could be limited by including a certain amount of redundant information in a data transmission, but it was unclear how that redundant data could actually be used.

Hamming's system, called Hamming codes, is based on the idea of parity. In its simplest form, parity involves adding an extra bit to each group of data bits being transmitted, so that the total group always has an odd or even number of one bits. For example, if "odd parity" is being used, the data bit group 10101010 (which has four one bits, an even number) another one bit would be added to make the number of ones odd: 101010101. Similarly, with the data bit group 11000111, which already has five ones (an odd number), a zero would be added, 110001110.

The receiving system then counts the number of one bits in each data group and checks to see whether it is still odd (if odd parity is being used). If a bit has been "flipped" from one to zero (or vice versa) during transmission, the count will be even, and the receiver knows that an error has occurred and can request that the transmitting system resend the data.

Hamming elaborated the simple parity idea by using multiplication and a special matrix, or group of digits. As a result of using two parity digits, Hamming codes can even detect when two separate errors have occurred—errors that otherwise might cancel each other out. Alternatively, Hamming codes can be limited to detect only single errors but also to determine which bit had the error. This means the data can be corrected without having to retransmit it.

In addition to his work on error correction, Hamming also made contributions to the development of higher-level, easier-to-use programming languages—work that would help lead to today's structured programming languages.

Hamming retired from Bell Labs in 1976, but his career was not over. He became head of the computer science department at the Naval Postgraduate School in Monterey, California. In his teaching, he tried to instill in his students the same flexibility and openness that he had demonstrated at Bell Labs. He believed that this was more important than imparting large amounts of detailed knowledge that was likely to be obsolete by the time his students were working in the real world. After a long, productive career, Hamming died on January 7, 1998.

Hamming received numerous awards including the Association for Computing Machinery (ACM) Turing Award in 1968 (he had served as that organization's president in 1958). The ACM also created the Hamming Medal in his honor. Hamming was also awarded the Emmanuel R. Piore Award of the Institute of Electrical and Electronic Engineers in 1979.

Further Reading

Hamming, Richard W. *The Art of Doing Science and Engineering: Learning to Learn*. New York: Taylor & Francis, 1997.

———. "How To Think About Trends" in Peter J. Denning and Robert M. Metcalfe, editors. *Beyond Calculation: The Next Fifty Years of Computing*. New York: Springer-Verlag, 1997, pp. 65–74.

Lee, J. A. N. *Computer Pioneers*. Los Alamitos, Calif.: IEEE Computer Society Press, 1995, pp. 360–366.

Hewlett, William Redington
(1913–2001)
American
Engineer, Entrepreneur

Most people have heard the story about the two guys who started a high-tech company in a garage and built it into a mega-corporation. Was this Steve Jobs and Steve Wozniak and Apple Computer in 1977? Perhaps, but the original "garage" story is much older. In 1939, William

Hewlett and DAVID PACKARD started an electronics company when the future Silicon Valley was mostly fruit orchards.

Hewlett was born on May 20, 1913, in Ann Arbor, Michigan, and the family moved to San Francisco when he was three. Hewlett's father, a physician and professor of medicine, died of a brain tumor when the boy was only 12. Later Hewlett speculated that if his father had lived longer, he might have chosen a career in medicine instead of engineering.

From a young age, Hewlett was interested in science and constructed elaborate home experiments. However, he did not do well in reading or writing because he suffered from dyslexia, a learning disability that was not well understood in the 1920s. Despite these problems, Hewlett graduated from the academically challenging Lowell High School and was then admitted to Stanford University, where he earned his B.A. degree in 1934. He would go on to earn a master's degree in electrical engineering at the Massachusetts Institute of Technology in 1936 and another engineering degree at Stanford in 1939.

In his freshman year, Hewlett met David Packard and they soon became friends. With the encouragement of Frederick Terman, their favorite electrical engineering professor, they decided to start their own electronics company. Their starting capital consisted of $538 and a Sears drill press. Their "factory" was the garage next door to Packard's, which today bears a historical placard proclaiming it the "birthplace of Silicon Valley." They flipped a coin to determine the order in which their names would appear, and the result was the Hewlett-Packard Corporation, soon to be known simply as HP.

Hewlett took the lead in invention and engineering, while Packard focused on business matters. Hewlett's first few inventions were of limited value—they included a harmonica tuner, an automatic urinal flusher, and a foul-line indicator for bowling alleys. Hewlett recalled to interviewers Rama Dev Jager and Rafael Ortiz that, "Our original idea was to take what we could get in terms of an order. Most of our initial jobs were contract jobs—ranging from just about anything."

However, Hewlett then turned to a design he had developed in graduate school for a resistance-capacitance audio oscillator. This device made it possible to accurately and less expensively test audio signals, a capability increasingly needed by the broadcast, sound recording, and defense industries. After determining that there was some interest in the product, they started to manufacture and sell it. (The Walt Disney Studios bought eight oscillators for use in producing a new animated film, *Fantasia*.)

As money came in, the company expanded rapidly, moving from the garage to a rented building in nearby Palo Alto. When World War II came along, Packard was exempted because of his role in managing the company's defense contracts, but Hewlett went into the army. He started in the Signal Corps but ended up as head of the army's New Development Division with the rank of lieutenant colonel.

After the war, Hewlett returned to HP to lead the company's research and development efforts. He would become vice president in 1947, executive vice president in 1957, president in 1964, and chief executive officer (CEO) in 1969. He would oversee many innovative products, including the scientific calculator, which burst upon the market in 1972 and quickly made the slide rule obsolete, and the laser printer, which went from an expensive accessory for the largest offices to a common desktop companion. (Indeed, although HP has produced computers and a variety of other kinds of systems, the company is best known today for its laser and inkjet printers. In 2002, however, the company seemed to be undertaking a new direction following its contentious acquisition of Compaq, a leading personal computer manufacturer.)

With Packard, Hewlett established a management style that became known as "the HP way." While it attracted much hype in management circles, Hewlett emphasized to Jager and Ortiz that "most of it is really common sense. The customer comes first. Without profits, the company will fail." To keep customers, HP emphasized building solid, dependable products that emphasized reliability over flashiness. However, independent-minded employees sometimes found the company's emphasis on teamwork to be stultifying. Meanwhile, Hewlett continued to take an interest in engineering problems, often brainstorming with researchers and uncovering hidden problems, a process that was dubbed "the Hewlett effect."

Beyond HP, Hewlett devoted much of his time and resources to a variety of philanthropic efforts, especially in the areas of education (serving as a trustee of Mills College in Oakland, California, as well as Stanford University) and medicine (as a director of what became the Stanford Medical Center as well as the Kaiser Foundation Hospital and Health Plan Board). He also served with scientific and technical organizations such as the Institute of Radio Engineers (which later became the Institute of Electrical and Electronics Engineers, or IEEE) the California Academy of Sciences, the National Academy of Engineering, and the National Academy of Arts and Sciences.

In 1966, Hewlett and his wife, Flora, formed the William and Flora Hewlett Foundation, a major nonprofit that funds a variety of projects with an emphasis on conflict resolution, education, environment, family and community development, performing arts, population, and U.S.–Latin American relations.

William Hewlett died on January 12, 2001. His achievements have been recognized by many awards including 13 honorary degrees, the National Medal of Science (1991), and the National Inventors Hall of Fame Award.

Further Reading

Allen, Frederick E. "Present at the Creation." *American Heritage* 52 (May–June 2001): 21ff.

Jager, Rama Dev, and Rafael Ortiz. *In the Company of Giants: Candid Conversations with the Visionaries of the Digital World*. New York: McGraw-Hill, 1997, pp. 225–232.

Packard, David, David Kirby, and Karen Lewis. *The HP Way: How Bill Hewlett and I Built Our Company*. New York: HarperBusiness, 1996.

Hillis, W. Daniel
(1956–)
American
Computer Scientist

Imagine a sprawling computer made only from Tinkertoy parts and fishing line that is smart enough to beat anyone at tic-tac-toe. Computer scientist W. Daniel Hillis built it while he was still in college, and today it is on display at the Boston Computer Museum. While this Babbage-like device might seem silly, it illustrates an important aspect of Hillis's success in "thinking outside the box" and bringing a radically different architecture to computing while showing his inventiveness in many startling ways.

Hillis was born on September 25, 1956, in Baltimore. His father was a doctor specializing in disease epidemics who frequently traveled, so the family lived in many different places and his parents generally taught him at home. As a boy, Hillis was an avid reader of science fiction who built toy robots and spaceships from whatever parts he could find around the house. His ambition was to attend the Massachusetts Institute of Technology (MIT), and upon acceptance there he started out studying neurophysiology, having become interested in the "hardware" aspects of the human mind. However, after Hillis received his bachelor's degree in mathematics in 1978, one of his advisers suggested that he learn about another kind of hardware—the computers being used at MIT's famous Artificial Intelligence Laboratory.

As Hillis worked toward his master's degree (received in 1981), he plunged into challenging artificial intelligence (AI) problems such as pattern recognition. By studying how a human baby learns to recognize its mother's face, AI researchers hoped to find algorithms that would allow computers to identify the significant features in pictures or allow a robot to better "understand" and interact with its environment. Hillis's Tinkertoy project and his general interest in toys also got him a summer job working on electronic toys for Milton Bradley.

After working with pattern recognition, Hillis concluded that if computers were going to perform as well as the human brain at such tasks, they would have to function more like the human brain. Conventional computers have a single processor that takes one instruction at a time and applies it to one piece of data at a time. Although the sheer speed of electronics makes computers far better at calculation than the human brain, the latter is much better at recognizing patterns. Just making the computer bigger and giving it more data will not solve the problem. As Hillis explained to a reporter: "If you try to make the computer smarter by giving it more information, it takes longer to process the answer, so it gets more stupid."

Unlike the single processor in most computers, the brain's millions of neurons simultaneously respond to stimuli (such as light) and signals from neighboring neurons, in turn firing off their own signals. Thus the recognition of a face by a human baby involves a sort of summing up of the work of many individual "processors." Researchers such as MARVIN MINSKY and SEYMOUR PAPERT had developed a way in which computers could simulate the operation of such a neural network, and it had shown considerable promise in tackling AI tasks.

However, Hillis believed that if a computer was really going to work like a brain, it too would have to have many separate processors working at the same time—a technique called parallel processing. Hillis decided to take this technique to its logical extreme. In 1986, he finished the design for a computer called the Connection Machine. It had 65,536 very simple processors that each worked on only one bit of data at a time. But since there were so many processors and they worked so fast, it meant that, for example, a pattern recognition program could deal with all the pixels of an image simultaneously. (Calculations on matrices or arrays of data would also be lightning fast, since all the bits would be calculated at once.) The first working Connection Machine was finished in 1986.

In 1988, having received his Ph.D., Hillis left MIT to work with Thinking Machines Corporation, a company he had founded in 1983 to develop parallel processing computers based on the Connection Machine principle. His machines were attractive to many users because while they had thousands of processors, the total cost of components was still much less than in regular mainframes or supercomputers, so a Connection Machine cost only about a third as much.

The exterior of the Connection Machine had a futuristic cube design, created by architect Maya Lin. Hillis admitted to an interviewer that while its blinking red lights "had some diagnostic use," the real reason for them was, "Who wants to spend his life working on something that looks like a refrigerator?"

Although the Connection Machine was impressive, serious hurdles had to be overcome before it could be brought into widespread use. Regular computer languages and operating systems did not have the facilities needed to coordinate multiple processors and allow them to communicate with one another. Therefore, "parallelized" versions of languages such as FORTRAN had to be developed. Programmers then had to be retrained to think in new terms, and a library of software tools and applications developed. Observers also noted that the original Connection Machine did not provide enough memory to allow each processor to work efficiently.

Hillis soon came out with an improved model, Connection Machine 2. While the commercial market was still slow to adopt the new design, officials at the Defense Department's Advanced Research Projects Agency (ARPA) saw a long-term advantage to parallel processing for dealing with the intensive calculation needs of applications such as nuclear weapons simulation. ARPA bought millions of dollars' worth of Connection Machines for its own use, as well as encouraging outside researchers to adopt them. (Eventually, the General Accounting Office investigated ARPA for limiting its grants to one company, and the agency began to allow other companies to participate in parallel computing contracts.) In addition to the military, Thinking Machines also sold Connection Machines to such researchers as earthquake scientists and airplane designers.

In 1991, Thinking Machines reached a high point of $65 million in sales, and had attracted much industry attention. Even IBM entered into an agreement to include some of the Connection Machine's technology in its own parallel computer designs.

Thinking Machines soon ran run into financial difficulties, however, losing $17 million in 1992. In part, this was due to a reduction of federal spending on computer technology. However, the commercial market for large-scale parallel computers also remained relatively small, in part because the ever-growing power of conventional computers enabled them to keep up with many demanding applications. Further, a competing architecture called "cluster computing," which uses networks of conventional computers to work together on problems, offered an attractive alternative path to greater computing power.

Thinking Machines experimented with building computers that used smaller numbers of more powerful processors that could be expanded as desired, but they sold only modestly. Although the company set a new sales record of $90 million in 1993, it had not really overcome either the challenges of the marketplace or the conflicts that had flared up between top executives. The company filed for bankruptcy in 1994, and after reorganizing changed its emphasis to developing software for "data mining," or the extraction of useful patterns from large databases. This business in turn was acquired by the large database company Oracle in 1999.

When Hillis founded his company he had stated that his ultimate goal was to "build thinking machines," and he continued to devote time to applying parallel computing to image recognition and other AI problems. In 1989, he developed a computer vision system based on a Connection Machine that could recognize objects. While it was far better than systems based on sequential computers, Hillis's computerized "eye" was disappointing: it still took several minutes to recognize objects that the human eye and brain can process in a fraction of a second.

Hillis has also been disappointed in the results of four decades of research in trying to get computers to "reason" in a way similar to human thinking. As he told one interviewer (quoted in *Current Biography Yearbook, 1995*), "I've been surprised at how little progress there's been. In retrospect, I think I was personally naive about how difficult common-sense reasoning was." Comparing AI to other sciences, Hillis says that "the science of intelligence resembles astronomy before Galileo invented his telescope. . . . I am like a telescope builder." Hillis's patient perspective can also be seen in a project called the Millennium Clock, being implemented by an organization called the Long Now Foundation. This clock is designed to "tick" only once a year and contains a "century hand" that would move one mark every 100 years.

Hillis has always taken a broad view of the significance of computing. His book *The Pattern on the Stone* provides a guide to "the simple ideas that make computers work," offering sophisticated insight into computer science even for readers who lack a mathematical or computing background. In this and other writings, Hillis suggests that rather than there being any one stupendous breakthrough, machine intelligence will gradually emerge or evolve just as biological intelligence has. And although the specific parallel computing architecture that Hillis and his company developed had some limitations, the general idea of computation through networked, cooperating objects has become in this age of the Internet a pervasive way of tackling applications.

Hillis's inventiveness extends well beyond parallel computing: He holds more than 40 U.S. patents including one for disk arrays, a technique that uses multiple hard disk drives working in tandem to speed up data storage or provide real-time "mirror" backups.

Hillis continues his work at MIT, often startling visitors who see him driving around in an old fire engine. Those who arrive at his office will find it filled with toys. Hillis serves as editor or advisory board member for a number of scientific organizations, including the Santa Fe Institute, a multidisciplinary research center dedicated to "emerging sciences." He has received the Association for Computing Machinery's Grace Murray Hopper Award (1989), the Ramanujan Award (1988), and the Spirit of American Creativity Award (1991).

Further Reading

Hillis, W. Daniel. "The Millennium Clock." *Wired*, 1995. Available on-line. URL: www.wired.com/wired/scenarios/clock.html. Downloaded on November 2, 2002.

———. *The Pattern on the Stone: The Simple Ideas That Make Computers Work*. New York: Basic Books, 1998.

Long Now Foundation. Available on-line. URL: http://www.longnow.org. Updated on October 10, 2002.

Moritz, Charles, ed. "Hillis, W. Daniel." *Current Biography Yearbook*. New York: H. W. Wilson, 1995.

Thiel, Timiko. "The Connection Machines CM-1 and CM-2." Available on-line. URL: http://mission.base.com/tamiko/cm/cm-text.htm. Downloaded on November 2, 2002.

⊠ Holberton, Frances Elizabeth (Betty Holberton)

(1917–2001)
American
Programmer, Mathematician

One of the ironies of computer history is that although the field was largely male-dominated at least until the 1990s, the majority of the first programmers to actually put the machines to work were women. Only today are historians starting to give wider public recognition to the group of female computing pioneers that includes JEAN BARTIK and Frances Elizabeth (Betty) Holberton.

Frances Elizabeth Holberton was born in 1917 in Philadelphia. Her father and grandfather were both astronomers, and they encouraged young Holberton to pursue her interest in science and particularly mathematics. Holberton's father had a liberal attitude about women pursuing careers and was scrupulous about treating his daughter and sons equally. This support helped her endure the teasing of classmates who saw her as a (literally) cross-eyed nerd.

Holberton went to the University of Pennsylvania with the intention of majoring in mathematics. At the time, however, women mathematicians were very rare, and one of her advisers, firmly believing that women belonged at home raising children, convinced Holberton to major in English and journalism—the kind of career suitable for a bright woman who would work a few years before marrying and settling down.

Holberton began her journalistic career as a writer for the *Farm Journal*, where she got a chance to use her mathematical knowledge in compiling and reporting on consumer spending and farming statistics. However, when the United States entered World War II an urgent call went out for mathematically trained women. The women (about 80 in number) were needed to perform massive, tedious computations for gun firing tables, using hand calculators at the Moore School at the University of Pennsylvania.

Shortly after she took her place in the ranks of human "computers," Holberton was offered an opportunity to join a secret project. Two researchers, J. PRESPER ECKERT and JOHN MAUCHLY, had been funded by the army to build an electronic digital computer called ENIAC. This machine would be able to calculate a firing table in a less than a minute, an operation that would take about 20 hours with a hand calculator. But people needed to learn how to set up the machine's intricate switches and wiring connections to enable it to perform a calculation. They needed programmers.

Holberton (along with Jean Bartik) became part of the first small group of women who learned to program ENIAC. At first, military security rules barred the women from actually entering the ballroom-size space containing the massive ENIAC. Another member of the programming team, John Mauchly's wife, Kay McNulty Mauchley Antonelli, later recalled in an interview for *Computerworld*.

> They gave us all the blueprints, and we could ask the engineers anything. We had to learn how the machine was built, what each tube did. We had to study how the machine worked and figure out how to do a job on it. So we went right ahead and taught ourselves how to program.

Finally the women were allowed entry and they could start to set up calculations—such as those needed by the scientists who were working on an even more secret project, the building of the first atomic bombs.

In 1947, Holberton left the ENIAC facility and went to work for the company that Eckert and Mauchly had set up to build the first commercial computers based on an improved ENIAC design. This new machine, Univac, was used in the 1950 census and then made a public splash when it correctly predicted Dwight Eisenhower's victory in the 1952 presidential election.

Having experienced the difficulty of programming computers by wires and switches and later, cryptic numeric instructions, Holberton felt a strong need to create a more user-friendly way to program Univac. She developed what amounted to one of the first programming languages, an instruction code called C-10. This code allowed abbreviated commands (such as "a" for add and "b" to bring a number into the processor) to be typed at a keyboard. She also persuaded the computer's designers to move the numeric keypad near the keyboard where it could be used more easily for entering numbers in to the machine. She even persuaded the company to change the computer's color from black to less intimidating beige.

In her later career Holberton worked on a variety of projects. She developed a number of data processing routines (such as a sorting program) that other programmers could then use to create applications such as payroll and inventory processing. She also wrote a "sort generator," helping pioneer the use of programs that could automatically generate customized data processing routines. Additionally, Holberton contributed to the development of the higher-level FORTRAN and COBOL programming languages (see also GRACE MURRAY HOPPER) while working at the navy's David Taylor Model Basin and the National Bureau of Standards (from which she retired in 1983).

Elizabeth Holberton died on December 8, 2001, in Rockville, Maryland. Fortunately, she lived long enough to become part of the new interest in the history of women in science and technology. She was honored by the Association for Women in Computing with its Lovelace Award. Her work is also recounted as part of the Univac exhibit at the Smithsonian's National Museum of American History.

Further Reading

Association for Women in Computing. Available on-line. URL: http://www.awc-hq.org. Downloaded on November 2, 2002.

Levy, Claudia. "Betty Holberton Dies; Helped U.S. Develop Computer Languages." *Washington Post*, December 11, 2001, p. B07.

McCartney, Scott. *ENIAC: The Triumphs and Tragedies of the World's First Computer*. New York: Berkeley Books, 1999.

Petzinger, Thomas, Jr. "Female Pioneers Fostered Practicality in Computer Industry." *Wall Street Journal*, November 22, 1996, p. B1.

Hollerith, Herman

(1860–1929)
American
Inventor

"Do not bend, fold, staple, or mutilate." This once famous phrase was often found on the punched cards that served as Social Security and payroll checks. The ubiquitous punch card transported data to and from the first generations of huge mainframe computers. Herman Hollerith invented the automatic tabulating machine, a device that could read the data on punched cards and display running totals. His invention would become the basis for the data tabulating and processing industry.

Hollerith was born on February 29, 1860, in Buffalo, New York. His father, George, a classical language teacher, died from an accident when the boy was only seven years old. His mother, Franciska, came from a family of skilled locksmiths that had immigrated to America and gone into the carriage-building business. After George died, Franciska kept the family going by starting a hat-making business.

As a boy, young Hollerith was an excellent student in every subject except spelling, which he apparently hated with such a passion that he once jumped out of a window to avoid taking a spelling test. When he was only 15, Hollerith won a scholarship to New York City College, and after studying there he went to the Columbia School of Mines in New York City, graduating in 1879. By then his main interests were mathematics, engineering, and drawing.

A portrait of Herman Hollerith next to one of his punch card tabulator machines. Such machines made it possible to complete the 1890 U.S. census in a fraction of the time needed by earlier methods. *(Photo courtesy Computer Museum History Center)*

After graduation, he went to work for the U.S. Census as a statistician. The 1880 census nearly broke down under the demands for counting the demographic and economic statistics being generated by what was rapidly becoming one of the world's foremost industrialized nations. Among other tasks, Hollerith compiled vital statistics for Dr. John Shaw Billings, who suggested to him that using punched cards and some sort of tabulator might keep the census from falling completely behind in 1890.

Hollerith left the Census Bureau in 1882 to teach mechanical engineering at the Massachusetts Institute of Technology (MIT). He continued to think about the tabulation problem with the aid of some of his MIT colleagues. One day, while riding on a train, Hollerith noticed the system that railroad conductors used to keep track of which passengers had already paid their fare. The tickets included boxes with descriptions of basic characteristics such as gender or hair color. When the conductor punched a passenger's ticket, he created a "punch photograph" by punching the appropriate description boxes. Hollerith decided that similar cards could be used to record and tabulate the census data for a person.

Hollerith began to design a punch card tabulation machine. His design was partly inspired by the Jacquard loom, an automatic weaving machine that worked from patterns punched into a chain of cards. He then got a job with the U.S. Patent Office, partly to learn the procedures he would need to follow to patent his tabulator. (Eventually he would file 30 patents for various data-processing devices.) He tested his machine with vital statistics in Baltimore, New York, and New Jersey.

Hollerith's system included a punch device that a clerk could use to record variable data in many categories on the same card (a stack of cards could also be prepunched with constant data, such as the number of the census district). The cards were then fed into a device something like a small printing press. The top part of the press had an array of spring-loaded pins that corresponded to tiny pots of mercury (an electrical conductor) in the bottom. The pins were electrified. Where a pin encountered a punched hole in the card, it penetrated through to the mercury, allowing current to flow. The current created a magnetic field that moved the corresponding counter dial forward one position. The dials could be read after a batch of cards was finished, giving totals for each category, such as an ethnicity or occupation. The dials could also be connected to count multiple conditions (for example, the total number of foreign-born citizens who worked in the clothing trade).

Hollerith received a contract to provide tabulators for the 1890 census, at the same time finishing his doctoral dissertation at the Columbia School of Mines, titled "The Electric Tabulating System." Aided by Hollerith's machines, a census unit was able to process 7,000 records a day for the 1890 census, about 10 times the rate in the 1880 count. As a result, the census was completed in two years as compared to the six years the previous count had taken.

The value of his machines having been conclusively demonstrated, Hollerith founded the Tabulating Machine Company in 1896. Soon his machines were being sold all over the world, as far as Russia, which was undertaking its first national census.

Starting around 1900, Hollerith brought out improved models of his machines that included such features as an automatic (rather than hand-fed) card input mechanism, automatic sorters, and tabulators that boasted a much higher speed and capacity. Hollerith machines soon found their way into government agencies involved with vital statistics, agricultural statistics, and other data-intensive matters, as well as insurance companies and other businesses.

Unfortunately, Hollerith was a much better inventor than he was a businessperson. He quarreled with business partners and family members. Facing vigorous competition and in declining health, Hollerith sold the Tabulating Machine Company, which merged with the International Time Recording Company and the Computing Scale Company of America. The new company, CTR (Computing-Tabulating-Recording) changed its name in 1924 to International Business Machines—IBM.

Hollerith continued to work as a consultant in the tabulating industry, and his stock and proceeds from the sale of the company ensured him a comfortable life. He built a new home and bought a farm. On November 17, 1929, Hollerith died of a heart attack.

The punched card, often called the Hollerith card, became a staple of business and government data processing. Companies such as IBM built more elaborate machines that could sort the cards and even perform basic calculations using the data. When the digital computer was invented in the 1940s, the punch card was a natural choice for storing programs and data. Gradually, however, high-speed tape and disk drives would take over the job, although there are still a few users of punch cards today.

Further Reading

Austrian, G. D. *Herman Hollerith: Forgotten Giant of Information Processing*. New York: Columbia University Press, 1982.

Kistermann, F. W. "The Invention and Development of the Hollerith Punched Card." *Annals of the History of Computing* 13, no. 3 (1991): 245–259.

Hopper, Grace Murray

(1906–1992)
American
Computer Scientist, Mathematician

Possibly the most famous woman in computing is Grace Hopper. Grace Brewster Murray Hopper was an innovator in the development of high-level computer languages in the 1950s and 1960s as well as the highest ranking woman in the U.S. Navy. She is best known for her role in the development of COBOL, which became the premier language for business data processing.

Hopper was born in New York City on December 9, 1906. As a child, she often accompanied her grandfather, a civil engineer, watching and asking questions as he worked with his maps and surveying equipment. The inquisitive girl also caused consternation when she dismantled all seven of the family's clocks in order to figure out how the alarms worked.

Hopper's family believed strongly in education for girls as well as boys. Young Hopper excelled at school, particularly in math classes. After graduating from high school, Hopper went to Vassar, a prestigious women's college, graduating in 1928 with a bachelor's degree and Phi Beta Kappa honors in mathematics and physics. She then went to Yale University for graduate study, receiving her M.A. degree in mathematics in 1930. Four years later, she became the first woman to receive a Ph.D. in mathematics from Yale. Meanwhile, she had married Vincent Foster Hopper, whom she would divorce in 1945.

Hopper taught at Vassar as a professor of mathematics from 1931 to 1943, when she joined the U.S. Navy at the height of World War II. As a lieutenant, junior grade, in the WAVES (Women Accepted for Volunteer Emergency Service), she was assigned to the Bureau of Ordnance, where she worked in the Computation Project at Harvard under pioneer computer designer HOWARD AIKEN. She became one of the first coders (that is, programmers) for the Mark I, a large, programmable computer that used electromechanical relays rather than electronics. She also edited and greatly contributed to the documentation for the Mark I, compiling it into a 500-page work, *Manual of Operations for the Automatic Sequence Controlled Calculator*.

After the war, Hopper worked for a few years in Harvard's newly established Computation

Grace Murray Hopper created the first computer program compiler and was instrumental in the design and adoption of COBOL. When she retired, she was the first woman admiral in U.S. Navy history. *(Photo courtesy of Unisys Corporation)*

Laboratory, helping with the development of the Mark II and Mark III. During this time, a famous incident occurred. One day the machine had failed, and when she investigated, Hopper found that a large moth had jammed a relay. She taped the moth to the system logbook and wrote, "First actual case of bug being found." Although the term *bug* had been used for mechanical glitches since at least the time of Edison, the incident did make *bug* a more common part of the vocabulary of the emerging discipline of computer programming.

In 1949, Hopper became senior mathematician at the Eckert-Mauchly Corporation, the world's first commercial computer company. J. PRESPER ECKERT and JOHN MAUCHLY had invented ENIAC, an electronic computer that was much faster than the Mark I and its electromechanical successors, and were now working on Univac, which would become the first commercially successful electronic digital computer. She would stay with what became the Univac division under Remington Rand (later Sperry Rand) until 1971.

While working with Univac, Hopper's main focus was on the development of programming languages that could allow people to use symbolic names and descriptive statements instead of binary codes or the more cryptic forms of assembly language. (This work had begun with John Mauchly's development of what was called Short Code and FRANCES ELIZABETH HOLBERTON's "sort generator.")

In 1952, Hopper developed A-0, the first compiler (that is, a program that could translate computer language statements to the corresponding low-level machine instructions.) She then developed A-2, a compiler that could handle mathematical expressions. The compiler was the key tool that would make it possible for programmers to write and think at a higher level while having the machine itself take care of the details of moving and manipulating the data. In her paper "The Education of a Computer," Hopper looked toward the development of increasingly capable compilers and other programming tools so that "the programmer may return to being a mathematician."

But Hopper knew that business computer users were at least as important as mathematicians and engineers. In 1957 she developed Flow-Matic. This was the first compiler that worked with English-like statements and was designed for a business data processing environment.

In 1959 Hopper joined with five other computer scientists to plan a conference that would eventually result in the development of specifications for a common business language. Her earlier work with Flow-Matic and her design input played a key role in the development of what would become the COBOL language. COBOL, in turn, would be the main business programming language for decades to come.

Hopper had retained her navy commission, and after her retirement in 1966 she was soon recalled to active duty to work on the navy's data processing needs. Even after Hopper reached the official retirement age of 62, the navy gave her repeated special extensions because they found her leadership to be so valuable. She finally retired in 1986 with the rank of rear admiral.

Hopper spoke widely about data processing issues, especially the need for standards in computer language and architecture, the lack of which she said cost the government billions of dollars in wasted resources. Admiral Hopper died January 1, 1992, in Arlington, Virginia, and was buried at Arlington Cemetery with full military honors. By then she had long been acknowledged as, in the words of biographer J. A. N. Lee, "the first lady of software and the first mother-teacher of all computer programmers." She has become a role model for many girls and young women considering careers in computing.

Hopper received numerous awards and honorary degrees, including the Computer Sciences Man (sic) of the Year (1969), the U.S. Navy Distinguished Services Medal (1986), and the National Medal of Technology (1991). The Association for Computing Machinery created the Grace Murray Hopper Award to honor distinguished young computer professionals. And the navy named a suitably high-tech Aegis destroyer after her in 1996.

Further Reading

Billings, C. W. *Grace Hopper: Navy Admiral and Computer Pioneer.* Hillfield, N.J.: Enslow Publishers, 1989.

"Grace Murray Hopper (1906–1992)." Available on-line. URL: http://www.unh.edu/womens-commission/hopper.html. Downloaded on November 2, 2002.

Lee, J. A. N. *Computer Pioneers*. Los Alamitos, Calif.: IEEE Computer Society Press, 1995.

Spencer, Donald D. *Great Men and Women of Computing*. Ormond Beach, Fla.: Camelot Publishing, 1999.

J

Jobs, Steve

(1955–)
American
Entrepreneur

As late as the 1970s, computers were large, forbidding-looking devices. The idea that people might actually have computers in their homes seemed absurd. Even when the microchip made desktop computers possible, it seemed unlikely that ordinary office workers or schoolchildren could learn to use them productively. One of the most important things that changed this perception was the creation of the colorful computer with the friendly name—Apple. Behind the Apple is the innovative, energetic, and often controversial Steve Jobs. Jobs was cofounder of Apple Computer and shaped the development and marketing of its distinctive Macintosh personal computer (PC).

Jobs was born on February 24, 1955, in Los Altos, California. He showed an enthusiastic interest in electronics starting in his years at Homestead High School and gained experience through summer work at Hewlett-Packard, one of the dominant companies of the early Silicon Valley. In 1974, he began to work for pioneer video game designer Nolan Bushnell at Atari. However, Jobs was a child of the counterculture as well as of the electronics age. He soon left Atari for a trip to India, and when he returned after a few months he went to work on a farm.

Jobs's technical interests soon revived. He became a key member of the Homebrew Computer Club, a group of hobbyists who designed their own microcomputer systems using early microprocessors. Meanwhile, Jobs's friend STEVE WOZNIAK had developed plans for a complete microcomputer system that could be built using a single-board design and relatively simple circuits. In it, Jobs saw the potential for a standardized, commercially viable microcomputer system. Jobs persuaded Wozniak to give up his job at Hewlett-Packard. They formed a company called Apple Computer (apparently named for the vanished orchards of Silicon Valley). To raise starting capital, Jobs sold his Volkswagen bus and Wozniak his programmable calculator. They built a prototype they called the Apple I, and when it seemed to work well they began to assemble more of them by hand. Although they could only afford to build a few dozen of the machines, they sold 50 to the Byte Shop in Mountain View, probably America's first home computer store. The machines made a favorable impression on the computer enthusiast community.

By 1977, Jobs and Wozniak were marketing a more complete and refined version, the Apple II. Unlike kits that could be assembled only by experienced hobbyists, the Apple II was ready to

Stephen Jobs cofounded Apple Computer, developing the pioneering Apple II with Steve Wozniak in 1977. In 1984, Jobs introduced the Macintosh with its attractive graphics and easy-to-use interface. After leaving Apple, Jobs returned in 1997 to help the company deal with slumping sales. *(© Douglas Kirkland/CORBIS)*

use "out of the box." It could be connected to an ordinary audiocassette tape recorder for storing programs. When connected to a monitor or an ordinary TV, the machine could create color graphics that were dazzling compared to the monochrome text displays of most computers. Users could buy additional memory (the first model came with only 4 kilobytes of RAM) as well as cards that could drive devices such as printers or add other capabilities.

The ability to run DANIEL BRICKLIN's spreadsheet program VisiCalc propelled the Apple II into the business world, and about 2 million of the machines were eventually sold. In 1982, when *Time* magazine featured the personal computer as its "man" of the year, Jobs's picture appeared on the cover. As he relentlessly pushed Apple forward, supporters pointed to Jobs's charismatic leadership, while detractors said that he could be ruthless when anyone disagreed with his vision of the company's future. However, by then industry giant IBM had entered the market. Its 16-bit computer was more powerful than the Apple II, and IBM's existing access to corporate purchasing departments resulted in the IBM PC and its clones quickly dominating the business market.

Jobs responded to this competition by designing a PC with a radically different user interface, based largely on work conducted during the 1970s at the Xerox PARC laboratory. The first version, called the Lisa, featured a mouse-driven graphical user interface that was much easier to use than the typed-in commands required by the Microsoft/IBM DOS. While the Lisa's price tag of $10,000 kept it out of the mainstream market, Jobs kept his faith in the concept. In 1984, Apple marketed Lisa's successor, the Macintosh, which provided much the same features at a much more competitive price. The machine quickly grabbed public attention when it was introduced in a Super Bowl commercial in which a rebellious woman runs out of a crowd of human drones and shatters a screen image of Big Brother. The message was that the Macintosh was the alternative to the stultifying, corporate IBM. The Mac was to be "the computer for the rest of us."

The Macintosh attracted millions of users, particularly in schools, although the IBM PC and its progeny continued to dominate the business market. Apple was now a major corporation, but Jobs kept his intense (many said "arrogant") attitude and his focus on implementing his vision of computing. Jobs had recruited John Sculley, former chief executive officer (CEO) of PepsiCo, to serve as Apple's CEO and take care of corporate considerations. But Sculley, consolidating his power base on Apple's board, essentially pushed Jobs aside, eventually leaving him with a tiny office that Jobs called "Siberia."

Jobs chafed at having no real work to do, and left the company in the fall of 1985. He then plunged into a number of innovative technologies. Using the money from selling his Apple stock, Jobs bought a controlling interests in Pixar, a graphics studio that had been spun off from LucasFilm. (Pixar would become well known in 1995 with the release of the film *Toy Story*, a tour de force of computer-generated imagery created from a cluster of 117 powerful Sun SPARCstations—computer workstations using specially designed processors.)

Jobs had his own ideas about graphics processing. He founded a new company called NextStep, or NeXT. The company focused on high-end graphics workstations that used a sophisticated object-oriented operating system. However, while NeXT's software (particularly its development tools) was innovative, the company was unable to sell enough of its hardware and closed that part of the business in 1993.

By 1996, Jobs had superior software but no hardware. Apple, meanwhile, had a Macintosh with an operating system more than 10 years old—one that no longer looked or worked significantly better than Microsoft's new Windows 95 for IBM-compatible PCs. Apple was thus in the market for a new operating system, and Jobs's NextStep won out over a competing, more expensive system called BeOS. As part of the deal, Jobs became an adviser to Apple's then CEO, Gil Amelio.

In 1997, with Apple's sales falling drastically, Amelio was removed by the board of directors. They appointed Jobs as interim CEO. Jobs quickly brought many of the executives from NeXT to Apple, and in 1998 Jobs was confirmed as CEO.

Jobs had some success in revitalizing Apple's consumer product line with the iMac, a colorful, slim version of the Macintosh. He promoted fresh interest in Apple's products with a new slogan, "Think Different." Jobs also focused on development of the new Mac OS X, a blending of the power of UNIX with the ease-of-use of the traditional Macintosh interface.

In April 2003 Jobs launched Apple's iTunes Music Store. The new service offered 200,000 music tracks for download at 99 cents apiece. The effort was backed by major record labels that, together with Jobs, are betting that consumers are willing to pay a modest amount for high-quality, reliable, legal music downloads as an alternative to the free file-sharing services.

Although the technical innovations were carried out by others, Jobs's vision and entrepreneurial spirit played a key role in bringing personal computing to a larger audience. The Macintosh interface that he championed would become the norm for personal computing, albeit in the form of Microsoft Windows on IBM PC-compatible systems. By maintaining Apple's presence in the market, he has also helped to maintain competition and innovation in PC hardware and software.

Jobs has been honored with the National Technology Medal (1985), and the Jefferson Award for Public Service (1987), and was named Entrepreneur of the Decade by *Inc*. magazine (1989).

Further Reading

Angelelli, Lee. "Steve Paul Jobs." Available online. URL: http://www.histech.rwth-aachen.de/www/quellen/Histcomp/Jobs.html. Downloaded on January 28, 2003.

Deutschman, Alan. *The Second Coming of Steve Jobs*. New York: Broadway Books, 2000.

Levy, Steven. *Insanely Great: The Life and Times of the Macintosh, the Computer That Changed Everything*. New York: Penguin, 2000.

⊠ Joy, Bill

(1954–)
American
Computer Scientist, Entrepreneur

Many of the most innovative software ideas were developed not in the business world of mainframes or IBM personal computers (PCs), but in college campuses and research labs where minicomputers and workstations run an operating system called UNIX. Bill Joy developed many of the key utilities used by users and programmers on UNIX systems. He then became one of the industry's leading entrepreneurs and later, a critic of some aspects of computer technology.

Joy was born in 1954 in Detroit, Michigan. He was an extremely precocious child, reading at age three and tackling advanced mathematics by age five. Because he was bored with school, his teachers thought he was learning disabled. However, when his intelligence was measured, it was literally off the scale. After skipping grades here and there, he graduated from Farmington High School at the minimum allowable age of 15.

Joy then attended the University of Michigan. Although he started out intending to be a math major, he soon became fascinated by computers and switched to computer science. After getting his undergraduate degree in 1975, Joy enrolled in the University of California, Berkeley in the computer science department.

UC Berkeley was a good choice for Joy. It was one of the two poles of the UNIX universe, the other being AT&T Bell Labs, where the operating system had been developed by KENNETH THOMPSON and DENNIS RITCHIE. Joy worked with Thompson, who was now at Berkeley, to add features such as virtual memory (paging), which allows a computer to use the disk as an extension of memory, giving it the capacity of a larger machine. He also added TCP/IP networking support to the operating system for use with Digital Equipment Corporation's (DEC) VAX minicomputers. TCP/IP would prove very important as more and more computers were connected to the ARPANET (later the Internet.)

These developments eventually led to the distribution of a distinctive version of UNIX called Berkeley Software Distribution (BSD), which rivaled the original version developed at AT&T's Bell Laboratories. The BSD system also popularized features such as the C shell (a command processor) and the text editors "ex" and "vi."

Bill Joy made key contributions to the UNIX operating system, including developing its network file system (NFS). As a cofounder of Sun Microsystems, Joy then helped develop innovative workstations and promoted Java as a major language for developing Web applications. However, Joy also warns of the dangers of runaway technology. *(Courtesy of Sun Microsystems)*

As opposed to the tightly controlled AT&T version, BSD UNIX development relied upon what would become known as the open source model of software development. This encouraged programmers at many installations to create new utilities for the operating system, which would then be reviewed and integrated by Joy and his colleagues. BSD UNIX gained industry acceptance and was adopted by DEC, makers of the popular VAX series of minicomputers.

In 1982, Joy left UC Berkeley and with SCOTT G. MCNEALY cofounded Sun Microsystems, a company that became a leader in the manufacture of high-performance UNIX-based workstations. With its powerful, expandable processors and high-end graphics, the Sun workstation became the tool of choice for scientists, engineers, designers, and even moviemakers, who were starting to use computer animation and imagery techniques. By the early 1990s, Sun had become a billion-dollar company and Joy had acquired a comfortable $10 million or so in personal net worth.

Even while becoming a corporate leader, Joy continued to refine UNIX operating system facilities, developing the Network File System (NFS), which was then licensed for use not only on UNIX systems but on VMS, MS-DOS, and Macintosh systems. Joy's versatility also extended to hardware design, and he helped create the Sun SPARC reduced instruction set (RISC) microprocessor that gave Sun workstations much of their power.

In the early 1990s, Joy turned to the growing world of Internet applications and embraced Java, a programming language created by JAMES GOSLING. He developed specifications, processor instruction sets, and marketing plans. Java became a very successful platform for building applications to run on Web servers and browsers and to support the needs of e-commerce. As Sun's chief scientist since 1998, Joy has led the development of Jini, a facility that would allow not just PCs but many other Java-enabled devices such as appliances and cell phones to communicate with one another.

Recently, however, Joy has expressed serious misgivings about the future impact of artificial intelligence and related developments on the future of humanity. Joy remains proud of the achievements of a field to which he has contributed much. However, while rejecting the violent approach of extremists such as Theodore Kaczynski (known as the Unabomber), Joy points to the potentially devastating unforeseen consequences of the rapidly developing capabilities of computers. Unlike his colleague RAYMOND C. KURZWEIL's optimistic views about the coexistence of humans and sentient machines, Joy points to the history of biological evolution and suggests that superior artificial life-forms will displace humans, who will be unable to compete with them. He believes that given the ability to reproduce themselves, intelligent robots or even "nanobots" might soon be uncontrollable.

Joy also expresses misgivings about biotechnology and genetic engineering, seen by many as the dominant scientific and technical advance of the early 21st century. He has proposed that governments develop institutions and mechanisms to control the development of such dangerous technologies, drawing on the model of the agencies that have more or less successfully controlled the development of nuclear energy and the proliferation of nuclear weapons for the past 50 years.

Joy received the Association for Computing Machinery's Grace Murray Hopper Award for his contributions to BSD UNIX before the age of 30. In 1993, he was given the Lifetime Achievement Award of the USENIX Association, "for profound intellectual achievement and unparalleled services to the UNIX community." He also received the *Computerworld* Smithsonian Award for Innovation in 1999.

Further Reading

Joy, Bill, ed. *The Java Language Specification, Second Edition*. Reading, Mass.: Addison-Wesley, 2000.

Joy, Bill. "Why the Future Doesn't Need Us." *Wired* 8, no. 4, April 2000. Available on-line. URL: http://www.wired.com/wired/archive/8.04/joy.html. Downloaded on December 2, 2002.

O'Reilly, Tim. *A Conversation with Bill Joy*. Available on-line. URL: http://www.openp2p.com/pub/a/p2p/2001/02/13/joy.html. Posted on February 13, 2001.

Williams, Sam. "Bill Joy Warns of Tech's Dangerous Evolution." *Upside.com*, January 18, 2001. Available on-line. URL: http://www.upside.com/Open_Season/3a648a96b.html. Downloaded on December 2, 2002.

K

Kahn, Philippe

(1952–)
French/American
Entrepreneur, Programmer

The growth of personal computing in the 1980s brought tremendous opportunities not only for developing software but for creating the compilers and other tools needed by programmers. Philippe Kahn was a pioneer in bringing innovative features and technical excellence to the personal computer (PC) industry.

Kahn was born in Paris in 1952 to a French mother and a German father. As a high school student, Kahn had many interests and talents. He wrote his first computer program at the age of 10 and was fascinated by the endless possibilities of software. He was an excellent science and math student, but had other interests as well. On the weekends, the teenage Kahn earned spending money by playing saxophone in a jazz group. And, as he had achieved his black belt in karate at age 16, it was unlikely that this particular nerd was picked on more than once.

As a graduate student at the Swiss Federal Institute of Technology during the 1970s, Kahn worked with NIKLAUS WIRTH, helping him with the development of Pascal, which became a popular computer language because of its clean structure and way of organizing programs. He also worked on the software for the Micral computer, which was developed by André Truong. Although the Micral was never sold, some historians consider it to be the first personal computer, predating the better-known Altair.

When Kahn came to California in 1982, he had $2,000, no job, and not even a green card allowing permanent U.S. residence. Nevertheless, he founded a company called Borland International which quickly became a market leader in programming languages and software tools. At the time, if someone wanted to develop "serious" software for a machine such as the IBM PC, he or she had to buy an expensive, hard-to-use BASIC or FORTRAN compiler from Microsoft.

Kahn put his intimate knowledge of Pascal implementation to work to create an alternative, Turbo Pascal, first released in 1983. Turbo Pascal put all the tools a programmer needed—editor, debugger, program window—together on the desktop. The compiler processed code at seemingly lightning-like speed, and the minimalist standard Pascal language was supplemented with graphics, file management, and other facilities. The whole package cost about $50, compared to the $500 or so for the Microsoft product.

Programmers flocked to Turbo Pascal and used it to create thousands of PC applications. Borland grew steadily through the 1980s until it had several thousand employees and sales of about $500 million a year. Meanwhile, Borland created similar programming environments for C, the language that was rapidly supplanting Pascal as the mainstream language for software development, and then moved on to C++, the object-oriented successor to C developed by BJARNE STROUSTRUP.

Another very successful Borland product was Sidekick. This program took advantage of an obscure feature of MS-DOS that allowed small programs to be tucked away in an unused portion of memory and then activated by a key combination. With Sidekick, a user running, for example, WordStar and needing an address for a letter could bring up Sidekick and retrieve the information from a simple database, read or make calendar entries, or even dial phone numbers—all without closing the WordStar program. While this ability to run more than one program at a time would become standard with Windows, at the time it was a programming tour de force that introduced a whole new category of software, the personal information manager, or PIM. Borland also developed the Quattro spreadsheet and Paradox database management system, both of which attracted a considerable following.

By the 1990s, however, Borland was facing increasing competition from Microsoft, which had developed its own low-cost integrated programming environments for C, C++, and Visual BASIC, all closely tied into the Windows operating system. Borland responded with Delphi, a Pascal-based system that also provided Windows-based visual programming. Although many programmers continued to prefer the Borland products, Microsoft's increasingly elaborate environments eventually took over most of the market for Windows software development tools.

Borland's spreadsheet and database products were also facing stiff competition from Microsoft, which introduced its Excel spreadsheet and Access database as part of its Microsoft Office software suite. Although Kahn tried to fight back with the acquisition of Ashton-Tate, developer of the venerable dBase program, Borland also had to deal with an expensive, time-consuming copyright infringement suit that accused its Quattro Pro spreadsheet of too closely mimicking the original Lotus 1-2-3 spreadsheet (now owned by IBM). Although Borland finally won in the U.S. Supreme Court in 1996, by then the market battle had been lost to Microsoft.

By the mid-1990s, the first "Internet boom" was underway, and Kahn wanted to revitalize Borland by taking the company in a new direction, specializing in networking and the sharing of information between computers. However, the company's board of directors was unwilling to support his plans. The year 1995 marked perhaps the lowest point in Kahn's career: He was essentially fired by Borland's board of directors and in his personal life was embroiled in a divorce that cost him his $9 million house. Meanwhile, the plunging value of Borland stock had wiped out most of Kahn's net worth.

However, Kahn showed the same resiliency he had displayed as a newly arrived immigrant. In 1995, together with his new wife, Sonia Lee, Kahn founded Starfish Software to specialize in systems for wireless linking and synchronization of "connected information devices" such as palmtop computers. One such device was Rolodex Electronics Xpress. This credit card-sized device has a built-in personal information manager, a system for keeping track of contacts, appointments, and notes. The device allows the user to make an Internet connection with a desktop or laptop PC to retrieve and display information needed "in the field." Kahn's efforts have led to the industry adoption of the TrueSynch standard for communicating and updating information between devices from different manufacturers. The growing capabilities of handheld computing devices and the new "smart" cell phones are likely to represent a continuing strong market for this type of software.

Kahn sold Starfish in 1998, but continued to run it as a division of the Motorola corporation. About the same time, Kahn founded a new company, LightSurf Technologies, which focuses on instant digital imaging, allowing pictures to be transmitted and received regardless of file format or platform. As he explained to an audience at the Vortex 2000 industry conference:

> What we want to do is what the Polaroid camera did, the Land camera 50 years ago, which is click, pull out, pass it on to the next person. Except that here, instead of sharing a little photograph, what you want is to have that image instantly share over the Internet, with the people you choose to share it with.

Kahn is famous for his physical energy and brashness. He has been known to jump into a swimming pool while dressed in a tuxedo, and he enjoys dirt biking, snowboarding, and surfing. Another sport in which he has been successful is sailboat racing, which requires paying attention to subtle changes while making long-term plans—not unlike the skills required of a successful CEO in a rapidly changing industry that has seen its share of stormy weather. While some observers noted a new sense of maturity in Kahn as he approached age 50, it seems clear that his brash, restless energy has not waned.

Together with Sonia Lee, Kahn has established the nonprofit Lee-Kahn Foundation to promote educational, health, and particularly environmental concerns. *Byte* magazine has named Kahn as one of the 20 most important people in the history of the computer industry.

Further Reading

Kahn, Philippe. "Philippe Kahn on the Wireless Internet." *Dr. Dobb's Tech Netcast*. November 10, 1999. Available online. URL: http://technetcast.ddj.com/tnc_program.html?program_id=64. Downloaded on December 2, 2002.

Kellner, Thomas. "Survivor: Philippe Kahn Making Comeback with LightSurf." *Forbes*, July 9, 2001, p. 128.

Metcalfe, Robert. Vortex 2000 ING Executive Forums. [Interview with Philippe Kahn.] Available on-line. URL: http://www.lightsurf.com/news/tech_speak/ts_052500.html. Posted on May 25, 2000.

Kahn, Robert

(1938–)
American
Engineer, Computer Scientist

In order for the Internet to reliably connect computers around the world, there must be a protocol, or agreement about how data will be formatted and handled. Robert Kahn played a key role in developing this protocol (called TCP/IP) and in the general architecture for the network that eventually became the Internet. One might say that he paved the Information Highway.

Kahn was born on December 23, 1938, in Brooklyn, New York. He attended Queens College and then City College of New York, receiving a bachelor's degree in electrical engineering in 1960. He then went to Princeton University for his advanced study. During his postgraduate years, Kahn also worked for Bell Laboratories, one of the nation's foremost private research laboratories.

After receiving his doctorate in electrical engineering from Princeton University in 1964, Kahn accepted a position as assistant professor at the Massachusetts Institute of Technology. In 1966, he took what he thought would be a brief sabbatical at Bolt, Beranek, and Newman (BBN), a pioneering company in computer networking and communications. Kahn quickly realized that BBN was at the heart of a coming revolution in how computers and their users would communicate, so he took the position of senior scientist to help BBN to design and implement new computer networks.

With more and more data being stored in government and business databases, the need to exchange data between different computer systems was becoming urgent. However, each type of computer had its own way of organizing the bits and bytes of data into meaningful words or numbers. If one computer system, such as a DEC PDP minicomputer, were to "understand" a message from another, such as an IBM mainframe, the data would have to be converted in real time to a format that the destination system could read. In 1969, Kahn led the team that developed the Interface Message Processor (IMP), which was essentially a specialized minicomputer. Once a properly configured IMP was provided for each computer, they could communicate over ordinary phone lines.

Meanwhile, Defense Department officials had become convinced that the nation needed a decentralized computer networking system that would allow facilities (such as radar stations and interceptor bases) to maintain communications even in the face of a nuclear attack. Kahn, together with VINTON CERF, LEONARD KLEINROCK, and LAWRENCE ROBERTS, worked out network protocols based on a system called packet-switching in which data could be broken into small, separately addressed pieces and relayed from one computer to another until they reached their destination, where they are reassembled into the original message. Such a system can take advantage of whatever connections are fastest, while also having the ability to reroute communications if a particular participant or "node" in the network is no longer functioning.

In 1972, Kahn took a key role in the leadership of this effort by joining the Information Processing Technology Office (IPTO) of the Defense Advanced Research Projects Agency (DARPA). Kahn became responsible for managing the program that developed ARPANET, ancestor of the Internet.

After Cerf joined Kahn in 1973, the two researchers built on an existing limited network protocol to develop the transmission control protocol/Internet protocol, or TCP/IP. The TCP part of the protocol handles error detection and transmission of packets containing errors, while the IP handles the overall addressing and processing of messages built from the packets.

With TCP/IP, any two computer systems (or entire networks) could communicate simply by following the rules for constructing packets and bundling them into messages. It did not matter what kind of data—word, number, picture, or something else—was in the packets. Once the message was transported to the destination, an appropriate program could process the data. It is this flexibility that makes it possible today for a variety of e-mail programs, Web browsers, database programs, and other kinds of software to work together over the Internet.

Kahn became director of the ITPO in 1979, continuing the development of ARPANET there until 1985. During this time, he also conducted research on extending and generalizing the concept of the packet-switched network to work with radio (wireless) and satellite communications (SATNET). In 1977, this interconnection had been shown by a now-famous demonstration in which messages were sent around the world using a mixture of satellite and ground line Internet links.

In 1986, Kahn left ITPO to found and head the nonprofit Corporation for National Research Initiatives (CNRI). This organization serves as a funding and coordination center for developing the technologies needed for what became known as the national information infrastructure—the high-capacity data networks essential for the Internet that would blossom with the coming of the World Wide Web in the 1990s. In this effort, Kahn and the CNRI worked together with the National Science Foundation (NSF) and DARPA, developing five "testbed" gigabit (billion-bit-per-second) networks. Kahn also developed the idea of the Knowbot, a program that could automatically link to different kinds of databases or other information sources over the Internet to retrieve requested information.

As it became clear that the 1990s would see an explosion in use of the Internet, Kahn also played a key role in the founding of the Internet Society, an organization that developed new technical standards for the network and sponsored seminars on social and other issues relating to the network. However, after an ongoing disagreement over some of these issues, Kahn left the board of directors of the Internet Society in 1996.

Like many complex developments, the Internet has no single inventor or "founding father." However, as Peter Kirsten wrote in the *Annals of the History of Computing*, "It was really Bob Kahn to which the major credit must go; it was his vision that started the main packet network technologies and their interconnection."

Robert Kahn has gotten used to the fact that his work has not brought him the celebrity enjoyed by other pioneers such as Netscape's MARC ANDREESSEN or even TIM BERNERS-LEE, creator of the World Wide Web. Indeed, Kahn remarked, in an interview: "It's amazing the number of people who don't know what the Internet really is. It's an architecture for interconnecting networks. It's not the World Wide Web. That's just an application on top of the Internet."

Despite the lack of fame, Kahn's work is well known within the technical community. He has received numerous awards including the PC *Magazine* Lifetime Award for Technical Excellence (1994), the Charles Stark Draper Prize of the National Academy of Engineering for 2001 (shared with Vinton Cerf, Leonard Kleinrock, and Lawrence Roberts), the Prince of Asturias Award for Scientific and Technical Research (2002) and the prestigious National Medal of Technology (1997).

Further Reading

Corporation for National Research Initiatives. Available on-line. URL: www.cnri.reston.va.us. Updated on October 17, 2002.

Hafner, Katie, and Matthew Lyon. *Where Wizards Stay Up Late: The Origins of the Internet*. New York: Simon and Schuster, 1996.

Kahn, Robert E. [Interview] *Omni*, December 1992, pp. 83ff.

Kapor, Mitchell

(1950–)
American
Entrepreneur, Programmer

Most of the computer pioneers had a heavy technical background—typically, college studies in fields such as mathematics, electrical engineering, and (starting in the 1970s) computer science. However, there are some individuals who bring a quite different set of experience and skills into the computer field. Such is the case with Mitchell Kapor, whose Lotus 1-2-3 program firmly established the spreadsheet as a business tool.

Kapor was born on November 1, 1950, in Brooklyn, New York, and attended public school in Freeport, Long Island. As a student at Yale University in the 1960s, he fashioned a customized degree program in cybernetics that that included diverse studies in psychology, cognitive science, linguistics, and computer science.

After receiving his bachelor's degree in 1971, Kapor worked at such odd jobs as radio disk jockey, teacher of transcendental meditation, and computer programmer. He then went to Campus-Free College (later called Beacon College) in Boston to study psychological counseling. He received his master's degree in counseling psychology in 1978 and worked for a short time as a counselor at the New England Memorial Hospital in Stoneham, Massachusetts.

While he was finishing his psychology studies Kapor had become intrigued by the personal computer (PC), which was just coming onto the market. He bought an Apple II and learned to program it so he could earn some extra money as a consultant. He then realized that

Mitchell Kapor founded Lotus, whose 1-2-3 integrated spreadsheet program became an early hit for the IBM PC. He then cofounded a group to help protect civil liberties in cyberspace. *(CORBIS SABA)*

he could adapt a statistical charting program called TROLL from the Massachusetts Institute of Technology (MIT) to the Apple II, making it available to more students and researchers. He called his version of the program "Tiny Troll."

Kapor then met DANIEL BRICKLIN, inventor of VisiCalc, the first PC spreadsheet program. VisiCalc had become the first true "hit" of the PC software market, leading thousands of business users to buy Apple II computers. Kapor worked with Bricklin at Personal Software to create VisiPlot, a program that creates charts and graphs from the data in VisiCalc spreadsheets. In 1980, Kapor sold all the rights to the new program to VisiCorp for $1 million.

Kapor then joined with business partner Jonathan Sachs to found Lotus Development in 1982, with the objective of creating a new spreadsheet program. This program, Lotus 1-2-3, integrated spreadsheet, graphing, and simple database functions into a single program with an easier-to-use, more consistent interface. Lotus 1-2-3 also provided on-line help screens to make it easier for businesspeople to master the unfamiliar technology. The program was written for the new IBM PC, which had rapidly overtaken the Apple II as the desktop computer of choice for business applications. Lotus 1-2-3 was a perfect fit for the booming PC market. In 1983, the year of its release, Lotus made $53 million in revenue, which jumped to $156 million the following year and $258 the third year, with the two-person company growing to more than 2,300 employees.

By 1986, however, Kapor was ready to move on to other ideas. As he told a *Newsweek* interviewer: "I was in a period of soul-searching, thinking about the immensity of changes [going from] two guys in a basement five years ago to the world's biggest software company. I felt the company was in great shape. . . . It was time the torch passed to the second generation of leadership." He therefore left the management ranks at Lotus, though he continued to work on a program called Lotus Agenda. This was one of the first programs to allow the user to keep track of personal information such as contacts, phone numbers, calendar entries, and notes. It started a new class of software called the personal information manager, or PIM. (See PHILIPPE KAHN for another approach to this type of software.)

In 1987, Kapor started a new company, ON Technology. Kapor's goal was to focus on developing technology that would allow groups of people to share documents and work together on projects. This emphasis on "groupware" was in keeping with the trend in the later 1980s and

1990s toward on networking and connectivity rather than focusing on the stand-alone PC. Further, Kapor wanted to achieve a new level of user-friendliness. He acknowledged in a "Design Manifesto" issued at a 1990 technical conference that existing operating systems just were not good enough: "Sooner or later, everyone that I know, except perhaps the most hard-core technical fanatic, wants to pick up the machine off the desk and throw it through the window."

Kapor recruited some top computer scientists from MIT, Harvard, and the Xerox Palo Alto Research Center (where the mouse and windows had been invented). However, by 1989 the ambitious attempt to create a new operating system had bogged down and many of the key programmers left for other projects. The company was able to develop only a few relatively minor applications.

In 1990, Kapor turned his attention to the social issues that were starting to emerge as more people used on-line communications such as bulletin boards, chat rooms, and Internet newsgroups. For example, that same year Steve Jackson Games was nearly driven out of business when the government seized its computers in connection with an investigation of hackers who were accused of accessing proprietary information. Although the game company had no real connection with the alleged hacker, the Secret Service refused to return the equipment, even though under the First Amendment publishers enjoy special protection. Eventually a federal judge ruled in favor of Jackson, awarding damages.

In response to the collision between law enforcement and new technology, Kapor and another "cyber-activist," John Perry Barlow, formed the Electronic Frontier Foundation (EFF). This group became known as a sort of American Civil Liberties Union for cyberspace, taking on the government and big corporations over issues such as privacy and free expression. Kapor served on the board of the EFF until 1994. During 1992 and 1993, Kapor also served as chair of the Massachusetts Commission on Computer Technology, which was set up to investigate computer crime and related civil liberties issues. As the 1990s continued, he also served on the Computer Science and Technology Board of the National Research Council and the National Information Infrastructure Advisory Council, as well as teaching courses in software design and computer-related social issues as an adjunct professor at the MIT Media Lab. He has also been active as a venture capitalist with Accel Partners and philanthropist, heading the Mitchell Kapor Foundation. Kapor received the 1996 Computer History Museum Fellow Award.

Further Reading

Cringely, Robert X. *Accidental Empires: How the Boys of Silicon Valley Make Their Millions, Battle Foreign Competition, and Still Can't Get a Date*. New York: HarperCollins, 1993.

Electronic Freedom Foundation. Available online. URL: http://www.eff.org. Downloaded on November 2, 2002.

Kapor, Mitchell. "Civil Liberties in Cyberspace." *Scientific American*, September 1991, p. X.

Kay, Alan C.
(1940–)
American
Computer Scientist

If Alan Kay had not done his pioneering work, personal computers (PCs) would still exist. But cryptic commands might have to be typed one line at a time, resulting in the operating system sending back text messages. There might never have been easy-to-click icons and a visual desktop for organizing work.

Alan Kay developed a variety of innovative concepts that changed the way people use computers. Because he devised ways to have computers accommodate users' perceptions and needs,

Kay is thought by many to be the person most responsible for putting the "personal" in personal computers. Kay also made important contributions to object-oriented programming (a term he coined), changing the way programmers organized data and procedures in their work.

Kay was born on May 17, 1940, in Springfield, Massachusetts. His family moved to Australia shortly after his birth, but the family returned to the United States because of the threat of a Japanese invasion of Australia. His father developed prostheses (artificial limbs) and his mother was an artist and musician. These varied perspectives probably contributed to Kay's interest in interaction with and perception of the environment.

An independent streak and a questioning mind also apparently characterized Kay from an early age, as he recalled later to Dennis Shasha and Cathy Lazere:

> By the time I got to school, I had already read a couple hundred books. I knew in the first grade that they were lying to me because I had already been exposed to other points of view. School is basically about one point of view—the one the teacher has or the textbooks have. They don't like the idea of having different points of view, so it was a battle. Of course I would pipe up with my five-year-old voice.

As he got older, Kay became a talented guitarist and singer and considered music as a career. However, after a brief stint at Bethany College in West Virginia, he was expelled for protesting the school's Jewish quota. After spending some time in Denver, Colorado, teaching music, he joined the U.S. Air Force.

In the early 1960s, Kay was working as a programmer at Randolph Air Force Base, near Denver. His job was to write the detailed data processing instructions, called assembly language, for a Burroughs 220 mainframe computer. His problem was that he needed to share this data with other types of computers owned by the air force, but each type of mainframe computer, such as Burroughs, Sperry Univac, or IBM, had its own completely different data format.

Kay then came up with an interesting idea. He decided to create "packages" that included data and a set of program functions for manipulating the data. By creating a simple interface, other users or programs could access the data. They did not need to know how the data was organized or understand the details of exactly how the data access function worked. Although his idea did not catch on at the time (Kay said to Cade Metz "I didn't get the big grok [realization] until 1966,") Kay had anticipated object-oriented programming, a concept that he and other computer scientists would bring to center stage in the 1970s and 1980s.

Meanwhile, later in the 1960s, while completing work for his Ph.D. at the University of Utah, Kay further developed his object-oriented ideas. He helped IVAN SUTHERLAND with the development of a program called Sketchpad that enabled users to define and work with onscreen graphics objects, while also working on the development of SIMULA, the first programming language specifically designed to work with objects. Indeed, Kay coined the term *object-oriented* at this time.

Kay viewed object-oriented programs as consisting of objects that contained appropriate data that could be manipulated in response to "messages" sent from other objects. Rather than being rigid, top-down procedural structures, such programs were more like teams of cooperating workers—or, to use one of Kay's favorite metaphors, like biological organisms interacting in an environment. Kay also worked on parallel programming, in which programs carried out several tasks simultaneously. He likened this structure to musical polyphony, in which several melodies are sounded at the same time.

Kay participated in the Defense Advanced Research Projects Agency (DARPA)-funded research that was leading to the development of the Internet. One of these DARPA projects was FLEX, an attempt to build a computer that could be used by nonprogrammers using onscreen controls. While the bulky technology of the late 1960s made such machines impracticable, FLEX incorporated some ideas that would be used in later PCs, including multiple onscreen windows.

During the 1970s, Kay worked at the innovative Xerox Palo Alto Research Center (PARC). Kay designed a laptop computer called the Dynabook, which featured high-resolution graphics and a graphical user interface. While the Dynabook was only a prototype, similar ideas would be used in the Alto, a desktop personal computer that could be controlled with a new pointing device, the mouse (invented by DOUGLAS ENGELBART). A combination of high price and Xerox's less-than-aggressive marketing kept the machine from being successful commercially, but STEVE JOBS would later use its interface concepts to design what would become the Macintosh.

On the programming side, Kay developed Smalltalk, a language that was built from the ground up to be truly object-oriented. Kay's worked showed that there was a natural fit between object-oriented programming and an object-oriented user interface. For example, a button in a screen window could be represented by a button object in the program, and clicking on the screen button could send a message to the button program object, which would be programmed to respond in specific ways.

After leaving PARC in 1983, Kay briefly served as chief scientist at Atari and then moved to Apple, where he worked on Macintosh and other advanced projects. In the early 1990s, Kay anticipated many of the trends in computer use that would be prominent at the start of the 21st century. He saw a transition from "personal computing" to "intimate computing," characterizing the latter as the use of mobile, wireless-networked computers that would use programs called "software agents" to help people find and track the data and daily communications (see also PATTIE MAES). As these agents become smarter and more capable, the idea of programming will be largely replaced by users "training" their agents much in the way that an executive might train a new assistant. In 1996, Kay became a Disney Fellow and Vice President of Research and Development at Walt Disney Imagineering. In 2002 Kay moved to a research environment reminiscent of PARC—Hewlett-Packard Laboratories. He told an interviewer that "the goal [of his new work] is to show what the next big relationship between people and computing is likely to be."

Kay retains a strong interest in how computers are being used (and abused) in schools. Reflecting on the success of the Macintosh, he noted later in an interview with Judy Schuster

> I think the thing that surprised me is that computers are treated much more like toasters, [with] predefined functions mainly having to do with word processing and spreadsheets or running packaged software, and less as an artistic material to be shaped by children and teachers.

Looking back at his father's work with prosthetics, he also noted in the same interview, "Put a prosthetic on a healthy limb and it withers." He believes that instead of being like cars, computers should be more like bicycles, amplifying the intellectual "muscles" of children and other learners.

Kay has won numerous awards, including an American Academy of Arts and Sciences fellowship, the Association for Computing Machinery Software Systems Award, and the J. D. Warnier Prize.

Further Reading

Gasch, Scott. "Alan Kay." Availale on-line. URL: http://ei.cs.vt.edu/~history/GASCH.KAY.HTML. Posted in 1996.

Kay, Alan. "The Early History of Smalltalk." In Thomas J. Bergin Jr. and Richard G. Gibson Jr., eds. *History of Programming Languages II*. Reading, Mass.: Addison-Wesley, 1996.

Shasha, Dennis, and Cathy Lazere. "Alan C. Kay: A Clear Romantic Vision." In *Out of Their Minds: The Lives and Discoveries of 15 Great Computer Scientists*. New York: Copernicus, 1995.

Metz, Cade. "The Perfect Architecture." *PC Magazine*, September 4, 2001, p. 187.

Schuster, Judy. "A Bicycle for the Mind, Redux." *Electronic Learning*, vol. 13, April 1994, p. 66.

Kemeny, John G.

(1926–1992)
Hungarian/American
Mathematician, Computer Scientist

In the early days of computing, learning to program was very difficult. Programs had to be written as a series of precise, detailed instructions, usually punched onto cards. The finished program was then handed over to the machine operator, who fed them to the computer. Since the programmer had no way to verify or test the instructions before submitting them, chances were good that one or more errors would crop up, halting the program or making its results useless. The programmer would then have to go through the whole tedious process of correcting and resubmitting the cards, often several times.

In the mid-1960s, however, a Dartmouth mathematician and professor named John Kemeny, together with a colleague, Thomas Kurtz, developed a new language called BASIC (for beginner's all-purpose symbolic instruction code). With BASIC, a student programmer could type simple commands at a Teletype terminal, receiving instant feedback from the computer. Any errors could be fixed on the spot, and the program run again to test whether it was correct.

Kemeny was born in Budapest, Hungary, on May 31, 1926. In 1940, the family fled to the United States because they believed that their country would soon fall to the Germans, who had already seized Czechoslovakia and Poland. The family settled in New York. When young Kemeny went to an American high school, he knew German and Latin as well as his native Hungarian, but not a word of English. Nevertheless, by the time he graduated in 1943, he led his class in academic honors.

Kemeny enrolled in Princeton University to study mathematics and philosophy, but in

John Kemeny, together with Thomas Kurtz, invented BASIC, an easy-to-use interactive programming language that let many students learn how to tap the power of computers. Kemeny then became president of Dartmouth College, leading that institution through the social tumult of the early 1970s. *(Courtesy Dartmouth College Library)*

1945 he received both his American citizenship and a draft notice. Because of his mathematical background, the army put him to work as a "computer"—that is, a person who performed complex calculations using the electromechanical calculating machines of the time. His workplace was Los Alamos, New Mexico, where the secret Manhattan Project was designing the atomic bomb. While at Los Alamos, Kemeny worked under JOHN VON NEUMANN, the prominent mathematician who was becoming increasingly involved in the design of electronic digital computers (see also J. PRESPER ECKERT and JOHN MAUCHLY). Kemeny soon shared von Neumann's enthusiasm for the possibilities of computing.

After the war Kemeny, returned to Princeton and he graduated with top honors in 1947, having served as president of the German Club and the Roundtable as well as having participated in the Court Club and the fencing team. He then studied for his doctorate, which he received only two years later. During his graduate studies, Kemeny worked with ALONZO CHURCH, the logician who had helped establish the theory of computability, as well as with Albert Einstein, who was working on his unified field theory.

After earning his Ph.D., Kemeny joined the Princeton faculty as an instructor in mathematical logic, though he later taught philosophy for several years. In 1953, he moved to Dartmouth College as a professor of mathematics and philosophy. He would remain there for the rest of his career. From 1954 to 1957, Kemeny served as chairman of the Dartmouth mathematics department, where he pioneered new methods of teaching mathematics. His emphasis on introducing advanced mathematics concepts into earlier stages of the curriculum and discarding rote drills and problem sets would help inspire the "New Math" movement that would later percolate down to the elementary school curriculum.

Dartmouth had no computer of its own at the time, so Kemeny frequently commuted 135 miles each way to use the computer at the Massachusetts Institute of Technology. He was therefore able to keep up with developments in computing, particularly the announcement of FORTRAN in 1957. Kemeny thought that FORTRAN represented an important step in making computers easier for scientists and engineers to use, because it accepted ordinary mathematical expressions rather than requiring the bit-by-bit specifications used by assembly language. "All of a sudden," Kemeny recalled in his 1972, book *Man and the Computer* "access to computers by thousands of users became not only possible but reasonable."

In 1959, Dartmouth finally got a computer, and Kemeny encouraged his students to use it just as they would the library or any other campus resource. However, existing computer operating systems allowed only one program to run at a time, and thus only one person at a time could have access to the expensive machine. Since the computer actually had enough memory to store several programs, Kemeny, sharing ideas also being developed by JOHN MCCARTHY and J. C. R. LICKLIDER, thought that it would make sense to have a computer load multiple programs and let them "take turns" running. Since the computer is much faster than human response time, the effect would be as though each user had his or her own dedicated machine.

Kemeny and a team of students and colleagues therefore developed the Dartmouth Time-Sharing System, or DTSS, which became available in 1964. The system was gradually expanded into a network that linked Dartmouth to 24 high schools and 12 other colleges in New England. Once time-sharing was available, it became practicable to think of making computing truly available to any student who might benefit from it. The problem was that the existing languages, even FORTRAN, were hard to use because programmers had to master their often complicated syntax.

Kemeny and a colleague, Thomas Kurtz, decided to take a different approach to teaching programming. They simplified the existing FORTRAN language and added simple PRINT and INPUT commands that made it easy to submit data, process it, and get the results. They then wrote an interactive compiler that could take program statements directly from students at terminals, compile them, and display the results almost instantly. Each instruction line began with a line number, and students could use a simple editing facility to recall and correct any line that contained an error. Kemeny and Kurtz also pioneered the idea of "beta testing"—they encouraged as many students as possible to use their prototype BASIC system and report any "bugs" they found.

Kemeny and Kurtz were interested in helping students, not making money, so they did not try to market their language. (Dartmouth owned the copyright, but made the BASIC system available for free to anyone who wanted to use it, such as General Electric, which distributed it with its own time-sharing system.)

In 1970, Kemeny became president of Dartmouth College. In that office, like many other college administrators, he was faced with the antiwar protests and social changes that had rocked the 1960s. Kemeny believed in that universities needed to reach out to the surrounding communities and in turn be more open to them. He made the college coeducational in 1972 and also worked to increase minority enrollment. (Kemeny himself was the first person of Jewish background to be appointed president of an Ivy League school.)

Kemeny also supported campus antiwar protests, particularly in response to the U.S. invasion of Cambodia in April 1970 and the killing of four students by National Guard troops at Kent State University in Ohio. Many liberals praised his progressive policies, but some of the more conservative alumni loudly objected, particularly when he banned the school's Indian sports mascot as racist.

Following the accident at the Three Mile Island nuclear power plant in 1979, Kemeny served as chair of an investigatory commission appointed by President Carter. The commission concluded that the nuclear industry had not paid sufficient attention to known safety problems and that regulators had been lax and biased in favor of the industry as opposed to public safety.

In 1980, Kemeny stepped down from the presidency of Dartmouth and his focus returned to the use of computers in education. One of the things both Kemeny and Kurtz were concerned about was what had happened to their BASIC language. Starting in the late 1970s, a variety of dialects, or versions of the language, had been written to run on the various models of personal computers. Because each version was somewhat different, programs written for an Apple II, for example, could not be run on an IBM PC without extensive modifications. Further, to the extent that a "standard" version of BASIC had been created by Microsoft, the language was poorly structured, particularly in lacking a proper mechanism for defining procedures and local variables, which could protect against data values being inadvertently changed.

Responding to this situation, Kemeny and Kurtz developed True BASIC in 1985, starting a company of the same name to sell it along with various educational software packages. They hoped that having a modern, consistent version of the language would make BASIC fully capable of becoming the language of choice for mainstream programming. However, although their language was well designed and well regarded by many computer scientists, their effort had come too late. C and particularly the object-oriented C++ would dominate the field, and BASIC, with the exception of Microsoft Visual BASIC (which gradually became better structured), would largely disappear.

Kemeny died on December 26, 1992. He was honored with the Computer Pioneer Award of the Institute of Electrical and Electronics Engineers

(1985). To acknowledge and carry on Kemeny's work in modernizing the mathematics curriculum, the Alfred P. Sloan Foundation funded the building of the Albert Bradley Center for mathematics study and research at Dartmouth.

Further Reading

Kemeny, John. *Back to BASIC: The History, Corruption and Future of the Language*. Reading, Mass.: Addison-Wesley, 1985.

———. *Random Essays on Mathematics, Education, and Computers*. Upper Saddle River, N.J.: Prentice-Hall, 1964.

Slater, Robert. *Portraits in Silicon*. Cambridge, Mass.: MIT Press, 1987.

Kernighan, Brian

(1942–)
Canadian/American
Computer Scientist, Writer

Many people assume that writing a computer program is something like writing a novel. The writer starts with a blank sheet of paper (or an empty screen), and writes every last instruction from scratch. Actually, however, it is more like designing a new car model. The designer uses many standard components and creates only those custom parts needed to define the new car's distinctive style and features. Similarly, programmers today generally rely on tested components, such as routines for sorting or formatting data, and write only the new code they need to complete the tasks needed for the particular application. Brian Kernighan demonstrated the effectiveness of this "software tools" approach to programming, and made major contributions to the development and growth of the UNIX programming environment.

Brian Kernighan was born in 1942 in Toronto, Canada. He received his bachelor's degree in engineering physics from the University of Toronto in 1964. Toward the end of his undergraduate years he saw his first computer, an IBM 650, and he gradually became interested in programming.

Kernighan came to the United States in 1965 and enrolled at Princeton University, earning a Ph.D. in electrical engineering in 1969. (He specialized in computer science, but at the time this was a subspecialty within electrical engineering, not a separate discipline.)

That same year, he joined Bell Laboratories in Murray Hill, New Jersey, and has spent his career in its Computing Sciences Research Center. He had arrived just in time for a revolution in computer operating systems. Two researchers, KENNETH THOMPSON and DENNIS RITCHIE, were developing UNIX. Besides being able to multitask, or run many programs at the same time, UNIX made it easy to write small programs that did useful things and then connect them together with software "pipes" that let the data flow from one program to the next in a stream of bytes.

Consider, for example, the task of finding and extracting selected lines of text, sorting them, and then printing them as a table. A user of a traditional operating system would have to write a program for each task. The "find" program would have to put the lines it found into a temporary data file. The sort program would then have to open this file, read in the lines, sort them, and write an output file. The table-making program would have to open the file and format it into a table, writing the results to still another file. With UNIX, however, the find, sort, and table programs could all be linked with the pipe symbol, and the data would flow through them, being processed by each program in turn and emerge in the final output.

To Kernighan and other UNIX pioneers, features like pipes and streams were more than neat technical ideas, they embodied a distinctive philosophy or approach to software design. Rather than writing programs, they were making tools others could use and build on. Together with P. J. Plauger, Kernighan wrote a book called *Software Tools* that illustrated how a set of basic tools for working with data, formatting text, and performing other tasks could be linked together to tackle formidable data processing tasks.

Ritchie (with some help from Thompson) was developing the C language at the same time Kernighan and his colleagues were designing UNIX tools. C has been criticized by some people as having too-cryptic syntax and for not giving programmers enough of a safety net in dealing with matters such as converting one type of data to another. But as he explained to Mihai Budiu, Kernighan believes that

> C is the best balance I've ever seen between power and expressiveness. You can do almost anything you want to do by programming fairly straightforwardly and you will have a very good mental model of what's going to happen on the machine; you can predict reasonably well how quickly it's going to run, you understand what's going on and it gives you complete freedom to do whatever you want.

Kernighan continues to work on a variety of projects that suit his needs and interests. For example, he has written a program to convert mathematical formulas from the format used by eqn (a UNIX program used for typesetting equations) to HTML or XML formats suitable for posting on Web pages. On a larger scale, together with colleagues David Gay and Bob Fourer, he is developing AMPL, a language designed for optimizing the solution to mathematical problems. The developers intend to have AMPL create objects that can be "plugged in" to a main program without the authors of the latter having to worry about how it works.

Kernighan has received the 1997 USENIX Lifetime Achievement Award for "books that educated us all, for tools we still use, and for insights in the use of language as a bridge between people and machines." He is also a Bell Laboratories Fellow (1995) and has been elected a member of the American Academy of Engineering.

Further Reading

"Brian Kernighan." Available on-line. URL: http://cm.bell-labs.com/cm/cs/who/bwk. Downloaded on December 3, 2002.

Budiu, Mihai. "An Interview with Brian Kernighan." Available on-line. URL: http://www-2.cs.cmu.edu/~mihaib/kernighan-interview. Posted on July 2000.

Kernighan, Brian W., and Bob Pike. *The Practice of Programming*. Reading, Mass.: Addison-Wesley, 1999.

———. *The Unix Programming Environment*. Upper Saddle River, N.J.: Prentice Hall, 1992.

Kernighan, Brian W., and P. J. Plauger. *Software Tools*. Reading, Mass.: Addison-Wesley, 1976.

Kernighan, Brian W., and Dennis M. Ritchie. *The C Programming Language*. Upper Saddle River, N.J.: Prentice Hall, 1988.

Kilburn, Thomas M.

(1921–2000)
British
Engineer, Inventor

Most accounts of computer history tend to focus on developments in the United States. Two leading candidates for the inventor of the electronic digital computer, JOHN VINCENT ATANASOFF and the team of J. PRESPER ECKERT and JOHN MAUCHLY, were American, and American companies would come to dominate the computer hardware and software industries. However, British researchers would also built important early computers and helped develop key concepts of modern computer architecture. A good example is engineer and computer designer Thomas Kilburn.

Thomas Kilburn was born on August 11, 1921, in Dewsbury, Yorkshire, in the United

Kingdom. As a boy, Kilburn showed considerable talent in mathematics, which he pursued almost to the exclusion of other interests. He then went to Cambridge University and graduated with first class honors in mathematics in 1942.

When World War II broke out in 1939, Kilburn served in the air reserve squadron at Cambridge, and hoped to become a pilot in the Royal Air Force. However, the authorities viewed his ability in mathematics as more valuable, and the noted scientist and philospher C. P. Snow recruited Kilburn into the secret Telecommunications Research Establishment (TRE) at Great Malvern. There he worked with Frederick C. Williams on improving radar systems, a technology that had provided a crucial edge in defending Britain against the German Luftwaffe. Kilburn's wartime work also gave him a solid background in electronics to complement his mathematical ability. That combination would be a crucial one for many computer pioneers.

After the war, Kilburn went to the electrical engineering department at the University of Manchester. There he worked again with Williams, helping him develop a way to use cathode-ray tubes (CRTs), similar to those found in television sets, as a form of computer memory. While Williams had come up with the basic idea, Kilburn worked out the precise way to direct the electron beams to store data bits on the tube surface. Williams and Kilburn thus created a fast random-access computer memory. ("Random access" meant that any piece of data could be stored or retrieved directly without having to go through intervening data.) By 1947, what had become known as the Williams tube could store 2,048 binary digits, or bits, on a single CRT.

The availability of a reasonably large, fast memory made it possible to build a new type of computer. Earlier electronic computers such as the American ENIAC could not store their programs in memory. This meant that programs either had to be preset (such as by throwing switches or plugging in cables) or read, one instruction at a time, from punched cards during processing.

In the new machine, which was christened the Baby, the entire set of program instructions could be read from cards or punched tape and stored using the Williams memory tube. This brought a new dimension of flexibility to programming. It meant that a program could be broken up into separate sections or subroutines and the computer could "jump" as required from one part of the program to another, and program "loops" could be run without having to punch duplicate cards. Given sufficient memory, data, too, could be stored and retrieved randomly, considerably speeding up processing. Although the Americans were also developing "stored program computers," historians generally credit the Baby of 1948 with being the first such machine.

The Baby was more of a demonstration than a complete computer, so the Manchester team proceeded to develop the Manchester Mark I (and a commercial version, the Ferranti Mark I). The latter machine, which came into service in 1951, not only had a CRT memory that could store 10,000 binary digits, it also had a secondary magnetic drum memory, the ancestor of today's hard disk drives. Its early operating software and program libraries were developed by ALAN TURING. The Mark I was technically successful, but the commercial market for computers in Britain was slow to develop, and the Mark I would eventually be overwhelmed by IBM and other American manufacturers.

When Williams decided to concentrate on noncomputer electronics in 1951, Kilburn was put in charge of computer engineering research at Manchester. Kilburn and his team then made a series of important developments. In 1954, they built a computer called the MEG that could handle floating-point decimal operations and ran about 30 times faster than the Mark I. A commercial version, the Mercury, replaced the

Williams CRT memory with a new, more reliable technology, an array of tiny doughnut-shaped magnetic cores.

Another interesting project by Kilburn's group was a small computer built in 1953 that used transistors instead of vacuum tubes for its processing elements. The first of its kind, the Transistor Computer proved that the newly invented transistor was a viable computer component that would be increasingly used as more reliable types of transistors were introduced.

In 1956, the Kilburn team concluded that with the availability of transistors, magnetic core memory, and other technologies it was feasible to build a much faster computer. This machine, called MUSE (for mu, or microsecond) was designed to execute 1 million instructions per second. This was about 1,000 times faster than the Ferranti Mark I. Like its counterparts in the United States, the LARC and IBM STRETCH, this new generation of computers had the speed and capacity to run large, complex software programs. The machine (renamed ATLAS) came into full service in 1964. For several years, the ATLAS was the world's most powerful and sophisticated computer, and the three full-sized and three smaller versions that were built gave unprecedented computing power to the British scientific community.

The power of this machine and the new generation of mainframes typified by the IBM System 360 made it both possible and necessary to develop new computer languages and compiler systems such as developed by JOHN BACKUS with FORTRAN and through the guiding inspiration of GRACE MURRAY HOPPER with COBOL.

Following the success of ATLAS, Kilburn took on a new task, turning his computer group within the Manchester electrical engineering department into a separate department of computer science. As head of this department, Kilburn was responsible for creating a complete curriculum for a field that was only beginning to be defined. The Manchester computer science department was the first of its kind in Britain and possibly in Europe.

Once the new department had been organized, Kilburn and his associates continued their research. They built a new machine, the MU5, which incorporated sophisticated new operating system features, including the ability to "segment" code so it could be shared by more than one running program.

Kilburn retired in 1981 and died on January 17, 2001, in Trafford, England. He received numerous honors, becoming a Fellow of the prestigious Royal Society in 1965 and a Commander of the British Empire (CBE) in 1973. He is also a recipient of the Computer Pioneer Award of the Institute of Electrical and Electronics Engineers (IEEE) Computer Society (1982) and the Eckert-Mauchly Award, given jointly by the IEEE and the Association for Computing Machinery (1983), "for major seminal contributions to computer architecture spanning a period of three decades [and for] establishing a tradition of collaboration between university and industry which demands the mutual understanding of electronics technology and abstract programming concepts."

Further Reading

Bellis, Mary. "Inventors of the Modern Computer: The Manchester Baby." About.com. Available on-line. URL: http://inventors.about.com/library/weekly/aa060998.htm. Posted on June 9, 1998.

Goldstine, Herman H. *The Computer: from Pascal to von Neumann*. Princeton, N.J.: Princeton University Press, 1972.

University of Manchester. "Tom Kilburn (1921–2001)." Available on-line. URL: http://www.computer50.org/mark1/kilburn.html. Downloaded on November 2, 2002.

Wilkinson, Barry. "Stored Program Computer." Available on-line. URL: http://www.cs.wcu.edu/~abw/CS350/slides1.pdf. Downloaded on November 2, 2002.

⊠ Kilby, Jack
(1923–)
American
Engineer, Inventor

Jack Kilby learned practical electronic engineering the hard way, building radios for U.S.-backed guerrillas fighting the Japanese in Burma during World War II. In 1959, he patented the integrated circuit, which allowed many components to be embedded in a single chip. Kilby and Robert Noyce (who patented a different process) share credit for the invention. *(Courtesy of Texas Instruments)*

Vacuum tubes made the electronic computer possible. Transistors made it smaller and more reliable. But it took the integrated circuit, invented virtually simultaneously by Jack St. Clair Kilby and ROBERT NOYCE, to make computers small enough to fit on people's desks or even in their pockets.

Kilby was born in Jefferson City, Missouri, on November 8, 1923. Kilby's father was an electrical engineer who became president of Kansas Power Company when the boy was four years old. The family thus moved to Great Bend, Kansas. Both parents loved books, and the family often recited memorized passages of Shakespeare.

Young Kilby had his heart firmly set on following in his father's footsteps and becoming an electrical engineer. He recalled to T. R. Reid that when his father used a ham radio to communicate with customers after the great blizzard of 1937, "I first saw how radio, and by extension, electronics, could really impact people's lives by keeping them informed and connected, and giving them hope." Kilby also learned from his father that engineering was not just what one could build, but whether one could build it cheaply enough. As he noted to Reid:

> You could design a nuclear-powered baby bottle warmer, and it might work, but it's not an engineering solution. It won't make sense in terms of cost. The way my dad always liked to put it was that an engineer could find a way to do for one dollar what everybody else could do for two.

Kilby also had his heart set on attending the Massachusetts Institute of Technology (MIT), one of the nation's premier engineering schools. Unfortunately, Kilby failed the MIT entrance exam with a score of 497, three below the minimum 500 points needed. Disappointed, he enrolled in the University of Illinois, which both of his parents had attended.

Kilby's studies were abruptly cut off by the Japanese bombing of Peal Harbor and the U.S. entry into World War II. Kilby promptly enlisted in the U.S. Army Signal Corps, where he gained practical electronics skills building and designing radio equipment. At the time, a portable radio for infantry use weighed 60 pounds and often broke down in the field. In modern war, soldiers, particularly those involved in scouting or special operations such as guerrilla attacks against the Japanese in Burma, desperately needed lighter, more compact communications. Kilby's engineering superiors sent him to Calcutta to buy radio parts on the black market. They then cobbled them together into improvised transmitters that were lighter yet more powerful than the official army issue.

After the war, Kilby returned to the University of Illinois and earned his bachelor's degree in electrical engineering in 1947. He then took a position with the Centralab Division of Globe Union, Inc., in Milwaukee, Wisconsin. This early work established what would become the overall goal for his career: building ever more compact and reliable electronic circuits and thus reducing the wiring and other manufacturing steps that added cost and complexity.

One way to do this is to use printing techniques to create the electrical conducting path directly on the surface of the circuit board. Kilby made two important early contributions to printed circuit technology, the use of silk-screening techniques and the printing of carbon resistors directly on a ceramic circuit base. Meanwhile, he also earned his master's degree in electrical engineering at the University of Wisconsin.

Kilby, like many other electronic engineers, had been inspired by the invention of the transistor in 1947. He attended a 1952 symposium at Bell Labs in Murray Hill, New Jersey, in which endless possibilities for using the compact devices were discussed. In 1958, Kilby moved to Dallas-based Texas Instruments (TI). That company had a U.S. Army contract to build "Micro-Modules," small, standardized components that already had the necessary wires built in and that could be simply snapped together to make circuits.

But Kilby believed that this approach was cumbersome, because the modules were still relatively bulky and large numbers of them would still have to be put together to make the more elaborate types of circuits. Kilby conceived the idea of eliminating the wiring entirely and packing the transistors, resistors, capacitors, and other components directly onto a tiny chip of semiconducting material. Because Kilby, as a new employee, did not qualify for vacation, he spent that summer using equipment borrowed from deserted labs to build a prototype using a tiny germanium chip. It was crude, but it worked.

In February 1959, TI filed a patent for Kilby's invention. Four months later, another researcher, Robert Noyce of Fairchild Semiconductor filed a patent for an integrated circuit that he had been developing independently. Although the two inventors respected one another, the companies became locked in a bitter legal battle. Eventually the courts awarded Kilby the patent for his integrated circuit design, but also upheld Noyce's patent for his process for manufacturing integrated circuits, which was different from that used by Kilby.

There was still the problem of getting this rather esoteric invention into a profitable consumer product. Transistor radios already were becoming popular. TI chairman Patrick T. Haggerty suggested that Kilby build a small calculator using an integrated circuit. Since a calculator circuit was considerably more complicated than that in a radio, it would nicely show off the benefits of the new integrated circuit technology. Kilby, together with Jerry D. Merryman and James H. Van Tassel, patented the first handheld electronic calculator in 1967. In order to provide printed output from the machine, Kilby also devised a printing technique that used heat to burn letters onto special paper.

Kilby left TI in 1970 and worked as a private consultant. He continued to develop a variety of inventions (he holds more than 60 U.S. patents). He also taught electrical engineering at Texas A&M University until 1984. As the personal computer and other electronics technology became pervasive in the 1980s and 1990s, one might have expected Kilby to embrace it. However, he prefers an analog rather than a digital watch, confesses a continuing fondness for his slide rule, and for a hobby takes black-and-white photographs with an old Hasselblad camera.

For many years Kilby's work, unlike that of people who invented computers, was largely unknown outside of the technical community. This changed in 2000, however, when Kilby got a call from the Royal Swedish Academy of Sciences. He had won the 2000 Nobel Prize in physics for his work on integrated semiconductor technology. Kilby has been the recipient of numerous other awards, including the Institute of Electrical and Electronics Engineers David Sarnoff Award (1966), the National Medal of Science (1969), induction into the National Inventors Hall of Fame (1982), and the National Medal of Technology (1990).

Further Reading

Reid, T. R. *The Chip: How Two Americans Invented the Microchip and Launched a Revolution*. Rev. Ed. New York: Random House, 2001.

Slater, Robert. *Portraits in Silicon*. Cambridge, Mass.: MIT Press, 1987.

Texas Instruments. Integrated Circuit Fact Sheets. Available on-line. URL: http://www.ti.com/corp/docs/company/history/firstic.shtml. Downloaded on November 2, 2002.

Kildall, Gary

(1942–1994)
American
Computer Scientist, Entrepreneur

To transform the first microcomputers from science projects to practical business tools required a way for ordinary users to create, organize, and manage files. Programmers, too, needed some means to instruct the computer how to control such devices as disk drives and printers. What was needed was an operating system complete enough to get the job done, versatile enough to allow for new kinds of hardware, and small enough to fit in the limited memory of early personal computers (PCs). Gary Kildall created the first widely used operating system for PCs. And if a crucial negotiation had worked out differently he, not BILL GATES, might have become the king of software.

Kildall was born on May 19, 1942, in Seattle, Washington. His family ran a navigation school for sailors. As a high school student, Kildall was more interested in tinkering with cars and other machines than in academic subjects. Among other things, Kildall built a burglar alarm and a Morse code training device (using a tape recorder and a binary flip-flop switch). After graduating from high school, Kildall taught navigation at his father's school, and this kindled a growing interest in mathematics.

Kildall went to the University of Washington, intending to become a mathematics teacher. However, his mathematics courses also gave him the opportunity to work with computers. After he took a couple of programming courses, his focus changed to computer science. Soon he was spending most of his time writing FORTRAN programs at the university's computer center.

In the mid-1960s, Kildall joined the naval reserve. This allowed him to continue his education without having to face the Vietnam War draft; he fulfilled his service requirements by attending Officer Candidate School during the summer. He earned his bachelor's degree in computer science in 1967 and continued with graduate studies. Working on the university's Burroughs 5500 mainframe gave him experience in file systems and storage techniques, while he also delved into the workings of an Algol language compiler.

Kildall received his master's degree in 1969. The navy then called him to active duty, giving him the choice of serving on a destroyer in the waters off Vietnam or teaching computer science at the Naval Postgraduate School in Monterey, California. Kildall recalled that to Robert Slater "It took me a couple of microseconds to make a decision about that."

While teaching naval students about computers, he earned his doctorate in 1972, writing a dissertation on "global flow analysis"—techniques for making computer code as efficient as possible. Together with his compiler studies, this work would give Kildall a solid background in the design and structure of operating systems and programming environments.

Shortly after getting his doctorate, Kildall was looking at a bulletin board at the University of Washington when he saw a notice offering to sell an Intel 4004 microprocessor chip for $25. This chip was the first of its kind, providing all the basic operations needed for computing. It thus offered the potential for building computers small and inexpensive enough for anyone to own.

Kildall did not have the parts or knowledge to actually turn the microprocessor into a working computer, so he did the next best thing. He wrote a program on the university's IBM System/360 that simulated the operation of the Intel 4004. It responded to the various instructions described in the 4004 manual in the same way the real chip would have, had it been connected to appropriate memory and input/output chips.

As he learned more about the 4004's repertoire of operations, Kildall began to think of possible applications for the chip. For example, he thought it might be possible to create a device that could automate some of the operations that navigators had to do by hand. The problem was that the chip only had built-in instructions for the simplest math operations, such as addition and subtraction. Multiplication and division could be done by carrying out sequences of addition or subtraction, but more complicated operations (such as the trigonometric functions needed for navigation) were difficult to program. Nevertheless, Kildall wrote code to carry out a number of useful mathematical operations.

When Kildall contacted Intel, their engineers were impressed by the repertoire of advanced math functions he had coded. Kildall thus became a part-time consultant for Intel while continuing his teaching duties as associate professor at the Naval Postgraduate School.

In 1973, Intel came out with a more advanced microprocessor, the 8008. Kildall rewrote his simulator to work with the new processor, and he also persuaded Intel that the new chip needed a systems implementation language—a tool that would let programmers more easily access the processor's particular features and develop programming utilities such as editors and assemblers. Kildall's new language, called PL/M (Programming Language for Microprocessors) became a valuable tool for the community that was starting to work together to develop the tools and applications that would make "serious" computing with microprocessor possible.

Now that the microprocessor had become more powerful, the main barrier to creating a truly useful desktop microcomputer was finding a faster way to store and retrieve programs and data. Punched paper tape was cheap but slow and prone to breakage. Magnetic tape was better, but still relatively slow, and in order to retrieve a specific piece of data on the tape, all the intervening tape had to be read through first.

One obvious solution was the magnetic disk drive, which had been pioneered by IBM and used on mainframes since the late 1950s. Hard drives were bulky and very expensive at the time, but in 1973 the floppy disk drive (which had also been invented by IBM) began to be sold on the general market by Shugart. Kildall suggested to Shugart that building an interface and controller board for floppy drives would enable them to be used with the new microcomputer systems that were starting to be developed.

Using his PL/M language and a simulation of the microcomputer system, Kildall wrote a set of routines for writing data to the drive, storing it as files, and reading it back. Eventually his friend John Torode was able to get a controller to work with the drive, and Kildall found that his basic file system worked pretty well. By 1973, he had added other utilities that users would need, such as a program for copying files. He now had the essentials for the operating system called CP/M (Control Program for Microcomputers).

By 1976, many other experimenters were building "homebrew" microcomputer systems, and Jim Warren, editor of the hobbyist magazine *Dr. Dobb's Journal,* suggested that Kildall sell CP/M directly to experimenters through an ad in the publication. Soon hundreds of orders were coming for the software at $75 a copy.

Sensing the growing business opportunity, Kildall and his wife, Dorothy McEwen, founded a company originally called Intergalactic Digital Research (the "Intergalactic" was soon dropped.) By 1977, about 100 companies had bought licenses to use CP/M for various microcomputer-based products, and that number grew to about 1,000 by the end of the 1970s. By 1984, the company would have $44.6 million in annual sales. But as McEwen later recalled to Slater, "There was never any thought of having a big company. It was just something that happened. It seemed like the right thing to do."

In 1977 Glen Ewing, a former student of Kildall's, suggested that he take the parts of CP/M that had to do with low-level operations such as disk access be separated out and organized into a module called BIOS, or Basic Input/Output System. Kildall immediately recognized the value of this idea. Once a BIOS was created, when someone wanted to use CP/M with a different hardware configuration, only the BIOS had to be customized, not the operating system as a whole.

Such smart technical decisions helped make CP/M very popular with many companies that marketed PCs in the late 1970s. Having so many machines running the same operating system meant that programs once written could then be run on many different machines. There were soon hundreds of software packages for CP/M systems.

In 1980, however, came a fateful turning point. IBM had decided to bring out is own personal computer. They naturally wanted to use the well-proven, versatile CP/M. Unfortunately for some reason Kildall was not informed that the IBM people were coming. According to one account, Kildall was out flying his private plane when the people from "Big Blue" arrived, but he later said that he was simply out on a business trip until late that afternoon.

At any rate, McEwen, who handled marketing for the firm, received the IBM representatives. According to Kildall's account to Robert Slater, they "threw a nondisclosure agreement at her, which is a typical IBM thing that scares everybody." McEwen was alarmed at the language in the agreement, including IBM's statement that they wanted to be able to use any ideas they heard while meeting with Digital. Further, negotiations soon bogged down because IBM wanted to pay a flat $200,000 for the right to use CP/M, while Kildall wanted either more money up front or a royalty to be paid for each IBM PC sold.

Meanwhile, however, IBM had turned to Bill Gates and Microsoft, asking him about possible alternatives. Gates quickly bought an obscure operating system from Seattle Computer Products, tweaked it, and renamed it PC-DOS. IBM then licensed that operating system for its forthcoming PC, and the rest was history. When Kildall learned about what had happened and then looked at PC-DOS, he bitterly complained that the latter operating system was virtually a clone of CP/M, using essentially the same names and syntax for commands. However, he decided that he did not have money to take on IBM in a protracted legal battle, and was mollified somewhat when IBM agreed to license CP/M 86 as an option for the IBM PC. Unfortunately, few users wanted to pay $250 for the CP/M option when PC-DOS already came with the machine.

Kildall was disappointed but did not give up. In 1984, Digital Research marketed DR-DOS, an operating system that combined the best features of CP/M and PC-DOS and could even multitask (run more than one program at a time). While a considerable technical achievement, it gained little market share in a world dominated by PC-DOS. Kildall also developed GEM (Graphical Environmental Manager), which offered a graphical user interface several years before the coming of Microsoft Windows, but this product fell afoul of Apple, which believed that it too closely copied the Macintosh user interface.

By the later 1980s, Kildall was broadening his interests. He pioneered the use of videodisks and later, CD-ROMS for providing multimedia content to PC users, and he also established two video production companies. In 1990, however, he reduced his activities and moved to a suburb of Austin, Texas, where he lived quietly and worked on fund-raising for children with AIDS.

Kildall died suddenly on July 11, 1994. On July 6, he had walked into a Monterey bar while wearing motorcycle leathers and biker-style patches and apparently got into a fight with bikers, suffering head injuries that would prove fatal. Although police investigated, no one was ever charged. Kildall was posthumously honored by the Software Publishers Association in 1995 and received the Excellence in Programming Award from *Dr. Dobb's Journal* in 1997.

Further Reading

Slater, Robert. *Portraits in Silicon*. Cambridge, Mass.: MIT Press, 1987.

Swaine, Michael. "Gary Kildall and Collegial Entrepreneurship." *Dr. Dobb's Special Report*, spring 1997. Available on-line. URL: at http://www.ddj.com/documents/s=928/ddj9718i/9718i.htm.

Wharton, John. "Gary Kildall, Industry Pioneer, Dead at 52." *Microprocessor Report* 8 (August 1, 1994): 10ff.

Kleinrock, Leonard

(1934–)
American
Engineer, Computer Scientist

When an email is sent, the message is turned into a series of small chunks of data, each individually addressed to the destination computer. Routing software then looks up that address to see how the packets should be relayed through a series of "nodes" or computer locations between the sending computer and the destination mail server. When the packets arrive at the server, sometimes over several different routes, they are reassembled into the complete message, which is left in the appropriate user mailbox to be picked up by the recipient's mail program. This system of packet-switching and routing was invented by Leonard Kleinrock, one of the pioneers whose work made today's Internet possible.

Kleinrock was born in 1934 and grew up in New York City. One day when he was only six years old, he was reading a Superman comic book when he became intrigued by a centerfold that explained how to build a crystal radio. It included a list of parts. The used razor blade could be obtained from his father (accompanied, one would hope, by a safety lecture). The piece of pencil lead and empty toilet paper roll were easy enough to find around the house. The earphone he apparently swiped from a public telephone. Only one part, a "variable capacitor," momentarily stumped the boy. However, he soon persuaded his mother to take him on the subway down to a radio/electronics store. The boy then confidently asked the store clerk for a variable capacitor, only to be asked in turn: "What

Leonard Kleinrock invented packet-switching, the process by which e-mail and other data is broken up into small pieces and routed across the Internet to its destination. This method of distribution makes the Internet robust, because if one machine fails, an alternate routing can be used. *(Photo by Louis Bachrach)*

size do you want?" He had no idea, but when he explained what he was building, the helpful clerk provided the correct part. After he went home he assembled the radio and was soon listening to music "for free," since the crystal radio used the power in the radio signal itself—no batteries required.

The crystal radio was only the first of young Kleinrock's many electronics projects, built from cannibalized old radios and other equipment. He attended the Bronx High School of Science, home of many of the nation's top future engineers. However, when it came time for college the family had no money to pay for his higher education, so he attended night courses at the City College of New York (then tuition-free) while working as an electronics technician and later as an engineer to help with the family finances. Because of his limited course time, he took more than five years to graduate, but in 1957 he graduated first in his class (day and evening) and had earned a fully paid fellowship to the Massachusetts Institute of Technology (MIT).

At MIT, Kleinrock had ample opportunity to work with computers, but unlike the majority of students who focused on programming, operating systems, and information theory, Kleinrock was still a "radio boy" at heart. Instead of following the pack, he became interested in finding ways for computers and their users to communicate with each other. The idea of computer networking was in its infancy, but he submitted a proposal in 1959 for Ph.D. research in network design.

In 1961, Kleinrock published his first paper, "Information Flow in Large Communication Nets." Existing telephone systems did what was called "circuit switching"—to establish a conversation, the caller's line is connected to the receiver's, forming a circuit that exists for the duration of the call. This meant that the circuit would not be available to anyone else, and that if something was wrong with the connection there was no way to route around the problem.

Kleinrock's basic idea was to set up data connections that would be shared among many users on an as-needed basis (called "demand access"). In Kleinrock's version of demand access (later to be called packet switching), instead of the whole call (or data transmission) being assigned to a particular circuit, it would be broken up into packets that could be sent along whatever circuit was most direct. If there was a problem, the packet could be resent on an alternate route. This form of "dynamic resource sharing" provided great flexibility as well as more efficient use of the available circuits. Kleinrock further elaborated his ideas in his dissertation, for which he was awarded his Ph.D. in 1963. The following year, MIT published his book *Communications Nets*, the first full treatment of the subject.

Kleinrock then joined the faculty at the University of California, Los Angeles (UCLA), where he started to work his way up the academic ladder. Meanwhile, the Defense Department, stunned by the apparent triumph of Soviet technology in Sputnik, had created an Advanced Research Projects Agency, or ARPA. ARPA had a particular interest in improving communications technology, particularly in finding ways to create communications networks that could survive in wartime conditions while linking researchers at different universities and laboratories together in peacetime. When they learned of Kleinrock's work on packet-switching, ARPA's researchers believed that they might have found the solution to their needs.

In 1968, ARPA asked Kleinrock to design a packet-switched network that would be known as ARPANET. The computers on the network would be connected using special devices called Interface Message Processors (IMPs), special-purpose minicomputers that would be designed and built by the Cambridge firm of Bolt, Beranek, and Newman (BBN). The overall project was under the guidance and supervision of one of Kleinrock's MIT office mates, LAWRENCE ROBERTS.

On October 29, 1969, they were ready to test the system. Kleinrock sat at a terminal at UCLA, connected to a computer with an Interface Message Processor. His assistant, Charles Kline, established a regular voice telephone connection to Stanford Research Institute (SRI), where the other computer and its IMP waited at the end of the data line.

For their first message, Leonard Kleinrock chose the word *login*, commonly used as a prompt for connecting to a remote computer. The receiving computer was cleverly programmed to finish the word by adding the "in" after the remote user had typed *log*.

Kline typed the *l*, and a few seconds later it appeared on the terminal as it was "echoed" back from Stanford. He then typed the *o*, which likewise appeared. He typed the *g*, at which point the remote computer was supposed to finish the word. Instead, the SRI host system crashed. After making adjustments for about an hour, however, they were able to complete the whole experiment. *Login* is hardly as dramatic as Samuel Morse's "What hath God wrought?" or Alexander Graham Bell's "Watson, come here, I need you!" Nevertheless, a form of communication had been created that in a few decades would change the world as much as the telegraph and telephone had done.

The idea of computer networking did not catch on immediately, however. Besides requiring a new way of thinking about the use of computers, many computer administrators were concerned that their computers might be swamped with users from other institutions, or that they might ultimately lose control over the use of their machine. Kleinrock worked tirelessly to convince institutions to join the nascent network. By the end of 1969, there were just four ARPANET "nodes": UCLA, SRI, the University of California, Santa Barbara, and the University of Utah. By the following summer, there were 10.

During the 1970s, Kleinrock trained many of the researchers who advanced the technology of networking. While Kleinrock's first network was not the Internet of today, it was the precursor to the Internet and an essential step in its development. In successfully establishing communication using the packet-switched ARPANET, Kleinrock showed that such a network was practicable. Lawrence Roberts, ROBERT KAHN, and VINTON CERF then created the TCP/IP (transmission control protocol/Internet protocol), which made it possible for many types of computers and networks to use the packet system. E-mail, Net news, the Gopher document retrieval system, and ultimately the World Wide Web would all be built upon the sturdy data highway that Kleinrock and other early ARPANET researchers designed.

In a UCLA press statement on July 3, 1969, (included in his "The First Days of Packet Switching") Kleinrock predicted, "We will probably see the spread of 'computer utilities,' which will service individual homes and offices across the country." This prediction would come true in the 1980s with the development of large on-line services such as CompuServe and America Online, and especially in the 1990s with the World Wide Web, invented by TIM BERNERS-LEE. However, Kleinrock believes that the Internet of today still has shortcomings and remains incomplete. It is largely "desk-bound," requiring that people sit at computers, and is often dependent on the availability of a particular machine.

Kleinrock looks toward a future where most network connections are wireless, and accessible through a variety of computerlike devices such as hand-held "palmtop" computers, cell phones, and others not yet imagined. In such a network, the intelligence or capability is distributed throughout, with devices communicating seamlessly so the user no longer need be concerned about what particular gadget he or she is using. Since the 1990s, Kleinrock has advocated this "nomadic computing" idea as energetically as he had spread the early gospel of the ARPANET. He is also affiliated with TTI/Vanguard, a forum for executive technology planning, and is chairman of the board of Nomadix, a company involved with mobile networking and Internet access.

Although his name is not well known to the general public, Kleinrock has won considerable recognition within the technical community. This includes Sweden's L. M. Ericsson Prize (1982), the Marconi Award (1986), and the National Academy of Engineering Charles Stark Draper Prize (2001).

Further Reading

Hafner, Katie, and Matthew Lyon. *Where Wizards Stay Up Late: The Origins of the Internet*. New York: Simon and Schuster, 1996.

Kleinrock, Leonard. "The First Days of Packet Switching." Available online. URL: http://www.lk.cs.ucla.edu/LK/Presentations/sigcomm2.pdf. Posted on August 31, 1999.

"Leonard Kleinrock, Inventor of the Internet Technology." Available on-line. URL: http://www.lk.cs.ucla.edu. Downloaded on November 2, 2002.

Knuth, Donald E.

(1938–)
American
Computer Scientist

In every field, there are a few persons whose work is recognized as "magisterial"—creating a comprehensive, authoritative view of the essential elements and developments in the field. Most computer scientists have tended to specialize in a particular area such as computer language structure, algorithms, operating system design, or networking. Donald Knuth, however, has combined penetrating insight into algorithms with a sweeping view of the art of programming.

Knuth was born on January 10, 1938, in Milwaukee, Wisconsin. His father worked as a bookkeeper and also ran a printing business. As a young boy, Knuth was thus exposed both to the algorithmic patterns of mathematics and the visual patterns of printed type, and these two worlds came together to form his lifelong interest.

As an eighth grader, Knuth entered a contest sponsored by a candy company. Participants tried to create as many words as possible from the letters in the phrase "Zeigler's Giant Bar." Knuth's list of 4,500 words far exceeded the contest's official list of only 2,500. Knuth's victory assured his school a new television set and a copious supply of candy bars.

In high school, Knuth became fascinated with algebra. Working without outside help, he devised a way to find the equation that described any pattern of connected straight lines. This type of system would become essential when Knuth later devised his computer typesetting system. Knuth also had a deep interest in music, becoming a capable pianist and later trying his hand at composition. In 1956, Knuth graduated from high school with a record-setting grade point average, and then enrolled in the Case Institute of Technology to study physics.

Knuth was a dedicated student who constantly pushed himself to achieve ever more ambitious goals, but he also had a wry sense of humor. His first published article as a college freshman was not a scientific paper but a humorous article for *Mad* magazine called "The Potrzebie System of Weights and Measures," a parody of the official standards for the metric system.

While working at a summer job preparing statistical graphs, Knuth learned about the school's IBM 650 computer. The machine was in daily use, but that was no problem—Knuth simply starting spending what he would later describe as "many pleasant evenings" working with the machine. Knuth's growing interest in computing led him to switch majors from physics to mathematics. He also applied his growing programming skills to writing a program that evaluated the performance of members of the Case basketball team, helping coaches to use the players more efficiently. Some observers credited Knuth with helping the team win the league championship in 1960. That year, he graduated, and the faculty was so impressed with his work that they awarded him a master's degree simultaneously with the expected bachelor's degree.

Knuth entered the computer industry by working for Burroughs, a business machine company that was trying to compete with IBM in the growing computer field. Besides helping Burroughs with software development, Knuth enrolled at the California Institute of Technology (Caltech) to study for his doctorate, which he received in 1963. Five years later, he became a professor of computer science at Stanford University, where he would spend most of his career. Knuth's first major contribution to computer science was his development of LR(k) or "left to right, rightmost" parsing, which proved to be the optimal way for a compiler to look through a program statement and process the individual tokens (words and numbers) according to the rules of the language.

Beyond particular technical challenges, Knuth was particularly interested in surveying and furthering the development of computer science as a discipline. During the 1960s, he served as editor of the journal *Programming Languages* of the Association for Computing Machinery (ACM). He introduced new course topics into the Stanford curriculum. His focus on data structures and algorithms would help propel the movement toward better structured programs in the 1970s.

As far back as the early 1960s, many computer scientists saw a need to bring together the many things that researchers were learning about algorithms, data structures, compiler design, and other topics. Although papers were being written for technical journals, there were few books that could serve as useful references. In 1962, the publisher Addison-Wesley asked Knuth (then a graduate student) if he would be willing to write a book about compilers.

Knuth began to work on the book, thinking at first that he would just have to compile existing ideas, organize them, and maybe write an overview. However, as he worked Knuth began to see how the theories of various computer scientists working on different topics fit into a larger picture, and he began to write large amounts of new material. Instead of a single book, Knuth ended up writing the first of a monumental series that would be called the Art of Computer Programming. The first volume, published in 1968, is titled *Fundamental Algorithms*. It deals with the basic "recipes" for processing data, starting with mathematical foundations, and uses an example computer processor called MIX to illustrate computer architecture. By the time the reader has worked with the various data structures and tackled the dozens of example problems, he or she has a solid understanding of how to solve a variety of kinds of problems with a computer.

Although the Art of Computer Programming was intended to have seven volumes in all, Knuth put it aside after the first volume to work on a new system of computerized typesetting. Bringing together his love of type design and his arsenal of programming tools, Knuth developed Metafont, a programming language designed for creating and modifying typefaces according to mathematical rules. He also developed TeX, a typesetting system that allowed users unprecedented control and precision in arranging text. Meanwhile, Knuth also championed the idea of "literate programming," developing formats for integrating program documentation and source code in a way that makes it much easier for people to understand how a program works.

Knuth eventually returned to working on the Art of Computer Programming series and has produced two more volumes thus far. In his spare time he also designed a 1,000-pipe church organ. Knuth has received many of the highest awards in the computer science field, including the ACM Turing Award (1974), the Institute of Electrical and Electronic Engineers (IEEE) Computer Pioneer Award (1982), the American Mathematical Society's Steele Prize (1986), and the IEEE John von Neumann Medal (1995).

Further Reading

Knuth, Donald E. *The Art of Computer Programming*. Vols. 1–3. 3rd ed. Reading, Mass.: Addison-Wesley, 1998.

———. *Computers and Typesetting*. Volumes A–E. Reading, Mass.: Addison-Wesley, 2000.

———. *Literate Programming*. Stanford, Calif.: Center for the Study of Language and Information, 1992.

"Don Knuth's Home Page." Available on-line. URL: http://www-cs-faculty.stanford.edu/~knuth. Downloaded on November 2, 2002.

Kurzweil, Raymond C.

(1948–)

American

Inventor, Entrepreneur, Writer

Today blind people can "read" regular books and magazines without having to learn Braille. Musicians have the realistic, yet synthetic, sound of dozens of different instruments at their fingertips. Tomorrow, perhaps intelligent computers and human beings will form a powerful new partnership. These inventions and forecasts and many more have sprung from the fertile imagination of Raymond Kurzweil.

Kurzweil was born on February 12, 1948, in Queens, New York, to an extremely talented family. Kurzweil's father, Fredric, was a concert pianist and conductor. Kurzweil's mother, Hanna, was an artist, and one of his uncles was an inventor.

Young Kurzweil's life was filled with music and technology. His father taught him to play the piano and introduced him to the works of the great classical composers. Meanwhile, he had also become fascinated by science and gadgets. By the time he was 12, Kurzweil was building his own computer and learning how to program. He soon wrote a statistical program that was so good that IBM distributed it. When he was 16, Kurzweil programmed his computer to analyze patterns in the music of famous composers and then create original compositions in the same style. His work earned him first prize in the 1964 International Science Fair, a meeting with President Lyndon B. Johnson in the White House, and an appearance on the television show *I've Got a Secret*.

In 1967, Kurzweil enrolled in the Massachusetts Institute of Technology (MIT), majoring in computer science and literature. Because he spent all of his spare time hidden away, working on his own projects, he became known as "the Phantom" to his classmates. One of these projects was a program that matched high school students to appropriate colleges, using a database of 2 million facts about 3,000 colleges. It used a form of artificial intelligence (AI) called an expert system, which applied a set of rules to appropriate facts in order to draw conclusions. The publisher Harcourt Brace paid $100,000 for the program, plus a royalty.

By the time Kurzweil received his B.S. degree from MIT in 1970, he had met some of the most influential thinkers in AI research, such as MARVIN MINSKY. Kurzweil had become fascinated with the potential of AI to aid and expand human potential. In particular, he focused on pattern recognition, or the ability to classify or recognize patterns such as the letters of the alphabet on a page of text. Pattern recognition was the bridge that might allow computers to recognize and work with objects in the world the same way people do.

Ray Kurzweil is one of America's most prolific and diversified inventors. His innovations include a reading machine for the blind, the flatbed scanner, and a music synthesizer. Kurzweil has also written about a future in which computers eventually exceed human intelligence. *(Photo © Michael Lutch)*

Early character recognition technology was limited because it could match only very precise shapes. This meant that such a system could only recognize one or a few character fonts, making it impractical for reading most of the text found in books, newspapers, and magazines. Kurzweil, however, used his knowledge of expert system and other AI principles to develop a program that could use general rules and relationships to "learn" to recognize just about any kind of text. This program, called Omnifont, could be combined with the flatbed scanner (which Kurzweil invented in 1975) to create a system that could scan text and convert the images into the actual alphabetical characters, suitable for use with programs such as word processors. This technology would be used in the 1980s and 1990s to convert millions of documents to electronic form. In 1974 Kurzweil established the Kurzweil Computer Products company to develop and market this technology.

Kurzweil saw a particularly useful application for his technology. If printed text could be scanned into a computer, then it should be possible to create a synthesized voice that could read the text out loud. This would enable visually handicapped people to have access to virtually all forms of printed text. They would no longer be limited to large print or Braille editions.

It would be a formidable challenge, however, because pronunciation is not simple. It is not enough simply to recognize and render the 40 or so unique sounds (called phonemes) that make up English speech, because the sound of a given phoneme can be changed by the presence of adjacent phonemes. Kurzweil had to create an expert system with hundreds of rules for properly voicing the words in the text. From 1974 to 1976, Kurzweil worked at the problem while trying to scrounge enough money to keep his company afloat. By the early 1980s, however, Kurzweil Reading Machines (KRMs) were opening new vistas for thousands of visually handicapped people.

Kurzweil sold his company to Xerox for about $6 million, which gave him plenty of money to undertake his next project. The blind musician Stevie Wonder had been one of the earliest users of the KRM, and he suggested to Kurzweil that he develop a music synthesizer. Using AI techniques similar to those used in the KRM, Kurzweil created an electronic synthesizer that can faithfully reproduce the sounds of dozens of different instruments, effectively putting an orchestra at a musician's fingertips.

In the 1980s and 1990s, Kurzweil applied his boundless inventiveness to a number of other challenges, including speech recognition. The reverse of voice synthesis, speech recognition involves the identification of phonemes (and thus words) in speech that has been converted into computer sound files. Kurzweil sees a number of powerful technologies being built from voice recognition and synthesis in the coming decade, including telephones that automatically translate speech. He believes that the ability to control computers by voice command, which is currently rather rudimentary, should also be greatly improved. Meanwhile, computers will be embedded in everything from eyeglasses to clothes, and since such computers will not have keyboards, voice input will be used for much of the activities of daily life.

During the 1990s, Kurzweil became a provocative writer on the significance of computing and its possible future relationship to human aspirations. His 1990 book *The Age of Intelligent Machines* offered a popular account of how AI research would change many human activities. In 1999 Kurzweil published *The Age of Spiritual Machines*. It claims, "Before the next century is over, human beings will no longer be the most intelligent or capable type of entity on the planet. Actually, let me take that back. The truth of that last statement depends on how we define *human*." Kurzweil suggests that the distinction between human and computer will vanish, and an intelligence of breathtaking capabilities will emerge from the fusion if a number of perils can be avoided.

Whatever the future brings, Raymond Kurzweil has become one of America's most honored inventors. Among other awards he has been elected to the Computer Industry Hall of Fame (1982) and the National Inventors Hall of Fame (2002). He has received the Association for Computing Machinery Grace Murray Hopper Award (1978), Inventor of the Year Award (1988), the Louis Braille Award (1991), and the National Medal of Technology (1999).

Further Reading

Kurzweil, Raymond. *The Age of Intelligent Machines*. Cambridge, Mass.: MIT Press, 1990.

———. *The Age of Spiritual Machines: When Computers Exceed Human Intelligence*. New York: Putnam, 1999.

KurzweilAI.net. Available on-line. URL: http://www.kurzweilai.net. Downloaded on November 2, 2002.

Kurzweil Technologies. Available on-line. URL: http://www.kurzweiltech.com/ktiflash.html. Downloaded November 2, 2002.

"The Muse (Inventor and Futurist Ray Kurzweil)." *Inc.*, March 15, 2001, p. 124.

L

Lanier, Jaron

(1960–)
American
Computer Scientist, Inventor

Many of the pioneers of computer science and technology grew up when there were no computers (or only largely inaccessible mainframe ones). They studied fields such as electrical engineering, mathematics, or physics, and later became intrigued by the potential of computing and became computer scientists or engineers. However, computers have now been around long enough for a new generation of innovators—people who came of age along with the personal computer (PC), powerful graphics workstations, and other technology. An example of this new generation is Jaron Lanier, who pioneered the technology of "virtual reality," or VR, which is gradually having an impact on areas as diverse as entertainment, education, and even medicine.

Lanier was born on May 3, 1960, in New York City, although the family would soon move to Las Cruces, New Mexico. Lanier's father was a Cubist painter and science writer and his mother a concert pianist (she died when the boy was nine years old). Living in a remote area, the precocious Lanier learned to play a large variety of exotic musical instruments and created his own science projects.

Lanier dropped out of high school, but sympathetic officials at New Mexico State University let him take classes there when he was only 14 years old. Lanier even received a grant from the National Science Foundation to let him pursue his research projects. Although he became fascinated by computers, he took an unconventional approach to their use. Inspired perhaps by the symbolic artistic world of his father, Lanier tried to create a computer language that would rely on understandable, "universal" symbols rather than cryptic commands and mathematical constructs.

After two years, however, Lanier left New Mexico State without completing his degree, having decided to go to Bard College in Annandale-on-Hudson, New York, to study computer music composition. He soon dropped out of Bard as well, helped organize protests against local nuclear power plants, and then returned to New Mexico, where he briefly supported himself by raising goats and acting as an assistant midwife.

By the mid-1980s, Lanier had gotten back into computing by creating sound effects and music for Atari video games and writing a commercially successful game of his own called Moondust. He developed a reputation as a rising star in the new world of video games.

It was at this time that Lanier decided that worlds depicted by games such as Moondust were

Jaron Lanier developed innovative virtual reality technology starting in the 1980s. Immersive graphics and sound place the user "inside" a computer-generated world, and special gloves and even body suits let the user interact with virtual objects. Today virtual reality is used for education, conferencing, and entertainment. *(Photo courtesy of Jaron Lanier)*

too limited, and he began to experiment with ways to immerse the player more fully in the experience. Using money from game royalties, he joined with a number of experimenters and built a workshop in his house. One of these colleagues was Tom Zimmermann, who had designed a "data glove" that could send commands to a computer based on hand and finger positions.

At age 24, Lanier and his experiments were featured in an article in *Scientific American*. One of the editors had asked him the name of his company. Lanier therefore decided to start one, calling it VPL (Visual or Virtual Programming Language) Research. As the 1980s progressed, investors became increasingly interested in the new technology, and Lanier was able to expand his operation considerably, working on projects for NASA, Apple Computers, Pacific Bell, Matsushita, and other companies.

Lanier coined the term *virtual reality* to describe the experience created by this emerging technology. A user wearing a special helmet with goggles, body sensors, and gloves, has a computer-generated scene projected such that the user appears to be "within" the world created by the software. The world is an interactive one: When the user walks in a particular direction, the world shifts just as the real world would. The gloves appear as the user's "hands" in the virtual world, and objects in that world can be grasped and manipulated much like real objects. In effect, the user has been transported to a different world created by the VR software.

That "different world" could be many things. It could be a simulated Mars for training NASA astronauts or rover operators, or a human body that could give surgeons a safe way to practice their technique. For many ordinary people, their first experience of VR would be in the form of entertainment. Although not full-fledged VR, many of the techniques developed by Lanier would be used to make "first person" video games more realistic. Virtual roller coaster and other rides, which combine projected VR and effects with physical movement, can provide thrills that would be impossible or too unsafe to build into a conventional ride.

Virtual reality technology had existed in some form before Lanier; it perhaps traces its roots back to the first mechanical flight simulators built during World War II. However, existing systems such as those used by NASA and the air force were extremely expensive, requiring powerful mainframe computers. They also lacked flexibility—each system was built for one particular purpose and the technology was not readily transferable to new applications. Lanier's essential achievement was to use the new, inexpensive computer technology of the 1980s to build versatile software and hardware that could be used to create an infinite variety of virtual worlds.

Unfortunately, the hippielike Lanier (self-described as a "Rastafarian hobbit" because of his dreadlocks) did not mesh well with the big business world into which his initial success had catapulted him. Lanier had to juggle numerous simultaneous projects as well as becoming embroiled in disputes over his patents for VR technology. In 1992, Lanier lost control of his patents to a group of French investors whose loans to VPL Research had not been paid, and he was forced out of the company he had founded. Lanier was philosophical about this turn of events, noting, "I think what we've done has permanently changed the dialogue and rhetoric around the future of media technology, and that's not bad."

Lanier did not give up after leaving VPL. During the 1990s, he founded several new companies to develop various types of VR applications. These include the Sausalito, California–based software company Domain Simulations and the San Carlos, California, company New Leaf Systems, which specialized in medical applications for VR technology. Another company, New York–based Original Ventures, focuses on VR-based entertainment systems.

Today Lanier also serves as lead scientist for the National Tele-Immersion Initiative, a coalition of universities that are studying possible advanced applications for the high-bandwidth Internet 2 infrastructure. Another of Lanier's recent research areas involves what he calls "phenotropics." This involves programs using AI techniques rather than strict protocols for connecting and cooperating with one another.

Like the World Wide Web, virtual reality technology has been proclaimed as the harbinger of vast changes in human existence, as can be seen in the writings of nanotechnologist K. ERIC DREXLER and inventor/futurist RAYMOND C. KURZWEIL. Surprisingly, Lanier has rejected the viewpoint that he calls "cybernetic totalism." Lanier does not deny that technology will change human lives in very important ways, but he objects to the idea that technology is some sort of inevitable and autonomous force beyond human control. As stated in his article "One Half of a Manifesto" he believes that "whatever happens will be the responsibility of individual people who do specific things."

Besides writing and lecturing on virtual reality, Lanier is active as both a musician and an artist. He has written compositions such as *Mirro/Storm* for the St. Paul Chamber Orchestra and the soundtrack for the film *Three Seasons*, which won an award at the Sundance Film Festival. Lanier's paintings and drawings have also been exhibited in a number of galleries.

Further Reading

"Jaron Lanier's Homepage." Available on-line. URL: http://people.advanced.org/~jaron. Downloaded on November 3, 2002.

Lanier, Jaron. *Information as an Alienated Experience*. New York: Basic Books, 2002.

———. "One Half of a Manifesto." Available on-line. URL: www.edge.org/3rd_culture/lanier/lanier_index.html. Downloaded on November 3, 2002.

Sherman, William R., and Alan B. Craig. *Understanding Virtual Reality: Interface, Application, and Design*. San Francisco: Morgan Kaufmann, 2002.

Lenat, Douglas B.

(1950–)
American
Computer Scientist

Computers are wonderful for computing, but teaching them to reason has been a long and often tedious process. Artificial intelligence researchers have created expert systems in which the program is given detailed knowledge about a subject in the form of rules or assertions. But

Douglas Lenat has tried to do something more—to give computers something like the common sense that most humans take for granted.

Douglas Lenat was born in Philadelphia on September 13, 1950, and also lived in Wilmington, Delaware. His parents ran a soda-bottling business. Lenat became enthusiastic about science in sixth grade when he started reading Isaac Asimov's popular nonfiction books about biology and physics.

About that same time, however, Lenat's father died suddenly, and economic problems meant that the family had to move frequently. As a result, Lenat was constantly being enrolled in new schools. Each school would put him in the beginning, rather than advanced, track because they had not evaluated him, and he often had to repeat semesters.

Lenat turned to science projects as a way of breaking out of this intellectual ghetto. In 1967, his project on finding the *n*th prime number got him into the finals in the International Science Fair in Detroit. At the fair, he and other contestants were judged by working scientists, and in many ways treated as scientists themselves. This experience confirmed Lenat's desire for a scientific career.

In 1968, Lenat enrolled in the University of Pennsylvania to study mathematics and physics, graduating in 1972 with bachelor's degrees in both disciplines plus a master's degree in applied math. However, he became somewhat disenchanted with both disciplines. He did not believe he could quite reach the top rank of mathematicians. As for physics, he found it to be too abstract and bogged down with its ever growing but incoherent collection of newly discovered particles.

Lenat became intrigued when he took an introductory course in artificial intelligence (AI). Although the field was still very young ("like being back doing astronomy right after the invention of the telescope," he told authors Dennis Shasha and Cathy Lazere), it offered the possibility of "building something like a mental amplifier that would make you smarter, hence would enable you to do even more and better things."

Artificial intelligence researcher Douglas Lenat programmed computers to use heuristics, or problem-solving techniques similar to those employed by human mathematicians. In 1984, he embarked on a 10-year project to create a giant database of "common sense" facts that would give computer programs broad knowledge about the world and human society. *(Courtesy of Cyc Corporation)*

At the time, the most successful approach to practical AI had been rule-based or expert systems. These programs could solve mathematical problems or even analyze molecules. They did it by systematically applying sets of specific rules to a problem. But while this approach could be quite effective within narrow application areas, it did not capture the wide-ranging, versatile reasoning employed by human beings. According to Shasha and Lazere, AI researchers had ruefully noted that "It's easier to simulate a geologist than a five-year-old." Human beings, even five-year-olds, were equipped with a large fund of what is called common sense. Because of this, humans approaching a given situation already know a lot about what to do and what not to do.

Inspired by JOHN MCCARTHY, who was trying to overcome such shortcomings, Lenat went to Stanford University for his doctorate, after first trying the California Institute of Technology for a few months. Lenat's adviser was Cordell Green, who had made considerable advances in what is known as "automatic programming." This is the attempt to set up a system that accepts a sufficiently rigorous description of a problem and then generates the program code needed to solve the problem.

Lenat became interested in applying this idea to mathematics. For his doctoral thesis, he wrote a program called AM (Automated Mathematician). The program "thinks" like a human mathematician. It applies heuristics (a fancy term basically meaning "guesses") that experience has shown to often be fruitful. For example, since many mathematical operations produce interesting results with certain values such as zero and one, the program will try these values and draw conclusions if they yield unusual results. Similarly, if several different techniques lead to the same result, the program will conclude that the result is more likely to be correct.

The AM program used 115 assertions or rules from set theory plus 243 heuristic rules. By exploring combinations of them, it was able to derive 300 mathematical concepts, including such sophisticated ones as "Every even number greater than three is the sum of two primes." (This was known to mathematicians as Goldbach's conjecture.) The AM program intrigued a number of mathematicians as well as other computer scientists such as DONALD E. KNUTH, who were interested in what paths the program took or did not take, and what happened as it moved farther from its mathematical moorings.

After receiving his Ph.D. in 1976, Lenat became an assistant professor at Carnegie Mellon University for two years, then returned to Stanford. He continued to explore the use of heuristics in programs. His new program, Eurisko, was an attempt to generalize the heuristic reasoning that had been surprisingly successful with AM, and to allow the program to not only apply various heuristics but to formulate and test new ones. Eurisko turned out to have some interesting applications to areas such as playing strategy games and designing circuits.

A continuing problem with heuristic programs is that they have very limited knowledge of the world—they are basically limited to a few hundred assertions provided by the programmer. Therefore, by the mid-1980s Lenat had turned his attention to creating a large "knowledge base"—a huge set of facts that ideally would be comparable to those available to an educated adult. AI pioneer MARVIN MINSKY had devised the concept of "frames," which are sets of facts about particular objects or situations (such as parts of a car or steps involved in taking an airline flight). Lenat got together with Minsky and ALAN C. KAY and together they did a literal "back of the envelope" calculation that about 1 million frames would be needed for the new program. Lenat dubbed it Cyc (short for "encyclopedia"). A consortium called the Microelectronics and Computer Technology Corporation, or MCC, agreed to undertake what was estimated to be at least a 10-year project, beginning in 1984.

Lenat pointed out to Shasha and Lazere that if this project succeeded

> This would basically enable natural language front-ends and machine learning front-ends to exist on programs. This would enable knowledge sharing among application software, like different expert systems could share rules with one another. It's clear that this would revolutionize the way computing worked. But it had an incredibly small chance of succeeding. It had all these

pitfalls—how do you represent time, space, causality, substances, devices, food, intentions, and so on.

On the website for Lenat's new company, Cycorp, which took over the project after it spun off from MCC in 1995, some examples of what Cyc can now do are highlighted:

> Cyc can find the match between a user's query for "pictures of strong, adventurous people" and an image whose caption reads simply "a man climbing a cliff." Cyc can notice if an annual salary and an hourly salary are inadvertently being added together in a spreadsheet.
>
> Cyc can combine information from multiple databases to guess which physicians in practice together had been classmates in medical school. When someone searches for "Bolivia" on the Web, Cyc knows not to offer a follow-up question like "Where can I get free Bolivia online?"

Cyc works by making and testing theories and checking them against the knowledge base. In a broad sense, it combines the heuristic approach that Lenat had pioneered with AM and Eurisko and the expert system and frames work by Minsky and EDWARD FEIGENBAUM.

The problem with applying expert system technology more widely and generally is that a system first needs access to a broad base of knowledge. To overcome this bottleneck, Lenat undertook the now 20-year-long CYC project: as an attempt to capture the culture's "consensus reality" into a single, functional knowledge base. It has been a grueling task, but by the late 1990s Lenat believed that the CYC system knew enough about the world that it would be able to start teaching itself more.

Meanwhile, Lenat has promoted knowledge technology widely and effectively. He has consulted with U.S. government agencies on national security-related technology, was a founder of Techknowledge, Inc., and has served on advisory boards at Inference Corporation, Thinking Machines Corporation, TRW, Apple, and other companies.

Although a decisive breakthrough had not occurred by the end of the 1990s, Cyc appeared to have a number of promising applications, including the creation of much smarter search engines and the analysis of patterns in large databases, often called "data mining." Lenat's research continues.

Lenat has received a number of awards for papers submitted to American Association for Artificial Intelligence (AAAI) conferences and became an AAAI Fellow in 1990. He has been a keynote or featured speaker at many conferences.

Further Reading

"Child's Play." *The Economist (U.S.)*, January 12, 1991, p. 80ff.

Cycorp. Available on-line. URL: http://www.cyc.com/ Downloaded November 3, 2002.

Lenat, Douglas B., and R. V. Guha. *Building Large Knowledge-Based Systems: Representation and Inference in the Cyc Project*. Reading, Mass.: Addison-Wesley, 1990.

Shasha, Dennis, and Cathy Lazere. *Out Of Their Minds: The Lives and Discoveries of 15 Great Computer Scientists*. New York: Copernicus, 1995.

Stork, David G., editor. *Hal's Legacy: 2001's Computer as Dream and Reality*. Cambridge, Mass.: MIT Press, 1996.

Licklider, J. C. R.

(1915–1990)
American
Computer Scientist, Scientist

Most of the early computer pioneers came from backgrounds in mathematics or engineering. This naturally led them to focus on the computer as a tool for computation and information processing. Joseph Carl Robnett Licklider, however, brought an extensive background in psychology to the problem of designing interactive computer systems that could provide better communication and access to information for users. Along the way he helped make today's Internet possible.

Licklider was born on March 11, 1915, in St. Louis, Missouri. His father was a proverbial "self-made man," having started as a farmer, then worked on the railroad to support his family when his own father died. Later he learned how to write and design advertisements, and finally ended up as a very successful insurance agent.

Young Licklider, called Rob by his friends, was caught up in the passion for flying that had been ignited by the exploits of Charles Lindbergh and other pioneer aviators. He became a highly skilled maker of hand-carved model airplanes. When he became 16, his interest switched to cars, and his parents bought him an old clunker on the condition that he not actually drive it off their property. He took the car completely apart and painstakingly put it back together until he understood how every part worked.

During the 1930s, he attended Washington University in St. Louis, his active mind flitting from one subject to another. By the time he was finished, he had earned not one B.A. degree but three: in psychology, mathematics, and physics. He concentrated on psychology for his graduate studies, earning an M.A. degree at Washington University and then receiving his Ph.D. from the University of Rochester in 1942.

The kind of psychology that intrigued Licklider was not the abstract or symbolic systems of Freud and the other psychoanalysts who were becoming popular, but the study of how the brain worked physically, how perception was processed, and of cognition. This study was coming into its own as researchers using a new tool, the electroencephalograph, or EEG, were stimulating different areas of the brain and measuring responses. They were thus creating the first maps of a planet less known than Mars—the human brain—pinpointing the motor center, the vision center, and other major regions.

In 1942, Licklider's knowledge of neuroscience was enlisted for the war effort when he went to work at the U.S. Army Air Corps' newly created Psycho-Acoustics Lab at Harvard University. (The lab was headed by physicist Leo Beranek, who would later play an important role in building the first computer networks). Their task was to understand how the very noisy environment of modern war, such as that found in a bomber, affected the ability of soldiers to function. Licklider's particular area of research was on how interference and distortion in radio signals affected a listener's ability to correctly interpret speech. Much to his surprise and that of his colleagues, Licklider discovered that while most forms of distortion made speech harder to understand, one particular kind, which he called "peak-clipping," actually improved reception. (In peak-clipping, the consonants ended up being emphasized in relation to the surrounding vowels.) This work also gave Licklider valuable experience in seeing how small things could greatly affect the quality of the interactions between persons and machines.

After the war, Licklider also participated in a study group at the Massachusetts Institute of Technology (MIT) led by NORBERT WIENER, pioneer in the new field of cybernetics, in the late 1940s. Wiener's restless, high-energy mind threw ideas at his colleagues like sparks coming off a Tesla coil. Some of those sparks landed in the receptive mind of Licklider, who through Wiener's circle was coming into contact with the emerging technology of electronic computing and its exciting prospects for the future. In turn, Licklider's psychology background allowed him a perspective quite different from the mathematical and engineering background shared by most early computer pioneers. This perspective could be startling to Licklider's colleagues. As William McGill recalled much later in an interview, cited by M. Mitchell Waldrop

> Lick was probably the most gifted intuitive genius I have every known. Whenever I would finally come to Lick with the mathematical proof of some relation, I'd discover that he already knew it. He hadn't worked it out in detail, he just . . . knew it. He could somehow envision the way information flowed, and see relations that people who just manipulated the mathematical symbols could not see. It was so astounding that he became a figure of mystery to all the rest of us . . .

Cybernetics emphasized the computer as a system that could interact in complex ways with the environment. Licklider added an interest in human-computer interaction and communication. He began to see the computer as a sort of "amplifier" for the human mind. He believed that humans and computers could work together to solve problems that neither could successfully tackle alone. The human could supply imagination and intuition, while the computer provided computational "muscle." Ultimately, according to the title of his influential 1960 paper, it might be possible to achieve a true "Man-Computer Symbiosis."

During the 1950s, Licklider taught psychology at MIT, hoping eventually to establish a full-fledged psychology department that would elevate the concern for what engineers call "human factors." (Psychology at the time was a section within the economics department.) From 1957 to 1962 he worked the private sector as a vice president for engineering psychology at Bolt, Beranek, and Newman, the company that would become famous for pioneering networking technology.

In 1962, the federal Advanced Research Projects Agency (ARPA) appointed Licklider to head a new office focusing on leading-edge development in computer science. Licklider soon brought together research groups that included in their leadership three of the leading pioneers in artificial intelligence (AI): JOHN MCCARTHY, MARVIN MINSKY, and ALLEN NEWELL. By promoting university access to government funding, Licklider also fueled the growth of computer science graduate programs at major universities such as Carnegie Mellon University, University of California at Berkeley, Stanford University, and MIT.

In his research activities, Licklider focused his efforts not so much on AI as on the development of interactive computer systems that could promote his vision of human-computer symbiosis. This included time-sharing systems, where many human users could share a large computer system, and networks that would allow users on different computers to communicate with one another. He believed that the cooperative efforts of researchers and programmers could develop complex programs more quickly than teams limited to a single agency or corporation.

Licklider's efforts to focus ARPA's resources on networking and human-computer interaction provided the resources and training that would, in the late 1960s, begin the development of what became the Internet. Licklider spent the last two decades of his career teaching at MIT, expanding the computer science program and forging connections with other disciplines such as the social sciences. Before his death on June 26, 1990, he presciently predicted that by 2000, people around the world would be linked in a global computer network. Licklider received the Franklin Taylor Award of the Society of Engineering Psychologists (1957), served as president of the Acoustical Society of America (1958), and received the Common Wealth Award for Distinguished Service in 1990.

Further Reading

"J. C. R. Licklider (1915–1990) [Biographical Timeline]." Available on-line. URL: http://www.columbia.edu~jrh29/years.html. Downloaded on November 3, 2002.

Licklider, J. C. R. "The Computer as Communication Device." *Science and Technology*, April 1968. Available on-line. URL: http://www.memex.org/licklider.pdf. Downloaded on December 2, 2002.

———. "Man-Computer Symbiosis." *IRE Transactions on Human Factors in Electronics* HFE-1 (March 1960): 4–11. Available on-line. URL: http://www.memex.org/licklider.pdf. Downloaded on December 2, 2002.

Waldrop, M. Mitchell. *The Dream Machine: J. C. R. Licklider and the Revolution That Made Computing Personal*. New York: Viking, 2001.

Lovelace, Ada (Augusta Ada Byron Lovelace; Ada, Countess of Lovelace)

(1815–1852)
British
Mathematician, Computer Scientist

It was remarkable enough for a woman in early-19th-century Britain to become recognized in the world of mathematics. But Ada, Countess of Lovelace would add to that the distinction of becoming arguably the world's first computer programmer and technical writer. Her work both helped explain the proposed Analytical Engine of CHARLES BABBAGE and recognized potential applications that would not be realized for more than 100 years.

Augusta Ada Byron was born on December 10, 1815, in London, England. Her father was the English Romantic poet Lord Byron. Byron was notorious for his various romantic affairs, and Lady Byron and her daughter separated from the poet only five weeks after the birth. Lady Byron tried to raise her daughter strictly to prevent her falling into the kind of dissolute life that she saw in her husband. According to biographer Joan Baum, "She arranged a full study schedule for her child, emphasizing music and arithmetic—music to be put to purposes of social service, arithmetic to train the mind."

Other children might have rebelled against such a program, but Ada showed both talent and passion for mathematics. Since schools beyond the elementary level were generally not available for girls, she was taught by a succession of tutors. Several of them were eminent mathematicians, such as William Frend of Cambridge and especially the logician August de Morgan of the University of London. The latter would describe his pupil as "an original mathematical investigator, perhaps of first-rate eminence."

In 1835, Ada Byron married William King, the eighth Baron King. When he became the first earl of Lovelace in 1838, this made his wife the countess of Lovelace. Unusual among men of his time, King supported his wife's mathematical studies and took pride in her accomplishments. However because of the aristocratic disdain for practical accomplishments, she would be pressured to sign her writings with only the initials "A. A. L."

A few years earlier, Byron had attended a party hosted by Mary Fairfax Somerville, one of the few prominent female scientists of her day. There she met Charles Babbage, who was intrigued by her mathematical interests and became her mentor and friend. In 1834, Babbage showed her some of the plans for his proposed calculating machines, the Difference Engine and the Analytical Engine.

In the late 1830s, Babbage embarked on an international tour to raise interest (and hopefully, funding) for the building of his machines. After a lecture in Italy in 1840, an Italian military engineer named Luigi Federico Menabrea wrote an article about the Analytical Engine that appeared in a French publication. In 1842, Babbage asked Lovelace to prepare an English translation.

However, Lovelace's article, which was published in *Taylor's Scientific Memoirs* in 1843, was much more than a translation. She added seven extensive notes that introduced new topics and many additional examples of the machine's operation. Lovelace's notes explained the use of the "store" (what would be called memory today), the organization and writing of programs, and even the use of loops to execute operations repeatedly and the breaking of instructions into sets of what today are called subroutines.

Lovelace also included a set of detailed instructions for calculating Bernoulli numbers, which suggests to 20th-century scholars that Lovelace is entitled to be called the world's first computer programmer. Looking at possible applications for the device, Lovelace suggested that the engine's speed and accuracy would make it possible to carry out calculations that would be impractical for human "computers."

Unfortunately, Lovelace's personal affairs and health both became problematic. Lovelace became involved in a number of sexual affairs, and as a result of one with John Crosse, possibly involving blackmail, she eventually pawned the Lovelace diamonds to pay his debts. As she fell further into debt, she turned to horse racing, possibly using a mathematical system in an attempt to recoup her losses, but she only fell farther behind.

Lovelace had always suffered from "delicate" health, having had measles and scarlet fever as a child and possibly migraine headaches and "fits" throughout her life. On November 27, 1852, she died of uterine cancer, her dreams of a scientific career and a new world of machine-woven mathematics unfulfilled.

Although his Analytical Engine would never be built, Babbage deeply appreciated Lovelace's work. Writing to her son Viscount Ockham, as quoted by Betty Toole, Babbage said that "In the memoir of Mr. Menabrea and still more in the excellent Notes appended by your mother you will find the only comprehensive view of the Anal. Eng. which the mathematicians of the world have yet expressed."

Historians have argued about the true extent of Lovelace's mathematical ability, but her remarkable presence and writings have been increasingly appreciated, both in their own right and as an inspiration for young women seeking scientific careers. A modern structured programming language, Ada, was named in her honor.

Further Reading

Baum, Joan. *The Calculating Passion of Ada Byron*. Hamden, Conn.: Archon Books, 1986.

Stein, Dorothy. Ada: *A Life and a Legacy*. Cambridge, Mass.: MIT Press, 1985.

Toole, Betty A., ed. *Ada: Enchantress of Numbers: A Selection from the Letters of Lord Byron's Daughter and Her Description of the First Computer*. Mill Valley, Calif.: Strawberry Press, 1992.

M

Maes, Pattie

(1961–)
Belgian/American
Computer Scientist

Someday, perhaps, computers will no longer give an error message or find nothing because the user misspelled a search term. Wading through websites and on-line catalogs to find good buys will be passé. Indeed, people will no longer have to type specific commands or click buttons. Rather, the computer will act like the perfect assistant who seems to know what you want before you want it. If that day comes, it will be due in large part to the pioneering work of Pattie Maes, developer of what has come to be called software agent technology.

Born on June 1, 1961, in Brussels, Belgium, Maes was interested in science (particularly biology) from an early age. She received bachelor's (1983) and doctoral (1987) degrees in computer science and artificial intelligence from the University of Brussels.

In 1989, Maes moved from Belgium to United States to work at the Massachusetts Institute of Technology (MIT), where she joined the Artificial Intelligence Lab. She worked with Rodney Brooks, the innovative researcher who had created swarms of simple but intriguing insectlike robots. Two years later, she became an associate professor at the MIT Media Lab, famed for innovations in how people perceive and interact with computer technology. At the Media Lab, she founded the Software Agents Group to promote the development of a new kind of computer program.

Generally speaking, traditional programming involves directly specifying how the computer is to go about performing a task. The analogy in daily life would be having to take care of all the arrangements for a vacation oneself, such as finding and buying the cheapest airline ticket, booking hotel rooms and tours, and so on. But since their time and expertise are limited, many people have, at least until recently, used the services of a travel agent.

For the agent to be successful, he or she must have both detailed knowledge of the appropriate area of expertise (travel resources and arrangements in this case) and the ability to communicate with the client, asking appropriate questions about preferences, priorities, and constraints. The agent must also be able to maintain relationships and negotiate with a variety of services.

Maes's goal has been to create software agents that think and act much like their human counterparts. To carry out a task using an agent, the user does not have to specify exactly how it is to be done. Rather, the user describes

Pattie Maes wants to promote the computer from tool to full-fledged assistant. She is creating "software agents" that can carry out general tasks without the user having to specify details or give commands. This could include making travel arrangements, shopping, handling routine e-mail, or watching the news for items of interest. *(Photo by Romana Vysatova)*

the task, and the software engages in a dialogue with the user to obtain the necessary guidance.

Today, many people obtain their airline tickets and other travel arrangements via websites such as Expedia.com or Travelocity.com. While these sites can be convenient and helpful for bargain hunters, they leave most of the overall trip planning to the user. With a software travel agent using the technology that Maes is developing, the program could do much more. It would know—or ask about—such things as the vacation budget and whether destinations involving nature, history, or adventure are preferred. The program might even know that a family member has severe asthma and thus the hotels should be within half an hour of a hospital.

The software agent would use its database and procedures to put together an itinerary based on the user's needs and desires. It would not only know where to find the best fares and rates, it would also know how to negotiate with hotels and other services. Indeed, it might negotiate with their software agents.

Software agents are often confused with expert systems such as those developed by EDWARD FEIGENBAUM and the "frames" approach to reasoning pioneered by MARVIN MINSKY. However, Maes explained in an interview with *Red Herring* magazine that "Rather than the heavily knowledge-programmed approach of strong AI proponents like Marvin Minsky, I decided to see how far simpler methods of statistically based machine learning would go."

The expert system relies upon a "knowledge base" of facts and rules, and uses a rather rigid procedure to match rules and facts to draw conclusions. Software agents, on the other hand, act more like people: They have goals and agendas, they pursue them by trying various techniques that seem likely to work, and they are willing to act on probability and accept "good enough" rather than perfect results, if necessary.

Some critics are worried that software agents might compromise privacy because they would know or be able to find out many intimate details about people. Other critics are concerned that if highly capable software agents take over most of the thinking and planning of people's daily lives, people may become passive and even intellectually stunted. Maes acknowledges this danger, quoting communications theorist Marshall McLuhan's idea that "every automation is amputation." That is, when something is taken over by automation, people "are no longer as good at whatever's been automated or augmented." As an example, she cites the pocket calculator's effect on basic arithmetic skills. However, Maes says that she is trying to emphasize designing software agents that help people more easily cooperate to accomplish complex tasks, not trying to replace human thinking skills.

In 1995, Maes cofounded Firefly Networks, a company that attempted to create commercial applications for software agent technology. Although the company was bought by Microsoft in 1998, one of its ideas, "collaborative filtering," can be experienced by visitors to sites such as Amazon.com. Users in effect are given an agent whose job it is to provide recommendations for books and other media. The recommendations are based upon observing not only what items the user has already purchased, but also what else has been bought by people who bought those same items. More advanced agents can also tap into feedback resources such as user book reviews on Amazon or auction feedback on eBay.

Maes started a new venture called Open Ratings in 1999. Its software evaluates and predicts the performance of a company's supply chain (the companies it relies upon for obtaining goods and services), which in turn helps the company plan for future deliveries and perhaps change suppliers if necessary. Since then, Maes has founded or participated in several other ventures related to e-commerce.

A listing of Maes's current research projects at MIT conveys many aspects of and possible applications for software agents. These include the combining of agents with interactive virtual reality, using agent technology to create characters for interactive storytelling, the use of agents to match people with the news and other information they are most likely to be interested in, an agent that could be sent into an on-line market to buy or sell goods, and even a "Yenta" (Yiddish for matchmaker) agent that would introduce people who are likely to make a good match.

Other ideas on Maes's drawing board include creating agents that get feedback from their user over time and adapt their procedures accordingly. For example, an agent can learn from what news items the user does *not* look at, and employ that information to predict how the user will respond in the future. Some of the applications most likely to arrive soon include agent-based Web search engines and intelligent e-mail filtering programs that can dig the messages most likely to be of interest or importance out of the sea of spam (unsolicited e-mail) and routine correspondence.

Maes has participated in many high-profile conferences, such as the (American Association for Artificial Intelligence and the Association for Computing Machinery (ACM) Siggraph, and her work has been featured in numerous magazine articles. She was one of 16 modern visionaries chosen to speak at the 50th anniversary of the ACM. She has also been repeatedly named by *Upside* magazine as one of the 100 most influential people for development of the Internet and e-commerce. *Time Digital* featured her in a cover story and selected her as a member of its "cyber elite." *Newsweek* put her on its list of 100 Americans to be watched in the year 2000. Also in 2000, the Massachusetts Interactive Media Council gave her its Lifetime Achievement Award.

Rather to Maes's amusement, a *People* magazine feature nominated her as one of their "50 most beautiful people" for 1997, noting that she had worked as a model at one time and that her striking looks have made her a "download diva." She deprecatingly notes that it is not hard to turn male eyes at MIT, because that institution still has a severe shortage of women.

Further Reading

D'Inverno, Mark and Michael Luck, editors. *Understanding Agent Systems*. New York: Springer-Verlag, 2001.

Maes, Pattie. "Intelligence Augmentation: A Talk with Pattie Maes." Available on-line. URL: http://www.edge.org/3rd_culture/maes. Downloaded on November 3, 2002.

"Pattie Maes' Home Page." Available on-line. URL: http://pattie.www.media.mit.edu/people/pattie. Updated on January 29, 1998.

"Pattie Maes on Software Agents: Humanizing the Global Computer." *Internet Computing Online*. Available on-line. URL: http://www.computer.org/internet/v1n4/maes.htm. Downloaded on November 3, 2002.

Mauchly, John
(1907–1980)
American
Engineer, Inventor

Together with J. PRESPER ECKERT, John William Mauchly codesigned the earliest full-scale digital computer, ENIAC, and its successor, Univac, the first commercially available computer. His and Eckert's work went a long way toward establishing the viability of the computer industry in the early 1950s.

Mauchly was born on August 30, 1907, in Cincinnati, Ohio. However, the family moved to Chevy Chase, Maryland. Young Mauchly loved to build electrical things: When he was only five, he improvised a flashlight. He also built intercoms and alarms and made pocket money installing electric bells and fixing broken wiring. But perhaps his most practical invention came when he was about 12 years old. By then, his love of late-night reading was colliding with his curfew. To solve the problem, he wired a pressure switch to a loose board in one of the stairs leading up to his bedroom and made it part of a circuit to which he attached a small light near his bed. Whenever a patrolling parent stepped on the board, the switch opened the circuit and the light went out, giving the boy ample time to shut off his room light. When the parent descended the stairs, the switch toggled again, turning the warning light back on.

Mauchly attended the McKinley Technical High School in Washington, D.C., where he was a top math and physics student and in his senior year edited the school newspaper. After graduation, his father wanted him to study engineering, which he considered to be a better paying and more reliable career than pure science. Mauchly won an engineering scholarship to Johns Hopkins University, but he soon became bored with engineering, which seemed to him to be mainly working by rote from a "cookbook." In his sophomore year, he changed his major to physics. (The University recognized his knowledge and talent, and he was allowed to take graduate courses.)

The spectral analysis problems he tackled for his Ph.D. (awarded in 1932) and in postgraduate work required a large amount of painstaking calculation. So, too, did his later interest in weather prediction, which led him to design a mechanical computer for harmonic analysis of weather data. Convinced that progress in calculation would be needed to prevent science from getting

John Mauchly is shown here with a portion of ENIAC. Completed in 1946, ENIAC is often considered to be the first large-scale electronic digital computer. *(Photo courtesy of Computer Museum History Center)*

bogged down, Mauchly would keep pondering the problem even after he left research for teaching.

Mauchly taught physics at Ursinus College in Philadelphia from 1933 to 1941. He had a definite knack for getting students' attention in his lectures. His last lecture before Christmas became a particular favorite—one year he used a skateboard to demonstrate Newton's laws of motion, and another time he wrapped Christmas presents in colored cellophane and showed students how to use the principles of spectroscopy to see through the wrapping to the goodies within.

In 1939, Mauchly decided to learn more about electronics, particularly the use of vacuum tubes that could be set up to count a rapid series of electronic pulses. He learned about binary switching circuits ("flip-flops") and experimented with building electronic counters, using miniature lightbulbs because they were much cheaper (though slower) than vacuum tubes. His experiments convinced him that electronic, not merely electrical circuits would be the key to high-speed computing.

On the eve of World War II, Mauchly went to the University of Pennsylvania's Moore School of Engineering and took a course in military applications of electronics. He then joined the staff and began working on contracts to prepare artillery firing tables for the military. Realizing how intensive the calculations would be, in 1942 he wrote a memo proposing that an electronic calculator be built to tackle the problem. The proposal was rejected at first, but by 1943 table calculation by mechanical methods was falling even farther behind. HERMAN HEINE GOLDSTINE, who had been assigned by the army to break the bottleneck, approved the project.

With Mauchly providing theoretical design work and J. Presper Eckert heading the engineering effort, the Electronic Numerical Integrator and Computer, better known as ENIAC, was completed too late to influence the outcome of the war. However, when the huge machine with its 18,000 vacuum tubes and rows of cabinets was demonstrated in February 1946, it showed that electronic logic circuits could be harnessed together to carry out calculations about 1,000 times faster than an electromechanical calculator, and with reasonable reliability.

ENIAC, however, was far from easy to use. In its original configuration, it was "programmed" by plugging dozens of wires into jacks to set up the flow of data. However, Mauchly and Eckert, working with the mathematician JOHN VON NEUMANN, gradually redesigned the machine. For example, a set of prewired logic modules were connected to a card reader so the machine could be programmed from punched cards, which was much faster than having to rewire the whole machine.

Mauchly and Eckert left the Moore School in 1946 after a dispute about who owned the patent for the computer work. They jointly founded what became known as the Eckert-Mauchly Computer Corporation, betting on Mauchly's confidence that there was sufficient demand for computers not only for scientific or military use but for business applications as well. By 1950, however, they were struggling to sell and build their improved computer, Univac, while fulfilling existing government contracts for a scaled-down version called BINAC. These computers included a crucial improvement: They were able to store their program's instructions in memory, allowing instructions to be repeated (looped) or even modified.

In 1950, they sold their company to Remington Rand, while continuing to work on Univac. In 1952, Univac stunned the world by correctly predicting the presidential election results on election night, long before most of the votes had come in.

Early on, Mauchly saw the need for a better way to write computer programs. Univac and other early computers were programmed through a mixture of rewiring, setting switches, and entering numbers into registers. This made programming difficult, tedious, and error-prone.

Mauchly wanted a way that variables could be represented symbolically: for example, Total rather than a register number such as 101. Under Mauchly's supervision, William Schmitt wrote what became known as Brief Code. It allowed two-letter combinations to stand for both variables and operations such as multiplication or exponentiation. A special program read these instructions and converted them to the necessary register and machine operation commands. While primitive compared to later languages such as Fortran or COBOL, Brief Code represented an important leap forward in making computers more usable.

Mauchly stayed with Remington Rand and its successor, Sperry Rand, until 1959, but then left over a dispute about the marketing of Univac. He continued his career as a consultant and lecturer. Mauchly and Eckert also became embroiled in a patent dispute arising from their original work with ENIAC. Accused of infringing on Sperry Rand's ENIAC patents, Honeywell claimed that the ENIAC patent was invalid, with another computer pioneer, JOHN VINCENT ATANASOFF, claiming that Mauchly and Eckert had obtained crucial ideas after visiting his laboratory in 1940.

In 1973, Judge Earl Richard Larson ruled in favor of Atanasoff and Honeywell. However, many historians of the field give Mauchly and Eckert the lion's share of the credit, because they built full-scale, practical machines.

Mauchly played a key role in founding the Association for Computing Machinery (ACM), one of the field's premier professional organizations. He served as its first vice president and second president. He received many tokens of recognition from his peers, including the Howard Potts Medal of the Franklin Institute, the National Association of Manufacturers Modern Pioneer Award (1965), the American Federation of Information Processing Societies Harry Good Memorial Award for Excellence (1968) and the Institute for Electrical and Electronics Engineers Computer Society Pioneer Award (1980). The ACM established the Eckert-Mauchly Award for excellence in computer design. John Mauchly died on January 8, 1980.

Further Reading

McCartney, Scott. *ENIAC: The Triumphs and Tragedies of the World's First Computer.* New York: Berkeley Books, 1999.

Stern, N. "John William Mauchly: 1907–1980," *Annals of the History of Computing* 2, no. 2 (1980): 100–103.

McCarthy, John

(1927–)
American
Computer Scientist, Mathematician

In the 1950s, popular articles often referred to computers as "giant brains." But while the computers of the time were indeed giant, and could calculate amazingly rapidly, they were not really brains. Computers could not reason the way people could, drawing conclusions that went beyond their precise instructions, or learn how to do things better. However, a number of researchers believed that this could change: the calculating power of the electronic computer could be organized in such a way as to allow for what came to be known as artificial intelligence, or AI. Starting in the 1950s, John McCarthy played a key role in the development of artificial intelligence as a discipline, as well as developing LISP, the most popular language in AI research.

John McCarthy was born on September 4, 1927, in Boston, Massachusetts. His father, an Irish immigrant, was active in radical labor politics. His mother, a Lithuanian Jew, was equally militant, involved in women's suffrage and other issues. Both were members of the Communist party, making young McCarthy what would be known as a "red diaper baby." McCarthy's father was also an inventor and

Starting in the mid-1950s, John McCarthy pioneered artificial intelligence (AI) research, organizing the key 1956 Dartmouth conference. He also developed the LISP programming language, a powerful tool for AI programming. *(Photo courtesy of John McCarthy)*

technological enthusiast, and the family placed a high value on science. These influences were translated in the young McCarthy into a passion for mathematics.

The family moved to California when the boy was still young, in part because his prodigal intellect was accompanied by poor health. He skipped much of grade school, attended Belmont High School in Los Angeles, and graduated when he was only 15. He entered the California Institute of Technology (Caltech) in 1944 to major in mathematics, but was called into the army, where he served as a clerk, returning to Caltech to get his bachelor's degree in 1948.

McCarthy then earned his Ph.D. at Princeton University in 1951. During the 1950s, he held teaching posts at Stanford University, Dartmouth College, and the Massachusetts Institute of Technology.

Although he seemed destined for a prominent career in pure mathematics, he encountered computers while working during the summer of 1955 at an IBM laboratory. He was intrigued with the potential of the machines for higher-level reasoning and intelligent behavior. He coined the term *artificial intelligence* (AI) to describe this concept. The following year, he organized a conference that brought together people who would become key AI researchers, including MARVIN MINSKY. In the Dartmouth AI Project Proposal, McCarthy suggested that

> the study is to proceed on the basis of the conjecture that every aspect of learning or any other feature of intelligence can in principle be so precisely described that a machine can be made to simulate it. An attempt will be made to find how to make machines use language, form abstractions and concepts, solve kinds of problems now reserved for humans, and improve themselves.

At the time, McCarthy was working on a chess-playing program. Because of its complexity, chess was an attractive subject for many early AI researchers. McCarthy invented a method for searching through the possible moves at a given point of the game and "pruning" those that would lead to clearly bad positions. This "alpha-beta heuristic" would become a standard part of the repertoire of computer game analysis.

McCarthy believed that the overall goal of AI required the incorporation of deep mathematical understanding. Mathematics had well-developed symbolic systems for expressing its ideas. McCarthy decided that if AI researchers were to meet their ambitious goals, they would need a programming language that was equally capable of expressing and manipulating symbols. Starting in 1958, he developed LISP, a language based on lists that could flexibly represent data of many kinds and even allowed programs to be fed as data to other programs.

According to one mathematician quoted in the book *Scientific Temperaments* (1982) by Philip J. Hilts:

> The new expansion of man's view of the nature of mathematical objects, made possible by LISP, is exciting. There appears to be no limit to the diversity of problems to which LISP will be applied. It seems to be a truly general language, with commensurate computing power.

LISP, rather than traditional procedural languages such as FORTRAN or Algol, would be used in the coming decades to code most research AI research projects, because of its symbolic, functional power. McCarthy continued to play an important role in refining the language, while moving to Stanford in 1962, where he would spend the rest of his career.

McCarthy also contributed to the development of Algol, a language that would in turn greatly influence modern procedural languages such as C. In addition, he helped develop new ways for people to use computers. Consulting with Bolt, Beranek, and Newman (the company that would later build the beginnings of the Internet), he helped design time-sharing, a system that allowed many users to share the same computer, bringing down the cost of computing and making it accessible to more people. He also sought to make computers more interactive, designing a system called THOR that used video display terminals. Indeed, he pointed the way to the personal computer in a 1972 paper titled "The Home Information Terminal."

McCarthy's life has taken him beyond intellectual adventures. He has climbed mountains, flown planes, and jumped from them, too. Although he came to reject his parents' marxism and what he saw as the doctrinaire stance of the radical left, since the 1960s McCarthy has been active in political issues. In recent years, he has been especially concerned that computer technology be used to promote rather than suppress democracy and individual freedom. He has suggested that the ability to review and, if necessary, correct all computer files containing personal information be established as a fundamental constitutional right.

In 1971, McCarthy received the prestigious A. M. Turing award from the Association for Computing Machinery. In the 1970s and 1980s, he taught at Stanford and mentored a new generation of AI researchers. He has remained a prominent spokesperson for AI, arguing against critics such as philosopher Hubert Dreyfus who claim that machines could never achieve true intelligence. As of 2001, McCarthy is Professor Emeritus of Computer Science at Stanford University.

Further Reading

"John McCarthy's Home Page." Available on-line. URL: http://www-formal.stanford.edu/jmc. Downloaded on November 3, 2002.

McCarthy, John. "The Home Information Terminal." *Man and Computer: Proceedings of the International Conference, Bordeaux, France, 1970*. Basel: S. Karger, 1972, pp. 48–57.

———. "Philosophical and Scientific Presuppositions of Logical AI" in H. J. McCarthy, and Vladimir Lifschitz, eds. *Formalizing Common Sense: Papers by John McCarthy*. Norwood, N.J.: Ablex, 1990.

McCorduck, Pamela. *Machines Who Think*. New York: W. H. Freeman, 1979.

McNealy, Scott G.
(1954–)
American
Entrepreneur

Today's world of software consists of two huge continents and islands of various sizes. The largest continent is controlled by Microsoft and its Windows operating system, used by the majority of business and personal computers. A decent-sized island is inhabited by the enthusiastic users of the Macintosh, particularly people involved in desktop publishing and video applications. But the second-largest continent is not well-known to most ordinary computer users. It is the land of UNIX, the powerful operating system developed by DENNIS RITCHIE in the early 1970s and given new popularity in the 1990s by LINUS TORVALDS, developer of a variant called Linux.

Computers running UNIX are the mainstay of many university computer science departments, research laboratories, scientists, and engineers. They include the powerful graphics workstations that visualize molecules or generate movie effects, made by Sun Microsystems, the multibillion-dollar company founded by entrepreneur Scott McNealy.

McNealy was born November 13, 1954, in Columbus, Indiana, but grew up in the Detroit suburb of Bloomfield Hills. His father was a successful business executive who became vice chairman of American Motors Corporation (AMC) in 1962. Young McNealy took a lively interest in his father's business and liked to help him figure out ways to improve AMC's competitive position against its larger competitors. He was also a good athlete, doing well in ice hockey, tennis, and golf.

After graduating from Cranbrook Kingswood preparatory school, McNealy enrolled in Harvard University, which had one of the nation's foremost programs in business and economics. He received his B.A. degree in economics in 1976. When he initially had trouble getting into a good graduate business school, he decided to experience the other side of industry by becoming a factory foreman at Rockwell International Corporation's plant in Ashtabula, Ohio. Because he had to work double shifts in anticipation of a pending strike, McNealy became exhausted, contracted hepatitis, and spent six weeks in the hospital.

He then managed to get into Stanford Business School, where he earned his MBA in 1980. Although intelligent, McNealy had little interest in theoretical studies. Rather, he focused on what he considered to be practical skills that

Scott McNealy cofounded Sun Microsystems, whose powerful workstations became the tool of choice for scientific visualization, engineering, and graphics. As Sun's chief executive officer, McNealy embraced network computing and the Java language as a way to revitalize the company's business during the 1990s. *(Courtesy of Sun Microsystems)*

would advance his career. He also preferred making things to thinking about them, resigning a position at FMC Corporation because, as he told *Fortune* magazine, "FMC put me on a strategy team, and I wanted to be a plant manager."

In 1981, McNealy got his first hands-on experience in Silicon Valley's burgeoning computer industry when he became director of management for Onyx Systems, a San Jose microcomputer manufacturer. But he had seen enough of the corporate lifestyle in his father's career, and would later tell *Industry Week* that his real ambition had been "to run a machine shop . . . my own little business with 40 or 50 people."

As it happened, McNealy would soon start making things, but not in the way he had imagined. In 1982, an engineer and former Stanford classmate named Vinod Khosla had gotten together with another Stanford graduate, Andreas V. Bechtolsheim, and software developer BILL JOY to start a company to manufacture computers. They invited McNealy to join them as director of manufacturing, even though, as he later noted, "I didn't know what a [computer] disk drive was. I couldn't explain the concept of an operating system."

The new company, Sun Microsystems, had a clear vision of what it wanted to build. At the time, desktop computers were not powerful enough to do much beyond word processing and spreadsheets, and they had only primitive graphics capabilities. Only mainframes or perhaps minicomputers were viable options for scientists and engineers who needed the best possible computing performance.

Sun's workstations brought together the powerful microprocessors (including the Sun SPARC with its innovative instruction set) with leading-edge high-resolution graphics capabilities and UNIX, the "serious" operating system already preferred by most scientific and engineering programmers. At a time when personal computers (PCs) generally stood alone, isolated on desktops, Sun's computers included built-in networking. The goal was to give each user both generous computing power and the ability to share data and collaborate.

The Sun business plan became a shining success as the 1980s advanced. In 1984, Sun had to raise money to be able to manufacture enough computers to meet the exploding demand, and McNealy was able to persuade Kodak, one of Sun's customers, to invest $20 million. At about the same time, Khosla, who had been Sun's chief executive officer (CEO), left the team. McNealy was tapped for a temporary appointment as president, and was confirmed as president and CEO a few months later.

The next milestone came in 1987, when Sun and AT&T signed an agreement under which Sun would develop a new version of AT&T's UNIX system, and then an agreement allowing AT&T to purchase a 20 percent interest in Sun. These events alarmed other computer and software developers who feared that the AT&T alliance would give Sun an inside track on UNIX development perhaps comparable to what Microsoft had with Windows. In May 1988, many of Sun's biggest competitors, including IBM, Digital Equipment Corporation, and Hewlett-Packard, formed the Open Software Foundation, dedicated to creating a new, nonproprietary version of UNIX. This effort did not pan out, and the UNIX alternative eventually came from a different direction—Linus Torvald's Linux. But Sun's competitors did enter the workstation market with their increasingly powerful desktop systems, and at the end of 1989 Sun had its first quarterly loss in earnings.

Industry analysts looked at Sun's faltering performance and the departure of some key executives and began to wonder if Sun had grown too large too fast. McNealy responded by changing the way the company was organized. Originally, the company had seven separate groups that competed internally for new projects, the idea being that this competition would sharpen their skills and keep their energy level high. But the competition had gotten out of hand, and often the groups were spending more time obstructing other groups' projects than finishing their own. McNealy drastically simplified the organization, combining the seven units into only two, and taking direct responsibility for engineering and manufacturing. He also cut costs while encouraging employee initiative.

For the first half of the 1990s, the new Sun seemed to shine brighter than ever, retaining its number-one position in the workstation market despite the increased competition. However, by mid-decade the workstation market had started to stagnate, and as desktop PCs with ever more powerful chips came out many tasks that formerly required a dedicated workstation could be now be accomplished with an ordinary PC. At the high end, Sun revitalized its workstations with the new UltraSPARC processor, creating machines powerful enough for the demanding movie industry, which was starting to produce movies entirely by computer, such as 1995's *Toy Story*.

Sun attempted a counterattack by starting to build its own low-cost PCs equipped with its Solaris UNIX system. However, these systems did not sell well, mainly because they were not compatible with Windows, Microsoft's dominant operating system in the PC market. Another Sun initiative, the "network computer" (a stripped-down, diskless PC that ran software from and stored data on a server) also largely failed, in part because the price of a full-fledged PC kept dropping.

Brighter prospects came from the explosive growth in the Internet and particularly the World Wide Web in the later 1990s. Sun's computers long had built-in networking capabilities (its slogan, in fact, was "the computer is the network"), and UNIX Web server software was powerful and stable, making it the preferred choice of the burgeoning Web development community. McNealy made a key decision that further strengthened Sun's position in the new Web world: He backed JAMES GOSLING in the development of his new language, Java. This language, something of a streamlined version of C++, included the key ability to run programs under virtually any operating system, including both UNIX and Windows. With Java, Web sites could download small programs called applets to a user's Web browser, and programmers could write larger, more traditional programs as well. While Java has not replaced Windows in the core office market, the freely distributed language has helped prevent Microsoft from taking over the Web.

McNealy became for many the champion for those who saw BILL GATES and Microsoft as stifling competition and innovation in the industry. Starting in 1998, along with other industry leaders McNealy testified against Microsoft, accusing the giant company of using its control of Windows to gain unfair advantage in the PC market. Although the judgment went against Microsoft, the consequences were not yet resolved as of 2002.

Sun's revenues rose healthily, with its stock going from $30 to $120 in 1999. However, since then the general downturn in the stock market and particularly in the tech sector has made life much tougher for Scott McNealy and Sun, as it has for other entrepreneurs. McNealy continues to work long hours, warning of more storms to come while proclaiming his faith in the long-term health of the industry.

Further Reading

Schlender, Brent. "The Adventures of Scott McNealy, Javaman." *Fortune*, October 13, 1997, pp. 70ff.

Southwick, Karen, and Eric Schmidt. *High Noon: The Inside Story of Scott McNealy and the Rise of Sun Microsystems*. New York: Wiley, 1999.

Metcalfe, Robert M.

(1946–)

American

Engineer, Entrepreneur, Writer

Having access to a computer network is now taken for granted, almost as much as having electricity, gas, and running water. Computers in most offices and schools and in an increasing number of homes are linked together into local networks, which in turn are connected to the Internet. Robert Metcalfe created Ethernet, the most widely used standard for wiring computers into a network.

Metcalfe was born in 1946 in Brooklyn, New York, and grew up nearby, on Long Island. From the age of 10 his ambition was to become an electrical engineer. For an eighth grade science project, young Metcalfe converted a model train controller into a simple electronic calculator. His interest in electronics would soon become focused on computers. In high school, he attended a Saturday morning science class sponsored by Columbia University, where students were given the opportunity to write programs and run them on an IBM mainframe.

In 1964, Metcalfe entered the Massachusetts Institute of Technology (MIT), and in 1969 he received bachelor's degrees in both electrical engineering and business management. The 1960s saw solid state electronics move from transistors to printed circuits to integrated circuit chips, and Metcalfe became part of a generation that was at home in both technology and business.

During his years at MIT, Metcalfe gained programming experience on mainframe computers, including an aging IBM 7094 and a Univac. However, the most exciting development was the arrival of a minicomputer on loan from the Digital Equipment Corporation (DEC). This sleek, refrigerator-sized machine cost "only" $30,000. However, shortly after Metcalfe was put in charge of the machine, a thief or thieves carted it away. Dismayed, Metcalfe screwed up his courage to tell DEC officials what had happened. To his relief, DEC decided to take advantage of the incident by proclaiming their machine the first computer small enough to steal.

After graduating from MIT, Metcalfe went to Harvard, where he earned an M.S. degree in applied mathematics (1970) and a Ph.D. in computer science (1973). Meanwhile, he had become deeply interested in computer networking. His doctoral dissertation on packet-switched networks (a concept first developed by LEONARD KLEINROCK) was written while working with the Advanced Research Projects Agency (ARPA), which was creating ARPANET which would eventually become the Internet. Metcalfe also designed the IMP, or Interface Message Processor, a special-purpose minicomputer that served as a data communications link between machines on the ARPANET. However, Metcalfe did more for ARPANET than just design hardware—he also "evangelized" for the network, promoting its possible applications in a pamphlet titled "Scenarios for the ARPANet." Later, referring to a popular line of handbooks, he would call this "the first *Internet for Dummies*."

Metcalfe became disillusioned with Harvard when its officials would not allow their computers to be connected to the ARPANET, so he moved back to MIT. There he worked with an early network called the Aloha Net, which linked together computers in MIT's Project MAC. A local network (computers wired together in an office or other facility) would complement the long-range ARPANET. Metcalfe began to sketch out a local network architecture that would eventually become Ethernet.

In 1972, Metcalfe went to work for Xerox. Although this company was primarily known to the public for its copying machines, it had established an innovative computer laboratory in California called the Palo Alto Research Center, or PARC. Researchers at this laboratory, such as ALAN C. KAY, would soon create the windowed, mouse-driven graphical computer user interface that is so familiar today.

Xerox was especially interested in networking because it had developed the laser printer, which then cost many thousands of dollars. If computers in an office (which at the time were minicomputers, since PCs had not yet been invented) could be connected into a network, they could share a laser printer, making it a more attractive purchase—if the network was fast enough to keep up with the printer. Metcalfe and his associate David R. Boggs decided that they could solve this problem by building a high-speed local area computer networking system.

Metcalfe's Ethernet (so called because it transmitted through a rather ethereal medium of wires) could link computers of any type as long as they had the appropriate interface card and followed the rules, or protocol. Ethernet adopted the packet-switching idea that had been pioneered with ARPANET, in which data is broken into small pieces, routed across the network, and reassembled at the destination. It also adopted a simple traffic control system. Before sending data, each computer waits until the line is clear. If a "collision" occurs because two computers inadvertently start sending at the same time, they stop and each waits a random time before transmitting again.

Importantly, with Ethernet data could flow at a fast 10 megabits (million bits) per second, meaning that printers would receive data fast enough to not stand idle and users working with other computers on the network would experience a prompt response. Ethernet was announced to the world in a 1976 paper titled "Ethernet: Distributed Packet-Switching for Local Computer Networks."

Meanwhile, Metcalfe moved to the PARC Xerox Systems Development Division in 1976, helping develop the Star, the innovative but commercially unsuccessful computer from which STEVE JOBS and Apple would largely take the design of the Macintosh. During this time, he also taught part time at Stanford.

Metcalfe thought that Ethernet had considerable commercial potential, but Xerox did not seem able to exploit it. Finally, in 1979 Metcalfe left to start his own company in Santa Clara, California. It was called 3Com, which stood for "computer communications compatibility." There he would wear many hats from division manager to president to chief executive officer (CEO), and he reveled as much in exercising his business skills as his technical ones, noting that most engineers did not understand the real importance of selling as a skill. He was able to persuade many companies both in the computer field (DEC, Intel, and Xerox) and outside (General Electric, Exxon) to adopt Ethernet-based networks as their standard. Looking back, Metcalfe told *Technology Review* that he is most proud of increasing the company's revenue "from 0 to $1 million a month."

Ethernet came into its own in the 1980s when PCs began to proliferate on desktops. The local area network, or LAN, enabled office users to share not only printers but also servers, computers with high-capacity hard drives and powerful applications. Novell and later Microsoft built networked operating systems on the foundation of Ethernet. If ARPANET and the Internet are the information superhighway, then Ethernet is the information boulevard, with speeds today of up to 1 billion bits a second. Meanwhile, 3Com became a Fortune 500 company.

Metcalfe's relations with the venture capitalists and board of directors who controlled 3Com were not very good, however. Even while he had done so much to increase the company's revenue he had been essentially demoted to vice president, and in 1990 the board chose Eric Benhamou, a young engineer, rather than Metcalfe, to be CEO.

Metcalfe decided to leave 3Com. More than that, he changed careers, becoming a publisher. He became publisher of *InfoWorld*, the flagship journal of the PC industry, and served as chief technology officer of its parent company, International Data Group (IDG), which would publish a popular series of how-to books "for Dummies." Metcalfe also became a journalist, writing a regular *InfoWorld* column called "From the Ether" until September 2000.

As a journalist and self-styled "technology pundit," Metcalfe tried to predict the future direction of the industry. His predictions were often controversial and sometimes wrong, as in 1996 when he warned of the imminent collapse of the Internet from traffic overload. However, his 1999 claim that Internet "dot-com" stocks were vastly overvalued and would soon collapse proved more prescient. Metcalfe continues to be active in a number of areas, including the venture capital firm Polaris Venture Partners.

Metcalfe has won a number of major awards, including the Association for Computing Machinery Grace Murray Hopper Award (1980) and the Institute for Electrical and Electronic Engineers Alexander Graham Bell Medal (1988) and its Medal of Honor (1996).

Further Reading

Metcalfe, Robert, and David R. Boggs. "Ethernet: Distributed Packet Switching for Local Computer Networks." *Communications of the ACM* 19 (July 1976): 395–404. Available on-line. URL: http://www.acm.org/classics/apr96. Downloaded on December 3, 2002.

———. "How Ethernet was Invented." *Annals of the History of Computing* 16 (1994): 81.

Spurgeon, Charles E. *Ethernet: The Definitive Guide*. Sebastapol, Calif.: O'Reilly, 2000.

———. "Invention Is a Flower, Innovation Is a Weed." *Technology Review* 102 (November 1999): 54.

Minsky, Marvin
(1927–)
American
Computer Scientist, Inventor

Starting in the 1950s, Marvin Minsky played a key role in the establishment of artificial intelligence (AI) as a discipline. Combining cognitive psychology and computer science, Minsky developed ways to make computers function in more "brainlike" ways and then offered provocative insights about how the human brain itself might be organized.

Minsky was born in New York City on August 9, 1927. His father was an ophthalmologist—in a memoir, Minsky noted, "Our home was simply *full* of lenses, prisms, and diaphragms. I took all his instruments apart, and he quietly put them together again." Minsky's father was also a musician and a painter, making for a rich cultural environment.

Minsky proved to be a brilliant science student at the Fieldston School, the Bronx High School of Science, and the Phillips Academy. He would later recall being "entranced" by machinery from an early age.

Before he could go to college, World War II intervened, and Minsky enlisted in the U.S. Navy, entering electronics training in 1945 and 1946. He then went to Harvard University, where he received a B.A. degree in mathematics in 1950. Although he had majored in mathematics at Harvard and then went to Princeton for graduate study in that field, Minsky was also interested in biology, neurology, genetics, and psychology, as well as many other fields of science. (His study of the operation of crayfish claws would later transfer to an interest in robot manipulators.)

Minsky recalled in a memoir published in the journal *Scanning* that

> perhaps the most amazing experience of all was in a laboratory course wherein a student had to reproduce great physics experiments of the past. To ink a zone plate onto glass and see it focus on a screen; to watch a central fringe emerge as the lengths of two paths become the same; to measure those lengths to the millionth part with nothing but mirrors and beams of light—I had never seen any things so strange.

This wide-ranging enthusiasm for the idea of science and of scientific exploration helped Minsky move fluidly between the physical and life sciences, seeing common patterns and suggestive relationships.

Minsky had become fascinated by the most complex machine known to humankind—the human brain. (He would later describe the brain as a "meat machine." However, calling the brain a machine did not mean that it was simple and "mechanical." Rather it meant that the brain was richly complex but accessible to scientific understanding.)

Minsky found the hottest thing in contemporary psychology, the behaviorism of B. F. Skinner, to be unsatisfactory because it focused only on behavior, ignoring the brain itself entirely. On the other hand, traditional psychologists who based their work on Freud and his followers seemed to deal only with ideas or images, not the process of thinking itself. He decided to explore psychology from a completely different angle.

In 1951, Minsky and Dean Edmonds designed SNARC, the Stochastic Neural-Analog Reinforcement Computer. At the time, it was known that the human brain contains about 100 billion neurons, and each neuron can form connections to as many as 1,000 neighboring ones. Neurons respond to electronic signals that jump across a gap (called a synapse) and into electrode-like dendrons, thus forming connections with one another. But what caused particular connections to form? How did the formation of some connections make others more probable? And how did these networks relate to the brain's most important task—learning? Minsky was surprised to find out how little researchers knew about how the brain actually did its work.

Since it was known that the brain used electrical signaling, Minsky decided to create an electrical model that might capture some of the brain's most basic behavior. SNARC worked much like a living brain. Its electrical elements responded to signals in much the same way as the brain's neurons. The machine was given a task (in this case, solving a maze), but unlike a computer, was not given a program that told it how to perform it. Instead, the artificial neurons were started with random connections. If a particular connection brought the machine closer to its goal, the connection was "reinforced" (given a higher value that made it more likely to persist). Gradually, a network of such reinforced connections formed, enabling SNARC to accomplish its task. In other words, SNARC had "learned" how to do something. Minsky then used the results of his research for his thesis for his Ph.D. in mathematics.

Another researcher, Frank Rosenblatt, would later build on this research to create the Perceptron, containing what became known as a neural network. Today, neural network software is used for applications such as the automatic recognition of characters, graphic elements, and spoken words.

In 1954, Minsky received his Ph.D. in mathematics. The following year, he invented and patented the confocal scanning microscope, an optical instrument that produced sharper images than the best existing microscopes. However, Minsky never promoted the invention, so later researchers would receive most of the credit for it.

The year 1956 was key in the development of artificial intelligence. That year, Minsky was an important participant in the seminal Dartmouth Summer Research Project in Artificial Intelligence, which set the agenda for this exciting new field for years to come. (There are many ways to define artificial intelligence, or AI. Minsky's common-sense definition is that artificial intelligence is "the field of research concerned with making machines do things that people consider to require intelligence.")

Not surprisingly, researchers tended to split into groups that emphasized different paths to the goal of creating an artificial intelligence. One group, including Rosenblatt and others pursuing neural networks, believed that developing more sophisticated neural networks was the key to creating a true computer intelligence. Minsky and his colleague SEYMOUR PAPERT, however, believed that the neural network or perceptron had limited usefulness and could not solve certain problems. They believed that AI researchers should pay more attention to cognition—the actual process of thinking. Minsky would describe this as switching from "trying to understand how the brain works to what it does." Thus, although exploring the biological roots of the brain had been fruitful, Minsky gradually moved to focusing on the structure and processing of information, a field that would become known as cognitive science (or cognitive psychology).

Minsky moved to the Massachusetts Institute of Technology (MIT) in 1957, serving as a professor of mathematics from 1958 to 1961, later switching to the electrical engineering department. During the same time, Minsky and JOHN MCCARTHY established Project MAC, MIT's first AI laboratory. In 1970, Minsky and McCarthy founded the MIT Artificial Intelligence Laboratory.

During the 1960s, Minsky and many other researchers turned to robotics as an important area of AI research. Robots, after all, offered the possibility of reproducing intelligent, humanlike behavior through interaction between a machine and its environment. But to move beyond basic perception to the higher order ways in which humans think and learn, Minsky believed that the robot (or computer program) needed some sort of way to organize knowledge and build a model of the world.

Minsky developed the concept of frames. Frames are a way to categorize knowledge about the world, such as how to plan a trip. Frames can be broken into subframes. For example, a trip-planning frame might have subframes about air transportation, hotel reservations, and packing. Minsky's frames concept became a key to the construction of expert systems that today allow computers to advise on such topics as drilling for oil or medical diagnosis. (This field is often called knowledge engineering.)

In the 1970s, Minsky and his colleagues at MIT designed robotic systems to test the ability to use frames to accomplish simpler tasks, such as navigating around the furniture in a room. The difficult challenge of giving a robot vision (the ability not only to perceive but "understand" the features of its environment) would also absorb much of their attention.

Although he is primarily an academic, Minsky did become involved in business ventures. He and Seymour Papert founded Logo Computer Systems, Inc., to create products based upon the easy-to-use but versatile Logo language. In the early 1980s, Minsky established Thinking Machines Corporation, which built powerful computers that used as many as 64,000 processors working together.

Minsky continued to move fluidly between the worlds of the biological and the mechanical. He came to believe that the results of research into simulating cognitive behavior had fruitful implications for human psychology. In 1986, Minsky published *The Society of Mind*. This book suggests that the human mind is not a single entity (as classical psychology suggests) or a system with a small number of often-warring subentities (as psychoanalysis asserted). It is more useful, Minsky suggests, to think of the mind as consisting of a multitude of independent agents that deal with different parts of the task of living and interact with one another in complex ways. What we call mind or consciousness, or a sense of self, may be what emerges from this ongoing interaction.

Since 1990, Minsky has continued his research at MIT, exploring the connections between biology, psychology, and the creations of AI research. One area that intrigued him was the possibility of linking humans to robots so that the human could see and interact with the environment through the robot. This process, for which Minsky coined the word "telepresence," is already used in a number of applications today, such as the use of robots by police or the military to work in dangerous areas under the guidance of a remote operator, and the use of surgical robots in medicine.

Minsky's wide-ranging interests have also included music composition (he designed a music synthesizer). His legacy includes the mentoring of nearly two generations of students in AI and robotics, as well as his seeking greater public support for AI research and computer science.

Minsky has received numerous awards, including the Association for Computing Machinery Turing Award (1969) and the International Joint Conference on Artificial Intelligence Research Excellence Award (1991).

Further Reading

Franklin, Stan. *Artificial Minds*. Cambridge, Mass.: MIT Press, 1995.

"Marvin Minsky Home Page." Available on-line. URL: http://www.media.mit.edu/people/minsky. Downloaded on November 3, 2002.

McCorduck, Pamela. *Machines Who Think*. New York: W. H. Freeman, 1979.

Minsky, Marvin. "Memoir on Inventing the Confocal Scanning Microscope." *Scanning*, vol. 10, 1988, pp. 128–138.

———. *The Society of Mind*. New York: Simon and Schuster, 1986.

Moore, Gordon E.

(1929–)
American
Entrepreneur

The microprocessor chip is the heart of the modern computer, and Gordon Moore deserves much of the credit for putting it there. His insight into the computer chip's potential and his business acumen and leadership would lead to the early success and market dominance of Intel Corporation.

Moore was born on January 3, 1929, in the small coastal town of Pescadero, California, north of San Francisco. His father was the local sheriff and his mother ran the general store. Young Moore was a good science student, and he attended the University of California, Berkeley, receiving a B.S. degree in chemistry in 1950. He then went to the California Institute of Technology (Caltech), earning a dual Ph.D. in chemistry and physics. He thus had a sound background in materials science that helped prepare him to evaluate the emerging research in transistors and semiconductor devices that would begin to transform electronics in the later 1950s.

After spending two years doing military research at Johns Hopkins University, Moore returned to the West Coast to work for Shockley Semiconductor Labs in Palo Alto. However, William Shockley, who would later share in a Nobel Prize for the invention of the transistor, alienated many of his top staff, including Moore, and they founded their own company, Fairchild Semiconductor, in 1958. (This company was actually a division of the existing Fairchild Camera and Instrument Corporation.)

Moore became manager of Fairchild's engineering department and the following year, director of research. He worked closely with ROBERT NOYCE, who was developing a revolutionary process for laying the equivalent of many transistors and other components onto a small chip.

Moore and Noyce saw the potential of this integrated circuit technology for making electronic devices, including clocks, calculators, and computers, vastly smaller yet more powerful. In 1965, he formulated what became widely known in the industry as Moore's law. This prediction suggested that the number of transistors that could be put in a single chip would double about every year (later it would be changed to 18 months or two years, depending on who one talks to). Remarkably, Moore's law still held true into the 21st century, although as transistors get ever closer together, the laws of physics begin to impose limits on current technology.

Moore, Robert Noyce, and ANDREW S. GROVE found that they could not get along well with the upper management in Fairchild's parent company, and decided to start their own company, Intel Corporation, in 1968, using $245,000 plus $2.5 million from venture capitalist Arthur Rock. They made the development and application of microchip technology the centerpiece of their business plan. Their first products were RAM (random access memory) chips.

Seeking business, Intel received a proposal from Busicom, a Japanese firm, for 12 custom chips for a new calculator. Moore and Grove were not sure they were ready to undertake such a large project, but then Ted Hoff, one of their first employees, told them he had an idea. What if they built one chip that had a general-purpose central processing unit (CPU) that could be programmed with whatever instructions were needed for each application? With the support of Moore and other Intel leaders, the project got the go-ahead. The result was the microprocessor, and it would revolutionize not only computers but just about every sort of electronic device. Intel's pioneering work in microprocessing would become especially valuable in the later 1970s when Japanese companies with advanced manufacturing techniques began to make memory chips more cheaply than Intel could. Intel responded by going out of the memory business and concentrating on building ever more powerful microprocessors.

Under the leadership of Moore, Grove, and Noyce, Intel established itself in the 1980s as the leader in microprocessors, starting when IBM chose Intel microprocessors for its hugely successful IBM PC. IBM's competitors, such as Compaq, Hewlett Packard, and later Dell, would also use Intel microprocessors for most of their PCs.

Gordon Moore and his Intel Corporation have provided generations of ever more powerful microprocessors for the computers found on most desktops. Moore also made a remarkably accurate prediction that processor power would double every 18 months to two years. *(Photo provided by Intel Corporation)*

As Moore recalled to *Fortune* magazine: "Andy [Grove] and I have completely different views of the early years. To me, it was smooth. We grew faster than I thought. But Andy thought it was one of the most traumatic periods of his life." Meanwhile, Moore would serve as vice president, president, chief executive officer, then chairman of the board at Intel, finally retiring from the company in 1995.

The public is often especially interested in what the most successful entrepreneurs see in the future. In an interview with the MIT *Technology*

Review at the start of the 21st century, however, Moore disclaimed such predictive powers:

> I calibrate my ability to predict the future by saying that in 1980 if you'd asked me about the most important applications of the microprocessor, I probably would have missed the PC. In 1990 I would have missed the Internet. So here we are just past 2000, I'm probably missing something very important.

Nevertheless, Moore does think that new developments in fields such as molecular biology are important and exciting. In computing, Moore looks toward reliable, widespread speech recognition technology as the key to many more people using computers in many new and interesting ways.

In his retirement, Moore enjoys fishing at his summer home in Hawaii while also becoming involved in environmental issues. He purchased thousands of acres of Brazilian rain forest in order to protect the species that are rapidly disappearing in the face of large-scale development. In 2002 Moore pledge $5 billion of his Intel stock to the newly created Gordon E. and Betty I. Moore Foundation, which emphasizes environmental research. Moore has also had a longtime interest in SETI, or the search for extraterrestrial intelligence.

Moore has been awarded the prestigious National Medal of Technology (1990), as well as the Institute of Electrical and Electronics Engineers Founders Medal and W. W. McDonnell Award, and the Presidential Medal of Freedom (2002).

Further Reading

Burgelman, Robert, and Andrew S. Grove. *Strategy Is Destiny*. New York: Simon and Schuster, 2001.

Intel Corporation. "Executive Bios: Gordon Moore." Available on-line. URL: http://www.intel.com/pressroom/kits/bios/moore.htm. Downloaded on November 3, 2002.

"Laying Down the Law." *Technology Review* 104 (May 2001): 65.

Mann, Charles. "The End of Moore's Law?" *Technology Review* 103 (May 2000): 42.

N

⊠ Nelson, Ted (Theodor Holm Nelson)
(1937–)
American
Computer Scientist, Inventor

The 1960s counterculture had an ambivalent relationship with science and technology. While embracing electronic music, laser shows, and psychedelic chemistry, it generally viewed the mainframe computer as a dehumanizing force and a symbol of the corporate establishment. However, a young visionary named Ted Nelson would appear on the scene, proclaiming to the flower children that they could and should understand computers. His pioneering ideas about using hypertext to link ideas via computer and his manifesto *Computer Lib/Dream Machines* helped spawn a sort of computer counterculture that would have far-reaching consequences.

Nelson was born in 1937 to actress Celeste Holm and director Ralph Nelson, but he was raised by his grandparents in the Greenwich Village neighborhood in New York City. Nelson attended Swarthmore College in Pennsylvania, earning a B.A. degree in philosophy in 1959. He then enrolled in Harvard University's graduate program in sociology and received his master's degree in 1963.

Nelson's first encounter with computers was by way of a programming course for humanities majors. He became intrigued not with the machine's calculating abilities but by its potential for organizing textual material. His master's project was an attempt to write a system that could help writers organize and revise their work. This ambitious project failed, largely because it was attempted long before the development of word processing, and most programs could do little more with text than store it in data fields and print it out. The mainframe languages of the time had little built-in support for text operations, so routines for text processing had to be written from scratch.

Although the programming bogged down, Nelson had hit on the concept that he dubbed "hypertext." The idea, familiar to today's Web users, is to create links that allow the reader to move at will from the current document to related documents. (VANNEVAR BUSH is usually given the credit for first conceptualizing hypertext, though he did not use that term. However, his Memex machine, proposed in the 1940s, would have been an electromechanical microfilm reading device, not a computer.)

Nelson formally introduced the idea of hypertext in a 1965 paper titled "A File Structure for the Complex, the Changing and the Indeterminate." This title hints at the implications of hypertext as a dynamic (changing) system of documents that combines the depth and detail of print with the ability to quickly update the content and to create endless potential paths for readers.

Nelson also contributed to the movement that led to the personal computer. In his 1974 book *Computer Lib/Dream Machines*, he made computers accessible to a new generation that was suspicious of a technology that was often associated with the oppressive behavior of governments and corporations. Along with LEE FELSENSTEIN and his pioneering computer bulletin board, and educators such as JOHN G. KEMENY and Thomas Kurtz (inventors of BASIC), Nelson helped create an interest in computers among students and activists at a time when "computer" meant a mainframe or minicomputer connected to time-sharing terminals. Thus, when the microprocessor made the personal computer practical in the later 1970s, there was already a culture of experimenters primed to use the new technology in imaginative ways.

In the late 1960s, Nelson started work on Project Xanadu, an ambitious hypertext system. The name intentionally evokes the exotic Mongolian city in Coleridge's poem "to represent a magic place of literary memory and freedom, where nothing would be forgotten." Nelson has drawn a clear distinction between the structure of Xanadu and that of the World Wide Web, invented by TIM BERNERS-LEE in the early 1990s. Nelson asserts on his home page that "The Web isn't hypertext, it's decorated directories." In his view, the Web is one-way hypertext: The reader can browse and follow links, but only the authorized webmaster of a site can create new links.

In Xanadu, writers can collaborate through "transclusion," or incorporation of existing documents (with their links intact) in new ones. Nelson suggested that one of the Internet's biggest current issues, the protection of intellectual property, could be built into such a system because the links and ownership would "follow" quoted material into new documents. At all times, readers could delve into the differences and branchings between versions of a document.

However, as decades passed and nothing reached the market, Xanadu began to be derided by many pundits as the ultimate in "vaporware," or software that is promised and hyped but never actually released.

Although there are still commercial possibilities and the Xanadu code was released as "open source" (available to any programmer for noncommercial use) in 1999, the legacy of Nelson's work on Xanadu is likely to remain subtle. By coining the term *hypertext* and showing how it might work, Nelson inspired a number of implementations, including especially Tim Berners-Lee's World Wide Web. And the vision of a hypertext system that truly allows both collaboration and the ability to automatically track the history and revision of text remains an intriguing possibility and challenge. Indeed, Xanadu is perhaps closer to what Berners-Lee originally envisioned for the Web than to today's Web with its limited hypertext, documents visually formatted like paper documents, and relatively passive reading.

Nelson continues his work today, teaching at Keio University, Fujisawa, Japan (where he received a Ph.D. in 2001), and the University of Southampton in England.

Further Reading

Nelson, Ted. *Computer Lib/Dream Machines*. Rev. ed. Redmond, Wash.: Microsoft Press, 1987.

Project Xanadu. Available on-line. URL: http://xanadu.com. Downloaded on November 3, 2002.

"Ted Nelson Home Page." Available on-line. URL: http://ted.hyperland.com. Downloaded on November 3, 2002.

von Neumann, John

(1903–1957)

Hungarian/American

Mathematician, Computer Scientist

Every field, it seems, is graced with a few strikingly original minds that seemingly roam at will through the great problems of science. In their wake, they leave theories and tools, each of which would be a major achievement to mark an entire career. Physics had Newton and Einstein and chemistry and biology had Pasteur. Mathematics and its infant offspring, computer science, had John von Neumann. Von Neumann made wide-ranging contributions in fields as diverse as pure logic, simulation, game theory, and quantum physics. He also developed many of the key concepts for the architecture of the modern digital computer and helped design some of the first successful machines.

John von Neumann developed automata theory as well as fundamental concepts of computer architecture, such as storing programs in memory along with the data. This scientific Renaissance man also did seminal work in logic, quantum physics, simulation, and game theory. *(Photo courtesy of Computer Museum History Center)*

Von Neumann was born on December 28, 1903, in Budapest, Hungary, to a family with banking interests who also cultivated intellectual activity. As a youth, he showed a prodigious talent for calculation and interest in mathematics, but his father opposed his pursuing a career in pure mathematics. Therefore, when von Neumann entered the University of Berlin in 1921 and the University of Technology in Zurich in 1923, he ended up with a Ph.D. in chemical engineering. However, in 1926 he went back to Budapest and earned a Ph.D. in mathematics with a dissertation on set theory. He then served as a lecturer at Berlin and the University of Hamburg.

During the mid-1920s, the physics of the atom was a dramatic arena of scientific controversy. Two competing mathematical descriptions of the behavior of atomic particles were being offered by Erwin Schrödinger's wave equations and Werner Heisenberg's matrix approach. Von Neumann showed that the two theories were mathematically equivalent. His 1932 book, *The Mathematical Foundations of Quantum Mechanics*, remains a standard textbook to this day. Von Neumann also developed a new form of algebra in which "rings of operators" could be used to describe the kind of dimensional space encountered in quantum mechanics.

Meanwhile, von Neumann had become interested in the mathematics of games, and developed the discipline that would later be called game theory. His "minimax theorem" described a class of two-person games in which both players could minimize their maximum risk by following a specific strategy.

In 1930, von Neumann immigrated to the United States, where he became a naturalized citizen and spent the rest of his career. He taught mathematical physics at Princeton University until 1933, when he was made a Fellow at the new Institute for Advanced Study in Princeton. He served in various capacities there and as a consultant for the U.S. government.

In the late 1930s, interest had begun to turn to the construction of programmable calculators or computers. The theoretical work of ALAN

TURING and ALONZO CHURCH had shown the mathematical feasibility of automated calculation, although it was not yet clear whether electromechanical or electronic technology was the way to go.

Just before and during World War II, von Neumann worked on a variety of problems in ballistics, aerodynamics, and later, the design of nuclear weapons. All of these problems cried out for machine assistance, and von Neumann became acquainted both with British research in calculators and the massive Harvard Mark I programmable calculator built by HOWARD AIKEN.

A little later, von Neumann learned that two engineers at the University of Pennsylvania, J. PRESPER ECKERT and JOHN MAUCHLY, were working on a new kind of machine: an electronic digital computer called ENIAC that used vacuum tubes for its switching and memory, making it about 1,000 times faster than the Mark I. Although the first version of ENIAC had already been built by the time von Neumann came on board, he served as a consultant to the project at the University of Pennsylvania's Moore School.

The earliest computers (such as the Mark I) read instructions from cards or tape, discarding each instruction as it was performed. This meant, for example, that to program a loop, an actual loop of tape would have to be mounted and controlled so that instructions could be repeated. The electronic ENIAC was too fast for tape readers to keep up, so it had to be programmed by setting thousands of switches to store instructions and constant values. This tedious procedure meant that it was not practicable for anything other than massive problems that would run for many days.

In his paper "First Draft of a Report on the EDVAC," published in 1945, and his more comprehensive "Preliminary Discussion of the Logical Design of an Electronic Computing Instrument" (1946) von Neumann established the basic architecture and design principles of the modern electronic digital computer.

Von Neumann's key insight was that in future computers the machine's internal memory should be used to store constant data and all instructions. With programs in memory, looping or other decision making can be accomplished simply by "jumping" from one memory location to another. Computers would have two forms of memory: relatively fast memory for holding instructions, and a slower form of storage that could hold large amounts of data and the results of processing. (In today's personal computers these functions are provided by the random access memory (RAM) and hard drive, respectively.) The storage of programs in memory also meant that a program could treat its own instructions like data and change them in response to changing conditions.

In general, von Neumann took the hybrid design of ENIAC and conceived of a design that would be all-electronic in its internal operations and store data in the most natural form possible for an electronic machine—binary, with 1 and 0 representing the on and off switching states and, in memory, two possible "marks" indicated by magnetism, voltage levels, or some other phenomenon. This logical design would be consistent and largely independent of the vagaries of hardware.

Eckert and Mauchly and some of their supporters later claimed that they had already conceived of the idea of storing programs in memory, and in fact they had already designed a form of internal memory called a mercury delay line. Whatever the truth in this assertion, it remains that von Neumann provided the comprehensive theoretical architecture for the modern computer, which would become known as the von Neumann architecture. Von Neumann's reports would be distributed widely and would guide the beginnings of computer science research in many parts of the world.

Looking beyond EDVAC, von Neumann, together with HERMAN HEINE GOLDSTINE and Arthur Burks, designed a new computer for the Institute for Advanced Study that would embody the von Neumann principles. That machine's design would in turn lead to the development of research computers for RAND Corporation, the Los Alamos National Laboratory, and in several countries, including Australia, Israel, and even the Soviet Union. The design would eventually be commercialized by IBM in the form of the IBM 701.

In his later years, von Neumann continued to explore the theory of computing. He studied ways to design computers that could automatically maintain reliability despite the loss of certain components, and he conceived of an abstract self-reproducing automaton, planting the beginnings of the field of cellular automation. He also advised on the building of Princeton's first computer in 1952 and served on the Atomic Energy Commission, becoming a full commissioner in 1954.

Von Neumann's career would be crowned with many awards reflecting his diverse contributions to science and technology. These include the Distinguished Civilian Service Award (1947), Presidential Medal of Freedom (1956), and the Enrico Fermi Award (1956). Von Neumann died on February 8, 1957, in Washington, D.C.

Further Reading

Aspray, William. *John von Neumann and the Origins of Modern Computing*. Cambridge, Mass.: MIT Press, 1990.

Heims, S. J. *John von Neumann and Norbert Wiener: From Mathematics to the Technologies of Life and Death*. Cambridge, Mass.: MIT Press, 1980.

Lee, J. A. N. "John Louis von Neumann." Available on-line. URL: http://ei.cs.vt.edu/~history/VonNeumann.html. Updated on February 9, 2002.

Von Neumann, John. *The Computer and the Brain*. New Haven: Yale University Press, 1958.

———. *Theory of Self-Reproducing Automata*. Edited and compiled by Arthur W. Burks. Urbana: University of Illinois Press, 1966.

Newell, Allen

(1927–1992)
American
Computer Scientist

According to pioneer artificial intelligence researcher Herbert Simon, there are four fundamental questions that science can ask: "the nature of matter, the origins of the universe, the nature of life, the workings of mind." In a moving tribute to his colleague, Simon says that Allen Newell devoted his long and productive career to pursuing that last question. As Newell would be quoted in a memoir by Herbert Simon:

> The scientific problem chooses you; you don't choose it. My style is to deal with a single problem, namely, the nature of the human mind. That is the one problem that I have cared about throughout my scientific career, and it will last me all the way to the end.

In the dawning computer age, a concern with the workings of the mind was very likely to bring a researcher into contact with what would become known as artificial intelligence research, or AI. In AI, two strands are inextricably intertwined: the growing capabilities of machines and their relationship to possible understandings of human mental functioning.

Newell was born on March 19, 1927, in San Francisco. His father was a distinguished professor of radiology at Stanford Medical School. In an interview with Pamela McCorduck for her 1979 book *Machines Who Think*, Newell describes his father as

> in many respects a complete man. . . . He'd built a log cabin up in the mountains. . . . He could fish, pan for gold, the whole bit. At the same time, he was the complete intellectual. . . . Within the environment where I was raised, he

> was a great man. He was extremely idealistic. He used to write poetry.

This example of a wide-ranging life and intellect would be carried into young Newell's early years as well. Spending summers at his father's log cabin in the Sierra Nevada range instilled in him a love of the mountains, and for a time he wanted to be a forest ranger when he grew up. The tall, rugged boy also naturally excelled at sports, especially football. At the same time Lowell High School in San Francisco offered a demanding academic program that encouraged his intellectual interests.

When World War II began, Newell enlisted in the navy. Following the war, he served on one of the ships monitoring the nuclear tests at Bikini Atoll, and was assigned the task of mapping the distribution of radiation in the area. This kindled an interest in science in general and physics in particular. When Newell left the navy, he enrolled in Stanford University to study physics. (He wrote his first scientific paper, on X-ray optics, in 1949.)

While at Stanford, Newell took a course from George Polya, a mathematician who had done important work in heuristics, or methods for solving problems. The idea that problem-solving itself could be investigated scientifically and developed into a set of principles would be a key to the approach to artificial intelligence later.

While still a graduate student, Newell also worked at RAND Corporation, a center of innovative research, in 1949–50. There he encountered game theory, the study of the resolution of competing interests. (This field had been established by JOHN VON NEUMANN and Oskar Morgenstern earlier in the 1940s and would become rather famous later through the life and work of John Nash.) Newell's work led him to research on organization theory, which in turn got him involved with experiments with individuals in groups who were given simulated problems to solve.

This effort eventually turned into an air force project at the Systems Research Laboratory at RAND that created a simulation of an entire air force early warning station—this at a time when such stations were the key to defense against an anticipated Soviet nuclear bomber attack. Running such a large-scale simulation required creating simulated radar displays, and that in turn meant Newell and his colleagues had to harness the power of computers. Using a primitive punch-card calculator, Newell devised a way to print out continuously updated positions for the blips on the simulated radar screen.

Fortunately, computer technology was rapidly advancing, with the Massachusetts Institute of Technology, IBM, and others developing computers that could store much more complex programs electronically. With stored programs, computers could modify data and even instructions in response to input conditions. In 1954, Newell attended a RAND seminar in which visiting researcher Oliver Selfridge described a computer system that could actually recognize and manipulate patterns, such as characters in strings. According to Simon's memoir, Newell experienced a "conversion experience" in which he realized "that intelligent adaptive systems could be built that were far more complex than anything yet done." He could do this by combining what he had learned about heuristics (problem-solving) with bits of simulation and game theory.

Newell decided to use chess as the test bed for his ideas. Researchers such as ALAN TURING and CLAUDE E. SHANNON had already made some headway in writing chess programs, but these efforts focused on a relatively mechanical, "brute force" approach to generating and analyzing possible positions following each move. Newell, however, tried to simulate some of the characteristics that a human player brings to the game, including the ability to search not for the theoretically best move but a "good enough" move, and the ability to formulate and evaluate short- and long-term goals. For example, a computer chess player might have short-term goals, such as clearing a file for its rook, as well as longer term goals such as taking control of the king side in preparation for an all-out attack on the opponent's king. In 1955, Newell presented his ideas in a conference paper titled "The Chess Machine: An Example of Dealing with a Complex Task by Adaptation."

That same year, Newell moved to Carnegie Mellon University (CMU) in Pittsburgh. The arena was changed from chess to mathematics, and Newell, working with Herbert Simon and Clifford Shaw, applied his ideas to writing a program called the Logic Theory Machine (LTM) which could prove mathematical theorems. By 1956, the program was running and demonstrated several proofs. Interestingly, unlike the usual procedure, it worked backward from a hypothesized theorem to the axioms from which it could be proven. In a paper, recounted in Simon's memoir, Newell described the LTM as "a complex information processing system . . . capable of discovering proofs for theorems in symbolic logic. This system, in contrast to the systematic algorithms . . . ordinarily employed in computation, relies heavily on heuristic methods similar to those that have been observed in human problem solving activity." An important part of the project was the development of an "information processing language," or IPL. Unlike regular programming languages that specify exact procedures, IPL was a higher-level logical language that could later be turned into machine-usable instructions. This way of outlining programs, called pseudocode, would become a standard tool for software development.

By 1960, Newell and his collaborators had created a more powerful program called the General Problem Solver, or GPS. This program could be given a specification for a "problem domain," a set of operators (ways to manipulate the elements of the problem domain), and guidelines about which operators were generally applicable to various situations. The program could then develop a solution to the problem using appropriate application of the operators. Then, in a further refinement, the program was given the ability to discover new operators and their appropriate use—in other words, it could learn and adapt.

Meanwhile, the Newell team had also created a chess-playing program called NSS (named for the last initials of the researchers). While NSS was not as strong a player as the "brute force" programs, it successfully applied automated problem-solving techniques, making it an "interesting" player. (Around this time, Newell predicted that within a decade the chess champion of the world would be a computer. If one accepts the 1997 victory of IBM's Deep Blue over Garry Kasparov as legitimate, Newell's prediction was off by about three decades.)

During the 1960s and 1970s, Newell delved more deeply into fundamental questions of "knowledge architecture"—how information could be represented within a program in a way such that the program might appear to "understand" it. Newell's "Merlin" program was an ambitious attempt to create a program that could understand AI research itself, being able to demonstrate and explain various other AI programs. Unfortunately, the program never worked very well.

Besides AI research, Newell became involved in a number of other areas. One was an attempt to build a simulation of human cognitive psychology called the Model Human Processor. It was hoped that the simulation would help researchers at the Xerox Palo Alto Research Center (PARC) who were devising what would become the modern computer user interface with mouse-driven windows, menus, and icons. The research was summarized in a book titled *The Psychology of Human-Computer Interaction* (1983).

Since the early 1970s, Newell had a lively interest in ARPANET, which gradually became the Internet. During the 1980s, Newell made important contributions to the CMU computer science curriculum and department, and to the establishment of Andrew, the campus computer network.

Finally, Newell attempted to draw together the models of cognition (both computer and human) that he and many other researchers had developed. His last problem-solving program, SOAR, demonstrated ideas that he explained in his book *Unified Theories of Cognition*. These techniques included learning by grouping or "chunking" elements of the problem, and the ability to break problems into subgoals or subproblems and then working back up to the solutions. Drawing on research from a number of different universities, the SOAR project continues today. It remains a tribute to Newell, one of whose maxims according to Simon was "Choose a project to outlast you."

Newell died on July 19, 1992. He had published 10 books and more than 250 papers and was the recipient of many honors. In 1975 he received the Association for Computing Machinery (ACM) Turing Award for his contributions to artificial intelligence. In turn the ACM, with sponsorship of the American Association for Artificial Intelligence, established the Newell Award for "contributions that have breadth within computer science, or that bridge computer science and other disciplines." Just before his death, Newell was awarded the National Medal of Science.

Further Reading

Card, Stuart K., et al. *The Psychology of Human-Computer Interaction*. Hillsdale, N.J.: Lawrence Earlbaum Associates, 1983.

McCorduck, Pamela. *Machines Who Think*. San Francisco: W. H. Freeman, 1979.

Newell, Alan. *Unified Theories of Cognition*. Cambridge, Mass.: Harvard University Press, 1994.

Rosenblum, Paul S., et al. editors. *The Soar Papers: Research on Integrated Intelligence*. Cambridge, Mass.: MIT Press, 1993.

Simon, Herbert A. "Allen Newell, March 19, 1927–July 19, 1992." Biographical Memoirs, National Academy of Science. Available online. URL: http://stills.nap.edu/readingroom/books/biomems/anewell.html.

Noyce, Robert

(1927–1990)
American
Inventor, Entrepreneur, Engineer

The region just south of San Francisco that became informally known as Silicon Valley was born in the 1960s when researchers discovered how to build thousands of miniature components into complex circuits inscribed on tiny bits of specially treated silicon—a valuable relative of ordinary sand. At the same time, a new corporate/technical culture was born, one driven largely by innovation and gauged by merit rather than traditional hierarchy. Robert Noyce played a key role in both the technology and the culture.

Noyce was born on December 12, 1927, in Burlington, Iowa. His father was a clergyman. From an early age, Noyce liked to tinker with a variety of machines. He enrolled in Grinnell College in 1946. As an undergraduate physics major, Noyce quickly acquired a reputation as a prankster and a bit of a troublemaker—a local farmer was enraged when he learned that Noyce had stolen one of his pigs for a college luau.

In 1948, one of Noyce's teachers had managed to obtain a couple samples of an exciting new invention, the transistor. The possibilities of the compact device helped steer Noyce away from pure science. When he graduated that same year with a B.S. degree in physics and mathematics, he went to the Massachusetts Institute of Technology (MIT) to study for his Ph.D. in electronics, which he received in 1953.

After his studies at MIT, Noyce worked at first as a research engineer at Philco Corporation, but in 1956 he joined William Shockley, one of the inventors of the transistor, at his new company, Shockley Semiconductor Laboratories.

Robert Noyce saw great possibilities for the transistor shortly after its invention in 1948. However, his desire for miniaturization soon went beyond separate transistors to integrated circuits that could place thousands of the devices on a single chip. In 1959, Noyce and Jack Kilby both filed patents for the new technology, and they share the credit for the invention that made the modern desktop computer possible. *(Photo provided by Intel Corporation)*

However, Shockley's abrasive style increasingly grated on Noyce and many other engineers, and he insisted on using germanium instead of a material many considered superior—silicon.

As writer Tom Wolfe would later remark:

> With his strong face, his athlete's build, and the Gary Cooper manner, Bob Noyce projected what psychologists call the halo effect. People with the halo effect seem to know exactly what they're doing and moreover make you want to admire them for it. They make you see the halos over their heads.

Thus when eight engineers decided to leave Shockley's company and strike out on their own, they naturally turned to the charismatic Noyce for leadership, even though he had not yet turned 30. In 1957, they started a new company, Fairchild Semiconductor, a division of Fairchild Camera and Instrument, which was seeking larger involvement in the burgeoning solid state electronics industry.

As engineering research manager at Fairchild, Noyce, unlike Shockley, did not try to micromanage. The corporate culture he created, with cubicles instead of offices and a minimum of hierarchy and "perks," would become familiar to the next generation of computer designers and programmers.

In this atmosphere, research proceeded remarkably quickly. The transistor had replaced the bulky, power-consuming vacuum tube as an electronic switch and amplifier. The next step was to form tiny transistors on the surface of a chip of semiconducting material such as germanium or silicon. This process could be compared to the difference between carving words on a page and using a printing press—if one also imagined printing a whole page of text into a space the size of the period at the end of a sentence!

The development of the miniature solid-state technology had also received an urgent boost: The same year Noyce and his colleagues had gone into business, the Soviets launched Sputnik. The Russian rockets were much larger than the American ones: to compete, the United States had to make its payloads much smaller, and the integrated circuit would be the key. Space and military applications would in turn create technology that fed into the civilian sector, resulting in (among other things) more compact, powerful calculators and computers.

In February 1959, Texas Instruments filed a patent for the work of JACK KILBY, who had developed an alternate approach to integrated circuits, and Noyce followed with his version in July of that year. From 1962 to 1967 the companies engaged in a protracted legal battle over whose patent was valid. The courts eventually ruled in favor of Noyce, and the companies agreed to license each other's technology. Meanwhile, Noyce had became Fairchild's general manager and vice president in 1959 and from 1965 to 1968 served as group vice president in the parent company.

The boom in what would soon become known as Silicon Valley was well underway, and investment money poured in as venture capitalists sought to start new semiconductor companies. Many of Fairchild's best engineers accepted these lucrative offers, and in June 1968 Noyce himself left. Joining GORDON E. MOORE and ANDREW S. GROVE, and funded by venture capitalist Arthur Rock, they founded Intel (short for either "integrated electronics" or "intelligence.")

Noyce at first concentrated Intel's efforts on developing integrated random access memory (RAM) chips, which would replace the older magnetic "core" memories. Intel's 1103 RAM chip launched a new generation of mainframes and minicomputers.

As the 1970s progressed, miniaturization efforts moved from the memory to the central processing unit, or CPU, the part of the computer that actually performs arithmetic and logical operations. In 1971, Intel had developed the first microprocessor, the Intel 4004, dubbed a "computer on a chip." By mid-decade, a more powerful chip, the Intel 8080, would become the most popular chip for the new personal computer (PC), and the 1980s would confirm Intel's market dominance when the IBM PC established the dominant PC architecture.

In 1978, Noyce retired from active involvement in Intel and became chairman of the Semiconductor Industry Association, devoting his efforts to lobbying on behalf of the now powerful industry as it struggled with Japanese competition, eventually ceding the memory market while maintaining dominance in microprocessors. In 1988, Noyce agreed to head Sematech, a government-funded semiconductor research consortium whose mission is to propel American microchip technology ahead of equally ambitious Japanese projects.

Noyce died on June 5, 1990. He received numerous medals, including the Ballantine Medal of the Franklin Institute (1966), the Institute of Electrical and Electronics Engineers (IEEE) Medal of Honor (1978) and Computer Society Pioneer Award (1980); the Faraday Medal of the Institution of Electrical Engineers (1979); the National Medal of Science (1979) and National Medal of Technology (1987).

In 2000, Jack Kilby received the Nobel Prize in physics for his work on integrated circuits. Most observers believe that Noyce would have shared in this award, except that Nobel prizes are not given posthumously.

Further Reading

Cortada, James W. *Historical Dictionary of Data Processing*. Westport, Conn.: Greenwood Press, 1987.

Leibowitz, Michael R. "Founding Father: Robert Noyce." *PC/Computing*, May 1989, pp. 94 ff.

Wolfe, Tom. "The Tinkerings of Robert Noyce." *Esquire*, December 1983, pp. 346ff.

Nygaard, Kristen
(1926–2002)
Norwegian
Computer Scientist

Many programmers today are familiar with object-oriented programming, which breaks programs into well-defined, modular parts that interact in specified ways. The very popular language C++, invented by BJARNE STROUSTRUP, uses object-oriented principles, as does Smalltalk, an innovative language created by ALAN C. KAY. But relatively few people know that it was a Norwegian computer scientist named Kristen Nygaard who created Simula, the first object-oriented language.

Nygaard was born on August 27, 1926, in Oslo, Norway. He earned his master's degree in mathematics at the University of Oslo in 1956. His master's thesis was on the theory of "Monte Carlo methods," an approach to simulation through the application of probability. By then he had been working as a programmer and operational research specialist for the Norwegian Defense Research Establishment, where he would continue until 1960. Nygaard had thus acquired both the theoretical and practical programming knowledge needed to develop computer simulations.

At the time, however, computer simulations were difficult to design and implement. This was largely because existing languages such as FORTRAN or even the better-structured Algol were designed to carry out a series of procedures, with the computer doing one thing at a time. With simulations, however, the researcher is trying to model the behavior of real-world objects, which can be everything from subatomic particles to customers waiting in a bank line.

Nygaard began therefore to think in terms of how to make the object rather than the procedure the building block of programming. From 1961 to 1965 Nygaard, working with Ole-Johan Dahl, developed a programming language called Simula, short for "Simulation Language." This language was built upon the procedural structure of Algol, but included the ability to have procedures or "activities" running at the same simulated time. Each procedure would keep track of where it had left off so it could be resumed, and the system as a whole would coordinate everything.

This scheme made it much easier to simulate situations in which many objects are involved, whether in a relatively simple structure such as a queue (that is, a waiting line) or more complex interactions. As Xerox and Apple researcher Larry Tesler was quoted in a *New York Times* obituary by John Markoff. "[Nygaard] understood that simulation was the ultimate application of computers. It was a brilliant stroke."

In 1967, Nygaard and Dahl announced an improved version of the language, Simula 67. The first Simula language had been essentially grafted onto Algol. In Simula 67, however, the concepts of the class and object were made explicit. A class is a description of the structure and properties of a thing, for example, a circle. It includes both internal information (such as the circle's radius) and procedures or capabilities that can be used by the program (such as to draw the circle on the screen). Once a class is defined, particular instances (objects) of that class are created and used by the program. For example, several actual circle objects might be created from the circle class and then drawn. Another interesting feature in Simula 67 is the ability to create a new class and have it "inherit" features from a previous class, much as children acquire features from their parents but then develop their own distinctive identity.

Although Simula never achieved widespread use, the papers describing the language were very influential in computer science circles. When Alan Kay, for example, read about Simula, he not only developed his own object-oriented language, Smalltalk, he and fellow researchers at the Xerox Palo Alto Research Center (PARC) began to create object-oriented operating systems and in particular, the user interface that would be adopted later by the Apple Macintosh and then Microsoft Windows. Thus the objects (menus, icons, windows) that personal computer (PC) users work with every day are descendants of the object-oriented ideas pioneered by Nygaard.

Meanwhile, Bjarne Stroustrup also encountered Nygaard's ideas, and for him they represented a way to improve the popular C programming language so that it could deal with increasingly complex modern applications. He incorporated Nygaard's key ideas of classes, objects, and inheritance into his new C++ language.

Nygaard served briefly as a professor in Århus, Denmark from 1975 to 1976, and then moved to the University of Oslo, where he would be a professor until 1996. Nygaard continued his theoretical work, developing Delta, a "general system description language" with Erik Holbaek-Hanssen and Petter Haandlykken.

Nygaard's interests increasingly expanded to include the political and social spheres. Starting in the early 1970s, he did research for Norwegian trade union organizations and explored the social impact of technology. He was also one of the country's first major environmental activists, heading the environment protection committee for the Norwegian Association for the Protection of Nature. Nygaard also became active in the leadership of the left-wing Venstre party, and also helped lead a successful movement opposing Norway's admission into the European Union, which activists feared would weaken labor rights and environmental protections.

As the founder of a key paradigm for programming and language design, Nygaard received numerous awards. In 1990, the Computer Professionals for Social Responsibility presented him with the Norbert Wiener Prize for his research on the social impact of computer technology, and he also became a member of the Norwegian Academy of Science. In 2001, Nygaard and Ole-Johan Dahl received the Association for Computing Machinery Turing Award, and in 2002 the two researchers were awarded the Institute of Electrical and Electronics Engineers von Neumann Medal for their key contributions to the development of object-oriented programming. Nygaard died on August 9, 2002.

Further Reading

"Home Page for Kristen Nygaard." Available on-line. URL: http://www.ifi.uio.no/~kristen. Downloaded on November 3, 2002.

Markoff, John. "Kristen Nygaard, 75, Who Built Framework for Modern Computer Languages." [Obituary] *New York Times*, August 14, 2002, pp. A19ff.

Wikipedia, the Free Encyclopedia. "Kristen Nygaard." Available on-line. URL: http://www.wikipedia.org/wiki/Kristen_Nygaard. Updated on October 25, 2002.

O

Olsen, Kenneth H.

(1926–)
American
Entrepreneur, Engineer

By the early 1960s, IBM had become the colossus of the computer industry, dominating sales of the large mainframe computers that were increasingly handling the data processing needs of large corporations and government agencies. However, the transistor and later, the integrated circuit microchip, would offer the possibility of building smaller, yet powerful computers that could be used by small companies, universities, and other users who could not afford a traditional mainframe. Kenneth Olsen and Digital Equipment Corporation (DEC) would pioneer these new minicomputers and Olsen would lead the company through the difficult decades to come.

Olsen was born on February 20, 1926, in Stratford, Connecticut, in a working-class community consisting largely of Norwegian, Polish, and Italian immigrants. His father designed factory equipment, including a machine to make safety pins and another to make universal joints for cars. He held several patents and later became a successful machine salesman.

As a boy, Olsen spent several summers working in a machine shop. A family friend, an electrical engineer, supplied the boy with technical manuals, which he was said to prefer to comic books. Olsen also fixed friends' and neighbors' radios for free, and a girlfriend was quite impressed when Olsen rigged a metal detector to find a watch she had lost at the beach.

Olsen's father taught a Bible class and his mother played the piano at the local church. The family's fundamentalist Christian background would play an important part in shaping Olsen's personality, and religion would always play a major role in his life. One of his high school teachers described Olsen as "quiet, dreary, and smart," yet in adulthood he would combine a solid, down-home manner with great energy, flexibility, and ability to innovate.

Olsen was still in high school when the United States entered World War II, but in 1944 he joined the navy, where he received his first formal training in electronics technology. After the war he entered the Massachusetts Institute of Technology (MIT), where his intelligence and drive enabled him to keep up with much better prepared classmates. He earned his bachelor's degree in electrical engineering in 1950 and master's in 1952.

The success of JOHN MAUCHLY and J. PRESPER ECKERT with ENIAC at the University of Pennsylvania had put electronic digital computing in the technological forefront. MIT had eagerly begun its own digital computer laboratory and Olsen joined its innovative research projects including the powerful Whirlwind, a computer designed as a real-time flight simulator for the air force. Besides its innovative architecture, Whirlwind also pioneered the use of interactive screen displays and controls. Olsen was put in charge of a group that built a small-scale test version of the machine's innovative ferrite core memory, a relatively fast, compact magnetic memory that replaced cumbersome memory tubes. This early work in making computers smaller and more interactive helped lay the foundation for his later development of the minicomputer industry. Meanwhile, according to Peter Petre's biography of Olsen in *Fortune* magazine, Jay Forrester, head of the lab, had found Olsen to be a "first-class practical engineer."

Olsen took a sudden six-month leave from MIT in 1950, when he went to Sweden, to court and marry Eeva-Liisa Aulikki Valve, a Finnish exchange student he had met earlier. To support himself while in Sweden, Olsen took a job as an electrician in a ball-bearing factory. This combination of decisiveness and persistence would become familiar to Olsen's colleagues and competitors in the decades to come.

When the U.S. Air Force began to build SAGE, a huge air defense control computer complex in 1951, the work took place at an IBM factory at Poughkeepsie, New York. Olsen was sent there to monitor the project's progress. In doing so, Olsen was shocked at what he found to be the IBM culture: as noted in his *Fortune* biography, "It was like going to a Communist state. They knew nothing about the rest of the world, and the world knew nothing about what was going on inside." Olsen confided to Norman Taylor, his MIT supervisor, that he had found IBM's production methods to be inefficient. "I can beat these guys at their own game," Olsen declared.

After returning to MIT, Olsen learned about another key technology when he led the project to build a small experimental computer that used transistors instead of vacuum tubes. By the late 1950s, engineers were well aware of the potential advantages of transistors, which were much more compact than vacuum tubes and consumed much less power. However, the reliability of the various types of transistors being developed was still in question. The "transistor computer" helped demonstrate the viability of using all solid-state technology.

Although Olsen had thought about the idea of starting his own company, it was Harold Ockenga, his pastor at the Park Street Church in Boston, who spurred him to take action. Although Ockenga was fundamentalist and conservative, he was quite willing to use the latest technology to carry on a radio ministry. Ockenga asked Olsen to revamp the church's Sunday school, which had become disorganized and neglected. Olsen, who had no formal management training, read all the management books he could find at the local library. He organized committees to plan the redevelopment of the school and applied marketing techniques to get the congregation to give the necessary financial support. As a result of this successful campaign, Olsen found not only that he had some management ability, but that he wanted to manage.

In 1957 Olsen, together with his brother Stan and a colleague, Harlan E. Anderson, founded the Digital Equipment Corporation (DEC). Their starting capital consisted of $70,000 from a venture capital company headed by Harvard professor Charles Doriot, who would also become a lifelong friend. DEC's business plan was to develop and market all-transistor versions of the various circuit modules used in engineering test equipment. The company rented space in an abandoned mill and had no office—just a battered desk next to the manufacturing equipment.

DEC had modest success with its component business, but Olsen decided he wanted to tackle the computer industry head on. As an engineer he realized that there were many tasks, such as monitoring scientific experiments, that would benefit greatly from computerization. However, a mainframe would be too expensive and unnecessarily powerful for such applications. Olsen conceived of building a simpler, more compact computer using the transistor component technology with which he had become familiar.

In 1960, DEC introduced the PDP-1. (PDP stood for "programmed data processor," a down-to-earth name that avoided much of the intimidating baggage around the word *computer.*) About the size of a large refrigerator, the PDP-1 consumed much less power than a mainframe and did not require a special air-conditioned computer room. The new "minicomputer" could thus go where computers had been unable to go before, such as scientific laboratories, engineering workshops, and factory floors.

The PDP-1 cost only $120,000, a fraction of the cost of an IBM mainframe. During the 1960s and early 1970s, subsequent models became more capable, yet cheaper. The PDP-8 became the first mass-produced minicomputer in 1965, and the PDP-11 of 1970 became a workhorse of laboratories and college computing departments where a new generation of computer scientists and engineers was busy developing such innovations as the UNIX operating system and the Internet. Further, many of the architectural features of the minicomputer, such as the use of a single "motherboard" to hold components, would be adopted in the later 1970s by the builders of the first desktop computers.

DEC prospered through the 1970s, growing from essentially a three-person company to one with 120,000 employees, second only to IBM in the computer industry. Olsen's management style, in which engineers formed more than 30 separate groups within the company, helped spur innovation and flexibility. However, the 1980s would bring new challenges.

According to writer David Ahl, Olsen had told the audience at a 1977 World Future Society meeting that "there is no reason for any individual to have a computer in his home." Olsen and DEC had thus virtually ignored the infant personal computer (PC) industry, while, ironically, stolid IBM in 1980 had decided to give its engineers free rein to build what became the IBM PC.

Throughout the 1980s, the PC would become more powerful and increasingly able to take over applications that had previously required a minicomputer. At the same time, another company, Sun Microsystems under SCOTT G. MCNEALY, had developed desktop workstations that used powerful new microprocessors. These workstations, which ran UNIX, began also to compete directly with DEC minicomputers. Starting in 1982, when a market downturn combined with falling earnings sent DEC stock plunging, the company essentially went into a decline from which it would never recover.

Despite having some success with its new line of minicomputers using VAX (virtual address extension, a flexible memory system), DEC found itself increasingly in debt. Olsen's once-praised management techniques were increasingly questioned. In particular, the many groups within the company were criticized as leading to incessant battles.

In 1983, the company was reorganized. Olsen had believed that by giving managers responsibility for profitability as well as production, he would make them mini-entrepreneurs and motivate them to new heights of success. Now, however, he concluded that the functions needed to be separated. However, the destruction of a system of relationships that had been built up over many years left many managers and engineers confused and demoralized.

That same year, DEC introduced the Rainbow, its belated entry into the PC market. However, the system was premium priced and worse, incompatible with the industry standard that had been established by IBM. DEC turned away from the desktop to focus on a new area, networking. It promoted Ethernet networks for offices. Networks were indeed a key trend of the later 1980s, and industry analysts began to proclaim DEC's recovery.

However, their optimism proved premature. By 1992, DEC was again heavily in debt, and Olsen retired in October of that year. The company was eventually bought by the PC maker Compaq.

Despite the difficulties and controversy over his management style, Olsen has been hailed by *Fortune* magazine as "America's most successful entrepreneur." DEC revolutionized the industry with its minicomputer technology, and helped create an alternative to the IBM corporate culture. Olsen has received many honors, including the Founders Award from the National Academy of Engineering (1982), the Institute of Electrical and Electronics Engineers Founders Medal (1993) and the National Medal of Technology (also 1993).

Further Reading

Petre, Peter. "America's Most Successful Entrepreneur." *Fortune*, October 27, 1986, pp. 24ff.

Rifkin, Glenn, and George Harrar. *The Ultimate Entrepreneur: The Story of Ken Olsen and Digital Equipment Corporation*. Chicago: Contemporary Books, 1988.

Omidyar, Pierre

(1967–)
French-Iranian/American
Entrepreneur, Inventor

In looking for the most important achievements of the information age, it is natural to focus on the technology—the chips and other components, the operating systems, the programming languages and software. But some of the most interesting and far-reaching developments are as much social as technical—bulletin boards, chat rooms, and e-mail, for example. They change the way people communicate and sometimes the way they buy and sell. One example of the latter is the e-commerce pioneered by JEFFREY P. BEZOS and Amazon.com. Another even more remarkable example is the on-line auction pioneered by Pierre Omidyar and his hugely successful eBay. On this website, millions of people bid for millions of items, while thousands of sellers run full- or part-time businesses selling art, antiques, collectibles, and just about every conceivable item.

Omidyar was born on June 27, 1967, in Paris. His family is of Iranian descent, and the name Omidyar means "he who has hope on his side" in Farsi. When he was six years old, Omidyar's family moved to Washington, D.C. His father, a doctor, accepted a position at Johns Hopkins University Medical Center. Omidyar's

Pierre Omidyar's eBay runs hundreds of thousands of auctions every day in which millions of users bid on virtually everything imaginable. Although e-commerce has been hyped as a remarkable business opportunity, so far only eBay has been a consistently profitable "dot-com." *(Corbis SABA)*

parents then divorced, but the family kept in close touch.

While working in his high school library, Omidyar encountered his first computer and it was love at first sight. He wrote a program to catalog books and continued to delve into programming, becoming what he later described as a "typical nerd or geek," often avoiding gym class in favor of extra computer time.

After graduating from high school, Omidyar enrolled at Tufts University to study computer science. However, after three years he became bored with classes and left school to go to Silicon Valley, the booming high-tech area just south of San Francisco. He helped develop a drawing program for the new Apple Macintosh, but after a year returned to Tufts finish his degree, which he received in 1988. He then returned to the valley to work for Claris, a subsidiary of Apple. There he developed MacDraw, a very popular application for the Macintosh.

By 1991, Omidyar had broadened his interest in computer graphics and drawing to an emerging idea called "pen computing." This involved using a special pen and tablet to allow computer users to enter text in ordinary handwriting, which would be recognized and converted to text by special software. Omidyar and three partners formed a company called Ink Development to work on pen computing technology. However, the market for such software was slow to develop. The partners changed their company name to eShop and their focus to e-commerce, the selling of goods and services on-line. However, e-commerce would not become big business until the mid-1990s when the graphical browser invented by MARC ANDREESSEN of Netscape made the Web attractive and easy to use. Meanwhile, Omidyar also did some graphics programming for the movie effects company General Magic.

Omidyar continued to think about the possibilities of e-commerce. The standard model was to take an existing business (such as a bookstore) and put it on-line. Although electronic, this kind of commerce essentially followed the traditional idea of "one to many"—one business selling goods or services to a number of customers who had no particular relationship to one another.

Omidyar, however, was interested in communities and the idea of decentralized control and access to the new technology. As he would later tell author Gregory Ericksen:

> The businesses that were trying to come onto the Internet were trying to use the Internet to sell products to people—basically, the more people, the more stuff that could be sold. Coming from a democratic, libertarian point of view, I didn't think that was such a great idea, having corporations just cram more products down peoples' throats. I wanted to give the individual the power to be a producer as well.

Omidyar learned more about Web programming and then created a site called AuctionWeb. (Popular lore has it that he did so to buy and sell collectible Pez dispensers for his fiancée, Pamela Wesley, but it was later admitted that the story had been created for publicity purposes.) AuctionWeb was based on a simple idea: Let one user put up something for bid, and have the software keep track of the bids until the end of the auction, at which point the highest bid wins.

At first, Omidyar made AuctionWeb free for both buyers and sellers, but as the site exploded in popularity his monthly Internet service bill shot up from $30 to $250. Omidyar then started charging sellers 10 cents per item listed, plus a few percent of the selling price if the item sold. Much to his surprise, Omidyar's mailbox started filling up with small checks that added up to $250 the first month, then $1,000, $2,000, $5,000 and $10,000. As recounted to Gregory

Ericksen Omidyar recalls his amazement: "I really started [what became] eBay not as a company, but as a hobby. It wasn't until I was nine months into it that I realized I was making more money from my hobby than [from] my day job."

Seeking money to expand his company, Omidyar received it from a surprising source—Microsoft, which bought his former company, eShop, with Omidyar receiving $1 million as his share. Omidyar then analyzed the auction business more closely. An auction is what economists call an "efficient market" in that it tends to create prices that satisfy both seller and buyer. However, traditional auctions have high overhead costs because the auction company has to get the goods from the seller (consigner), accurately describe and publicize them in a catalog, and pay for the auctioneer, related staff, and shipping the goods to the winners.

Omidyar's key insight was that the essence of an auction was bringing seller and potential buyers together. Thanks to the Web, it was now possible to run an auction without cataloger, auctioneer, or hotel room. The job of describing the item could be given to the seller, and digital photos or scanned images could be used to show the item to potential bidders. The buyer would pay the seller directly, and the seller would be responsible for shipping the item.

Because eBay's basic overhead costs are limited to maintaining the Web site and developing the software, the company could charge sellers about 2 percent instead of the 10–15 percent demanded by traditional auction houses. Buyers would pay no fees at all. And because the cost of selling is so low, sellers could sell items costing as little as a few dollars, while regular auction houses generally avoid lots worth less than $50–$100.

With the aid of business partner and experienced Web programmer Jeff Skoll, Omidyar revamped and expanded the site, renaming it eBay (combining the "e" in electronic with the San Francisco Bay near which they lived). Unlike the typical e-commerce business that promised investors profit sometime in the indefinite future, eBay made money from the first quarter and just kept making more.

When Omidyar and Skoll sought venture capital funding in 1996, it was not that they needed money but that they needed help in taking the next step to become a major company. Through their relationship with a venture capital firm, Benchmark, they gained not only $5 million for expansion but the services of Meg Whitman, an experienced executive who had compiled an impressive track record with firms such as FTD (the flower delivery service), the toy company Hasbro, Procter & Gamble, and Disney. Although she was not initially impressed by eBay, Whitman was eventually persuaded to become the company's vice president in charge of marketing. eBay's growth continued: By the end of 1997 about 150,000 auctions were being held each day.

In 1998 they decided to take the company public. The timing was not good, since the Dow Jones Industrial Average dropped several hundred points while they were preparing their initial public offering. But in the "road show" in which Omidyar and Whitman met with prospective investors, they were quietly effective. When eBay stock hit the market it began to soar. As writer Randall Stoss reported: "Down in San Jose, eBay employees abandoned their cubicles and formed a giant conga line, a snake of conjoined, singing delirious adults that wound through an ordinary-looking office in an ordinary smallish office building in an ordinary-looking business park." By the time the trading day ended, Omidyar's stock was worth $750 million and Whitman and the other key players had also done very well.

Despite a worrisome system crash in June, 1999 was another year of impressive growth for eBay. The company kept adding new auction categories and special programs to reward high volume "power sellers." However, the company had its critics. One possible weakness in the eBay model was that it relied heavily on trust by the seller and especially the buyer. What if a buyer won an item only to receive something that was not as described or worse, never received anything at all? But while this happened in a small number of cases, Omidyar through his attention to building communities for commerce had devised an interesting mechanism called feedback. Both sellers and buyers were encouraged to post brief evaluations of each transaction, categorized as positive, neutral, or negative. A significant number of negative feedbacks served as a warning signal, so both sellers and buyers had an incentive to fulfill their part of the bargain. As Omidyar explained to Ericksen, "We encourage people to give feedback to one another. People are concerned about their own reputations, and they are very easily able to evaluate other people's reputations. So it turns out that people kind of behave more like real people and less like strangers." The system was not perfect, but the continued patronage of several million users suggested that it worked. (An escrow system was also made available for more expensive items.)

As the 21st century dawned, Omidyar became less personally involved with eBay. In 1998, he stepped down as chief executive officer, and, the post passed to Whitman. Omidyar and his wife, Pam, turned their attention toward running a foundation for helping community education programs, particularly with regard to computer literacy. The couple also moved back to Omidyar's birthplace, Paris, to live in a modest home.

Further Reading

Cohen, Adam. "Coffee with Pierre: Creating a Web Community Made Him Singularly Rich." *Time*, December 27, 1999, pp. 78ff.

———. *The Perfect Store: Inside eBay*. New York: Little, Brown & Company, 2002.

Ericksen, Gregory K. *Net Entrepreneurs Only: 10 Entrepreneurs Tell the Stories of Their Success*. New York: Wiley, 2000.

Sachs, Adam. "The Millionaire No One Knows." *Gentlemen's Quarterly*, May 2000, p. 235.

P

Packard, David

(1912–1996)
American
Entrepreneur, Engineer

The technological powerhouse known today as Silicon Valley can be said to have begun in a Palo Alto garage where two young engineers, David Packard and WALTER REDLINGTON HEWLETT, started Hewlett-Packard (HP) with a treasury of $538. In 2000, the company and its affiliates would have $60 billion in sales. In the intervening decades, Packard created a new form of corporate management for a new industry.

Packard was born on September 7, 1912, in Pueblo, Colorado. His father was an attorney; his mother, a teacher. Packard's childhood was well balanced, including the many outdoor activities available in the small rural community, as well as plenty of books. His favorites were those about science and electrical technology. He built his own radio while still in elementary school. However, the tall, strapping Packard also excelled in football, basketball, and track in high school.

After high school, Packard enrolled in Stanford University to study electrical engineering. At Stanford, he was president of his fraternity, Alpha Delta Phi and a star on the track team. He earned his bachelor's degree with high honors in 1934.

After taking some graduate courses at the University of Colorado, Packard worked at General Electric in Schenectady, New York. He started out on the traditional electrical engineering side, where a supervisor replied to his interest in electronics by informing Packard that the field would never amount to anything. Undeterred, Packard managed to get assigned to the vacuum tube electronics department.

Packard then decided to resume his graduate studies, returning to Stanford for his master's degree, which he earned in 1939. Meanwhile, however, he had begun to talk with a college friend, another electrical engineer named William R. Hewlett, about the possibility of their going into business together. While another famous pair, STEVE JOBS and STEVE WOZNIAK, would make a similar decision four decades later, they would do so in the relatively healthy economy of the mid-1970s. Packard and Hewlett, however, were still in the throes of the Great Depression—by the late 1930s the economy had improved somewhat, but was still very uncertain.

In 1939, Packard and Hewlett set up shop in Packard's garage, with that $538 in capital and a smattering of secondhand equipment. Hewlett won a coin toss, so the firm was dubbed Hewlett-Packard. Packard seemed to gravitate toward the business side, leaving most of the technical work to Hewlett. They tinkered with a variety of inventions, including an electronic harmonica tuner, a weight reduction machine, and a bowling alley foul-line indicator.

Their first successful product, however, was an audio oscillator that Hewlett had written about for his college thesis. This device offered a superior and much less expensive way to measure and generate signals needed for broadcasting, the recording industry, and the defense industry that was beginning to tool for war.

One of their first sales of the audio oscillator was to a new movie production studio run by Walt Disney, who bought eight of the devices at $71.50 each for use in the production of the sound track of the groundbreaking animated film *Fantasia* (1940). All in all, Hewlett-Packard made $1,653 in profit their first year. They then began a practice that the company would follow for three decades: They promptly invested their profit into expanding their production facilities near Stanford, having outgrown the garage. (The cordial relationship between the company and Stanford, which had owned the land, was a prototype for the close cooperation between major universities and industry that would grow in the 1950s and 1960s.)

During the 1950s, HP became a major producer of electronic test instruments and components, including parts for the growing computer industry. In the 1960s, they increasingly innovated, rather than just supplying the industry. By the 1970s, the company had become well-known for pioneering the electronic calculator (first desktop, then hand-held), which soon banished the slide rule from engineering offices.

But David Packard's greatest contributions to the industry would be managerial, not technical. He was one of the first to establish the "open" style that is characteristic of high-tech companies today. Workers had flexible cubicles and an open floor plan—not even Packard and Hewlett had private offices. Product development was not directed from above, but carried out by small groups that included their own engineers, research and development, and marketing components. In effect, the increasingly large company was run as a group of smaller companies. Employees were encouraged to make suggestions and even drop in on a project to work awhile if they felt they could make a contribution.

Another aspect of what became known as the "HP way" was regard for employees' welfare. In an economic downturn, instead of laying off some employees, everyone's working hours and thus salaries were cut, even those of Packard and Hewlett. The level of employee benefits also set an industry standard. Meanwhile, Packard roamed freely through all the groups and projects—so much so that his staff awarded him an honorary "degree" of M.B.W.A., or "Master by Wandering Around."

Packard's high profile as an executive made him an increasingly important player in American business. He served on the boards of such companies as National Airlines, General Dynamics Corporation, and United States Steel, and financial firms such as Crocker-Citizens National Bank and Equitable Life Insurance of America.

Business led to politics and Packard, a Republican, was increasingly courted by party officials and officeholders. When Richard Nixon became president in 1968, he appointed Packard as deputy secretary of defense under Defense Secretary Melvin Laird. Nixon supporters touted Packard's proven business success as bringing a necessary discipline to the Pentagon's $80 billion budget. However, critics pointed out that HP was one of the nation's biggest defense contractors, to the tune of about $100 million a year. They suggested that this would create a conflict of interest. Packard responded by agreeing to put his HP stock in a charitable trust that would benefit from the proceeds and then return the shares to him after he had left office.

Frustrated by Washington politics, Packard left the post after only two years. In his autobiography *The HP Way,* he wrote that working with the Washington bureaucracy was "like pushing one end of a 40-foot rope and trying to get the other end to do what you want." In 1972 Packard returned to HP as its chairman and chief executive officer, although he and Hewlett retired from active involvement in 1977.

During the 1980s, HP made a name for itself in the area for which it is best known today—the computer printer. The company's ink jet and laser printers continue to be very popular today. Packard and Hewlett made a brief reappearance at HP in 1990 when the company ran into financial and organizational trouble. After reorganizing the company without layoffs, Packard retired for good in 1993. He died on March 26, 1996. In 2000, the company and its spinoff, Agilent, had $60 billion in sales.

The significance of Packard and Hewlett's achievement can be found on a placard on that original garage door, naming it as "the birthplace of Silicon Valley." By creating a culture that fostered creativity and innovation, Packard did much to shape the astonishing growth of the electronics and computer industry in the last decades of the 20th century.

Packard received numerous awards, including the Gandhi Humanitarian Award and Presidential Medal of Freedom, both awarded in 1988. Packard also created an ongoing gift to the community in the form of the David and Lucille Packard Foundation, which upon Packard's death was endowed with his entire fortune of $6.6 billion. The foundation has a particular interest in helping further the education of minority children. Separate bequests also went to Stanford and to historically black colleges.

Further Reading

Allen, Frederick E. "Present at the Creation." *American Heritage*, vol. 52, May–June 2001, p. 21.

Moritz, Charles, ed. *Current Biography Yearbook 1969*, pp. 318–321.

Packard, David. *The HP Way: How Bill Hewlett and I Built Our Company*. New York: HarperBusiness, 1996.

Papert, Seymour

(1928–)
South African/American
Computer Scientist

Lists, pattern-matching, control structures, recursion . . . back in the 1960s when the discipline of computer science was getting off the ground, such esoteric topics were viewed as being only suitable for professionals or at best, college students. But a decade or so later, one could find elementary school students exploring these and other concepts by typing at a keyboard and moving a "turtle" across the floor or the screen. They were using LOGO, a special computer teaching language invented by Seymour Papert, an artificial intelligence pioneer and innovative educator.

Papert was born in Pretoria, South Africa, on March 1, 1928. He attended the University of Witwatersrand, earning his bachelor's degree in mathematics in 1949 and Ph.D. in 1952. As a student he became active in the movement against the racial apartheid system, which had become official South African policy in 1948. His unwillingness to accept the established order and his willingness to be an outspoken activist would serve him well later when he took on the challenge of educational reform.

Papert went to Cambridge University in England and earned another Ph.D. in 1952, then did mathematics research from 1954 to 1958. During this period artificial intelligence, or AI, was taking shape as researchers began to explore the possibilities for using increasingly powerful computers to create or at least simulate intelligent behavior. In particular, Papert worked closely with another AI pioneer, MARVIN MINSKY, in studying neural networks and perceptrons. These devices made electronic connections much like those between the neurons in the human brain. By starting with random connections and reinforcing appropriate ones, a computer could actually learn a task (such as solving a maze) without being programmed with instructions.

Papert's and Minsky's research acknowledged the value of this achievement, but in their 1969 book *Perceptrons* they also suggested that this approach had limitations, and that researchers needed to focus not just on the workings of brain connections but upon how information is actually perceived and organized.

This focus on cognitive psychology came together with Papert's growing interest in the process by which human beings assimilated mathematical and other concepts. From 1958 to 1963, he worked with Jean Piaget, a Swiss psychologist and educator. Piaget had developed a theory of learning that was quite different from that held by most educators. Traditional educational theory tended to view children as being incomplete adults who needed to be "filled up" with information.

Seymour Papert created LOGO, a programming language that makes sophisticated computer science concepts accessible even to young children. He believes that the key to education is "constructivism," or learning by doing, not the passive reception of information. *(Photo courtesy of Seymour Papert)*

Piaget, however, observed that children did not think like defective adults. Rather, children at each stage of development had characteristic forms of reasoning that made perfect sense in terms of the tasks at hand. Piaget believed that children best developed their reasoning skills by being allowed to exercise them freely and learn from their mistakes, thus progressing naturally. (This idea became known as constructivism.)

In the early 1960s, Papert went to the Massachusetts Institute of Technology (MIT), where he cofounded the MIT Artificial Intelligence Laboratory with Minsky in 1965. He also began working with children and developing computer systems better suited for allowing them to explore mathematical ideas. "Somewhere around that time," Papert recalled to *MIT News*, "I got the idea of turning technology over to children to see what would happen as they explored it." The tool that he created to enable this exploration was the LOGO computer language.

First implemented in 1967, LOGO was based on Lisp, a language developed by JOHN MCCARTHY for AI research. LOGO has many of the same powerful capabilities for working with lists and patterns, but Papert simplified the parenthesis-laden syntax and added other features to make the language more "friendly."

LOGO provided a visual, graphical environment at a time when most programming resulted in long, hard-to-read printouts. At the center of the LOGO environment is the "turtle," which can be either a screen cursor or an actual little robot that can move around on the floor, tracing patterns on paper. Young students can give the turtle simple instructions, such as FORWARD 50 or RIGHT 100 and draw everything from squares to complicated spirals. As students continued to work with the system, they could build more complicated programs by writing and combining simple procedures. As they work, the students are exploring and grasping key ideas such as repetition and recursion (the ability of a program to call itself repeatedly).

As computer use became increasingly prevalent in daily life in the 1980s, a debate rose over the concept of "computer literacy." What would the next generation need to learn about computing to be sufficiently prepared for the coming century? Some critics, such as CLIFFORD STOLL, proclaimed that computers were overrated and that what was needed was a rededication to quality interactions between teachers and students. Many pragmatic educators, on the other hand, decided that students needed to be prepared to use software such as word processors and spreadsheets (and in the later 1990s, the World Wide Web), but they did not really need to *understand* computers.

In creating LOGO, Papert believed that he had demonstrated that "ordinary" students could indeed understand the principles of computer science and explore the wider vistas of mathematics. But when he saw how schools were mainly using computers for rote learning, he began to speak out more about problems with the education system. Building on Piaget's work, Papert called for a different approach as described in *MIT News:*

> When we say we educate children, it sounds like something we do to them. That's not the way it happens. We don't educate them. We create contexts in which they will learn. The goal of our research is to find ways of helping even the youngest children take charge of their own learning.

Papert often makes a distinction between "instructivism," or the imparting of information to students, and "constructivism," or a student learning by doing. Using LOGO, a child learns by trying things out, observing the results, and making modifications.

Papert also helped with the development of LEGO LOGO, a product that uses the popular building toy to create robots that can then be programmed using the LOGO language. LOGO itself has also continued to evolve, including versions that allow for parallel programming (the creation of many objects that can operate simultaneously) and can be used to create artificial life simulations.

Papert retired from MIT in 1998, but remains very active, as can be seen from the many websites that describe his work. Papert lives in Blue-Hill, Maine, and teaches at the University of Maine. He has also established the Learning Barn, a laboratory for exploring innovative ideas in education. Papert has worked on ballot initiatives to have states provide computers for all their students, as well as working with teenagers in juvenile detention facilities. Today, educational centers using LOGO and other ideas from Papert can be found around the world.

Papert has received numerous awards including a Guggenheim fellowship (1980), Marconi International fellowship (1981), the Software Publishers Association Lifetime Achievement Award (1994), and the *Computerworld* Smithsonian Award (1997).

Further Reading

Abelson, Harold, and Andrea DiSessa. *Turtle Geometry: The Computer as a Medium for Exploring Mathematics*. Cambridge, Mass.: MIT Press, 1981.

"The Connected Family Website." Available on-line. URL: http://www.connectedfamily.com. Downloaded on November 3, 2002.

Harvey, Brian. *Computer Science Logo Style*. 3 vols. 2nd ed. Cambridge, Mass.: MIT Press, 1997.

Papert, Seymour. *The Children's Machine: Rethinking School in the Age of the Computer*. New York: Basic Books, 1993.

———. *The Connected Family: Bridging the Digital Generation Gap*. Marietta, Ga.: Longstreet Press, 1996.

———. *Mindstorms: Children, Computers and Powerful Ideas*. 2nd ed. New York: Basic Books, 1993.

"Papert Describes his Philosophy of Education." *MIT News*. Available on-line. URL: http://web.mit.edu/newsoffice/++/1990/may16/23181.html. Posted on May 16, 1990.

"Professor Seymour Papert." Available on-line. URL: http://www.papert.org. Downloaded on November 3, 2002.

Perlis, Alan J.

(1922–1990)
American
Computer Scientist, Mathematician

The first computers were generally built by electrical engineers and programmed by mathematicians. As the machines became larger and more capable and more extensive programs were developed, the pioneer programmers and computer designers began to feel a need to develop a systematic body of theory and knowledge. They gradually realized that if computing was going to play a central part in so many human activities, it would have to be understood and taught as a science even while remaining more of an art for many of its most skilled practitioners. Perhaps no one played a greater part than Alan Perlis in the establishment of computer science as a discipline.

Perlis was born on April 1, 1922, in Pittsburgh, Pennsylvania. He attended the Carnegie Institute of Technology (now Carnegie Mellon University) and received a B.S. degree in chemistry in 1943. He then switched his focus to mathematics, taking graduate courses at both the California Institute of Technology (Caltech) and the Massachusetts Institute of Technology (MIT); he received an M.S. degree (1949) and a Ph.D. (1950) in mathematics from the latter institution.

The power of the new digital computing technology was drawing mathematicians and scientists into new research tracks, and Perlis was no exception. Following his graduation he worked as a research mathematician at the U.S. Army's Aberdeen Proving Grounds, and in 1952 he briefly joined Project Whirlwind, an ambitious MIT program to develop a computer fast and powerful enough to provide real-time simulation of air defense operations.

Perlis then went to Purdue, where he was an assistant professor of mathematics from 1952 to 1956. The director of the Statistical Laboratory had decided that the engineering department needed a computer. In the early 1950s, however, a computer was not an "off the shelf" item. Most were custom built or early prototypes of later commercial machines. Perlis recalled later in a talk on computing in the Fifties reprinted in J. A. N. Lee's *Computer Pioneers* that early computers could be strange indeed:

> At Purdue we searched for one year for a computer—the university having given us $5,000 and a Purdue automobile to travel through the eastern and middle-western part of the country. We saw some very strange computers. One I will never forget was being developed by Facsimile in New York City in a loft, in the Lower East Side, on the third floor that one reached by going up on a rickety freight elevator. The machine was called the Circle Computer. It had one important attribute: there was a socket on the top of the computer into which one could plug an electric coffee pot. The Circle Computer showed me that machines really do have a personality . . . the Circle Computer would only run when one of its designers was in the same room.

Eventually they found something called a Datatron 205, which cost $125,000. Perlis and his colleagues then had to design programming courses around it, which was difficult because the very concept of a programming language was still quite tenuous. (GRACE MURRAY HOPPER, for example, had just started to work with a compiler that could handle symbolic instruction codes.) Perlis began working on a compiler of his own, and he continued this work when he moved to Carnegie in 1956.

Carnegie obtained an IBM 650, a machine that marked IBM's entry into computing, and Perlis finished his compiler, which he called IT, for Interpretive Translator. Perlis said that he "will never forget the miracle of seeing that first program compiled, when an eight-line Gaussian elimination [mathematical procedure] program was expanded into 120 lines of assembly language."

In 1957, Perlis designed and taught what might well be the first true course in programming. Earlier courses had started either with mathematics and then a machine implementation, or with the study of the circuitry and operation of the machine itself. This new course talked about how programs were organized and structured, and then gave examples in both the higher-level compiler language and assembly language.

Perlis's courses were very popular with the engineers who wanted to learn to program. Of course there was only one computer, and thus, as Perlis recounts in his historical talk

> the way one used the computer in those days was to line up on Monday morning to sign up for time in blocks half an hour to an hour or even two hours after midnight. A cot was always kept next to the computer, and in the air-conditioning unit of the computer there was a small chamber in which one kept beer and Coca-Cola.

In the late 1950s, pioneer computer scientists began to work together to create a new higher-level structured programming language, one that would eventually be called Algol. (Algol in turn would inspire the Pascal and C languages in the 1970s.) Perlis was appointed chairman of the committee that created Algol. Meanwhile, Perlis served as chairman of the Carnegie mathematics department and, from 1960 to 1964, director of the computation center.

In 1962, Perlis became cochairman of an interdisciplinary program in systems and communications sciences. This led to support from the Defense Department's Advanced Research Projects Agency (ARPA), which in 1965 funded a graduate department of computer science. Perlis served as its first chairman, until 1971.

In 1971, Perlis moved to Yale, where he built that university's computer science department and served as its chairman in the latter half of the decade. He also taught at the California Institute of Technology. During this time he wrote two authoritative textbooks, *Introduction to Computer Science* (1972) and *A View of Programming Languages* (1970), coauthored with B. A. Galler.

Moving into the 1980s, Perlis served on a number of national committees and advisory boards in the National Science Foundation, the Pennsylvania Council on Science and Technology, and the National Research Council Assembly. He was always an advocate for the importance of computer science and an explainer of its relationship to other disciplines and interests.

Perlis also kept his eye on the future. When hearing talks about how chip technology was progressing, he commented, in his historical talk: "Isn't it wonderful that we can now pack 100 ENIACs in one square centimeter?" He predicted that computers would soon become as ubiquitous (and thus effectively invisible) as electric motors. However, he also left his profession with a challenge: "Computers really are an explanation of what *we* are. Our goal, of course, ought to be to use this tool of our age . . . to help us understand why we are the way we are, today and forever." Perlis died on February 7, 1990.

As a pioneer, Perlis achieved a number of "firsts"—he was the first president of the Association of Computing Machinery (ACM), the first winner of the ACM Turing Award, and the first head of the Computer Science Department at Carnegie Mellon University. He was also elected to the American Academy of Arts and Sciences (1973) and the National Academy of Engineering (1976).

Further Reading

"Alan J. Perlis, 1922–1990: A Founding Father of Computer Science as a Separate Discipline." *Communications of the ACM*, 33 (May 1990): 604ff.

Lee, J. A. N. *Computer Pioneers*. Los Alamitos, Calif.: IEEE Computer Society Press, 1995.

Perlis, Alan. "Epigrams in Programming." Available on-line. URL: http://www.cs.yale.edu/homes/perlis-alan/quotes.html. Downloaded on November 3, 2002.

Postel, Jonathan B.
(1943–1998)
American
Computer Scientist

When a Web user types a URL (uniform resource locator) into a browser, what ensures that the request will get to the website serving that page? Just as a modern postal system could not work without each building having a unique address, the Internet could not work without a system to assign names and numbers to sites so that software can properly direct Web requests, e-mail, and other messages. The aptly named Jonathan Postel was largely responsible for setting up the Internet address system, and he served as its manager and steward for nearly 30 years.

Postel was born on August 6, 1943, in Altadena, California. After attending Van Nuys High School and a local community college, he enrolled in the University of California, Los Angeles (UCLA), where he earned a bachelor's degree in engineering in 1966 and a master's degree in 1968. He went on to receive his Ph.D. from UCLA in 1974.

In the late 1960s, the Defense Department's Advanced Research Projects Agency (ARPA) funded the development of a new kind of decentralized network that could be used to connect many different kinds of computers. At UCLA, Postel and his former high school classmate VINTON CERF worked on the software that in October 1969 first connected a UCLA computer with three other facilities to create the beginnings of ARPANET, ancestor of the Internet.

At UCLA's Network Management Center, Postel helped run tests to monitor the performance of the network. He also coauthored or edited many of the documents called RFCs (request for comments) that announced proposed specifications for everything from TCP/IP (the basic protocol for routing packets on the Internet) to SMTP (simple mail transfer protocol, the basis for virtually all e-mail used today). Resisting all pressures from the government agencies that funded the work, he insisted that all RFCs be unclassified and publicly available.

In the early days, the numbers needed to tell the software where to find each computer on the network were assigned more or less arbitrarily—for a time, Postel kept track of them on a piece of paper. But as the number of machines on the network grew to tens and then hundreds, such an ad hoc system became impracticable. After all, users could not be expected to remember a number such as 63.82.155.27 to log onto a computer or send someone e-mail.

Postel and his colleagues devised the domain name system (DNS) to translate between the cryptic numbers used by the network and names such as ucla.edu or well.com. By using standard suffixes such as .com for commercial (business) or .edu for educational institutions, locations on the Internet could be better identified as well as separated into manageable groups or domains. This division made it possible to allocate blocks of numbers to allow for a smoother expansion of the network as new users came on-line. Postel was also responsible for a small but important feature—the "dot," or period, used to separate parts of a domain name.

Postel served as director of the Internet Assigned Numbers Authority (IANA), which managed the domain system until its role was taken over in 1998 by the nonprofit Internet Corporation for Assigned Names and Numbers (ICANN). (Postel was also a founding member of this organization.)

Postel's involvement with the Internet went well beyond the narrowly technical. He was always interested in the overall architecture of the network and in its growing impact on society as a whole. Postel was the first individual member of the Internet Society, and founded the Internet Architecture Board.

Postel joined the UCLA faculty in 1977 and remained there for the rest of his life. His achievements may not have been as exciting as the invention of a new kind of computer or the meteoric rise of an entrepreneur such as Bill Gates. After his death, colleague Stephen Crocker was quoted in the *New York Times* obituary as saying

> Jon's greatest achievement was operating IANA. It was a very central part of the infrastructure. As mundane and as simple as it seemed, he set policies that made it very easy for the network to grow. He minimized bureaucratic delay and at the same time kept silly and nonsensical things to a minimum.

Rather more poetically, Postel's long-time colleague Vinton Cerf said that "Jon has been our North Star for decades. He was the Internet's Boswell [chronicler] and its technical conscience." In the last years of his life, the bearded Postel, often described as looking like an Old Testament prophet, quietly advocated for the expansion of the Internet and the seamless addition of new services such as shared applications and teleconferencing.

Postel died on October 16, 1998, following emergency heart surgery. Postel's honors include the SIGCOMM award (1998) and the Silver Medal of the International Telecommunications Union (1998). The Postel Center for Experimental Networking (PCEN) has been established to carry on Postel's work by supporting visiting scholars and research. The Internet Society has also established the Jonathan B. Postel Service Award.

Further Reading

Cohen, Danny. "Remembering Jonathan B. Postel: Working with Jon." Available on-line. URL: http://www.postel.org/remembrances/cohen-story.html. Posted on November 2, 1998.

Hafner, Katie. "Jonathan Postel Is Dead at 55; Helped Start and Run Internet." *New York Times*, October 18, 1998, p. 139.

Internet Assigned Numbers Authority (IANA) Available on-line. URL: http://www.iana.org. Updated on August 25, 2002.

Postel Center for Experimental Networking (PCEN). Available on-line. URL: http://www.postel.org/jonpostel.html. Updated on January 7, 2002.

Rader, Ross William. "One History of DNS." Available on-line. URL: http://www.byte.org/one-history-of-dns.pdf. Posted on April 25, 2001.

R

Rabin, Michael O.

(1931–)
German/American
Mathematician, Computer Scientist

To most people, the chief virtues of the computer are speed and accuracy. If a program is working properly, it should give the right answer every time. Indeed, most of the functions of business and government assume and depend on the reliability of their data processing systems. However, a mathematician and computer scientist named Michael Rabin has shown how previously insurmountable problems can be tackled if one allows for randomness and is willing to accept results that are highly probable but not absolutely guaranteed.

Rabin was born to a Jewish family in Breslau, Germany (today Wroclaw, Poland) in 1931. His father was a rabbi and professor at Breslau's Theological Seminary, who had come from Russia. Rabin's mother was also a scholar, with a Ph.D. in literature, and was a writer of children's stories.

Rabin's father remembered the bitter history of anti-Jewish pogroms in his native Russia. When the Nazis took power in Germany in 1933, he went to the board members of the Theological Seminary and urged them to move the institution to Jerusalem. However, they, like many German Jews, considered themselves to be patriotic Germans and believed that their nation needed them to provide a liberal, humanist voice. They refused to leave, but in 1935 Rabin and his family moved to Haifa in what was then called Palestine.

In elementary school in Haifa, young Rabin became fascinated with microbiology, having read the classic book *Microbe Hunters* by Paul de Kruif. However, when he was about 12, Rabin encountered two older students sitting in the corridor trying to solve geometry problems. They had found a problem they could not solve, and challenged the younger boy to tackle it. Rabin solved the problem and as he later told an interviewer, "the fact that by pure thought you can establish a truth about lines and circles by the process of proof, struck me and captivated me completely."

Rabin then came into conflict with his father, who wanted him to attend religious school, while he wanted to pursue science at the secular Reali School. He won the argument, and after graduating from the Reali School entered Hebrew University in Jerusalem, earning a master's degree in mathematics in 1953.

At the time, people in Israel were just starting to read about digital computing. Rabin became interested in some of the field's deepest and most abstract questions. He read a paper by ALAN TURING about computability, or the question of what kinds of mathematical problems were theoretically solvable by a computer. He also learned about the Turing Machine, a very simple computer that can carry out only two operations (making a mark or skipping a space), which Turing had shown was fundamentally equivalent to any future digital computer. The idea that mathematics could not only solve problems but determine whether problems *could* be solved seemed amazingly powerful.

Israel did not yet have a computer center, so Rabin decided to move to the United States. He studied mathematics briefly at the University of Pennsylvania but then went to Princeton to earn a Ph.D. in logic, which he received in 1956. There he studied under ALONZO CHURCH, who had created a theory of computability that was different from, but functionally equivalent to, that of Turing. His doctoral thesis discussed the computability of mathematical structures called groups. By representing groups using a structure similar to that of programs, he showed that for many groups it was not possible to demonstrate certain properties, such as whether the group is commutative. (Ordinary numbers are commutative for multiplication because, for example, 3×5 is equal to 5×3.)

In 1957, Rabin and another young mathematician, Dana Scott, were given summer jobs by IBM Research and allowed to pick any mathematical topic of interest. What they chose to develop was a fundamentally different approach to computing, which they called nondeterminism. Computers normally followed defined steps such that each time the same input was given, the result should be the same. This can be modeled by what is called a "finite state machine." A very simple example is a traffic light: If the light is green, the next state will always be yellow, the next state after that red, and so on.

Rabin and Scott created a finite state machine in which the next state after a given input was not forever fixed, but instead might be one of several possibilities. Any path that leads to the defined goal (or end state) is considered acceptable. They then used set theory to show that any given indeterminate state machine could be converted to an equivalent determinate one.

Published in 1959, their paper proved to have fundamental applications in many areas of computer science, where programmers could chose between deterministic and nondeterministic algorithms depending on which seemed to better represent the problem or application.

Meanwhile, Rabin continued his research for IBM. Artificial intelligence pioneer JOHN MCCARTHY gave him a problem in cryptography, namely how to make it hard for an enemy to determine the cipher key from an intercepted message. Rabin realized that there were mathematical functions that are easy to do "forward" but very hard to do backward. Using such a function would make a code hard to crack. For example, as JOHN VON NEUMANN had pointed out, a 100-digit number squared results in a 200-digit number. If the middle 100 digits are then extracted, the result is easy to calculate from the original number (with a computer, anyway), but would take virtually forever for someone else to determine, since that person would have to square all the possible original numbers, extract the middle 100 digits, and compare them to the key.

This problem and solution interested Rabin in another issue, computational difficulty or complexity. That is, once a problem has been shown to be theoretically solvable, how is the minimum amount of work (mathematical operations) measured that it inherently requires? Working with Michael Fischer of Yale University, Rabin developed tools that could be used to determine the complexity of a given algorithm, and thus its practicality given the available computing resources.

Rabin's work on complexity or difficulty also had clear applications for cryptography, since it provided tools for estimating how "strong" (hard to crack) a particular code was. One of the most common ways to create cryptosystems (code schemes) in the mid to late 20th century involved multiplying two very large prime numbers. This is because factoring such a number (determining what numbers multiply together to get that number) requires a huge amount of computation—hopefully more than the would-be code breaker can muster.

The problem of getting those large primes is also difficult, though. For the past four centuries or so, mathematicians have tried to come up with a way to quickly test whether some large number is prime, rather than having to painstakingly examine large numbers of possible factors. One test called the Reimann hypothesis seemed to work well, but mathematicians had failed to prove that it would always work.

In 1975, however, Rabin took a different approach to the problem. Building on previous work by Gary Miller, he developed a "primality" test where randomly chosen numbers between 1 and the number to be tested were tested for a certain relationship. If any of the random numbers showed that relationship, the number in question was not prime. But Rabin's insight was that if 150 of these random numbers were tested and the relationship was not found, the number was almost certainly a prime. (The probability that it is not prime is less than one in 1 trillion trillion trillion trillion trillion trillion trillion!) Building his procedure into a program, Rabin's colleague Vaughan Pratt was able to quickly generate huge prime numbers such as $2^{400} - 593$.

Besides finding an easy way to generate primes for the public key cryptography, first developed in the mid-1970s, Rabin's method had shown the power of randomization and allowing for a small possibility of error. Randomization is used today in computer networks. If two machines try to send a message at the same time, the result is an electronic "collision." If each machine is programmed simply to wait some fixed time *n* after a collision, they will promptly collide again. However, if the program calls for each machine to generate a random number and wait that long, chances are a new collision will be avoided.

In 2001, Rabin came up with a new way to apply randomization to create what he believes would be an uncrackable code. The inherent weakness in most codes is that the longer a given key is used, the more likely it is that "eavesdroppers' can crack the code by applying various statistical techniques to the encrypted text. Traditionally, therefore, intelligence agencies used a "one-time pad" for their most sensitive communications. As the name implies, it contains a random key that is used only for one message, making it virtually impossible to crack. The disadvantage is that perhaps thousands of different one-time pads must be generated, stockpiled, and distributed for use. Besides being cumbersome and expensive, such a system is vulnerable to surreptitious diversion or copying of the pads.

Rabin has found an ingenious way to create an electronic version of the one-time pad. Using a set of rules, each of the two users selects random digits from a variety of sources, such as signals from TV satellites, fluctuating values such as stock averages, even the contents of websites. This creates a unique key known only to the two users, who can then use it to encode their communications. Because a large number of random bits of information have been selected from a virtually infinite universe of ephemeral data, the only way someone would break the code would be to know when the key was being made and record all those data streams—a virtually impossible task even given future ultrapowerful supercomputers. If Rabin's scheme achieves a practical implementation, it would, of course, represent a mortal threat to codebreakers such as the National Security Agency.

In addition to his work on nondeterministic algorithms, complexity, and randomization, Rabin has also worked on parallel computing and the MCB software environment for coordination of computers in "clusters." He has developed the IDA (Information Dispersal Algorithm) as a way to prevent data loss in case of system failure, and has also applied his randomization techniques in developing the protocols that maintain communication between the machines. Rabin and his doctoral student Doug Tygar also developed ITOSS (Integrated Toolkit for Operating System Security), a facility that could be added to current or future operating systems to make them harder to hack.

Today Rabin divides his time between his posts as the Albert Einstein Professor of Mathematics and Computer Science at Hebrew University, Jerusalem, and as Thomas J. Watson Sr. Professor of Computer Science at Harvard University. In 1976, Rabin and Dana Scott shared the Association for Computing Machinery Turing Award for their work on nondeterministic state machines.

Further Reading

Brooks, Michael. "Your Secret's Safe: The Ephemeral Signals That Flood the Airwaves Are the Key to Creating an Uncrackable Code." *New Scientist*, 173, no. 2324 (January 5, 2002): 20ff.

Kolata, Gina. "Order Out of Chaos in Computers; Computer Scientists are Controlling Decision-making with Probabilistic Methods, Which Work in Cases Where Determinism Does Not." *Science*, 223 (March 2, 1984): 917ff.

"Dr. Michael O. Rabin." Available on-line. URL: http://www.sis.pitt.edu/~mbsclass/hall_of_fame/rabin.htm. Downloaded on November 3, 2002.

Shasha, Dennis, and Cathy Lazere. *Out of Their Minds: The Lives and Discoveries of 15 Great Computer Scientists*. New York: Copernicus, 1995.

Raymond, Eric S.

(ca. 1958–)
American
Programmer, Writer

The first generation of computer pioneers came from traditional backgrounds in mathematics, engineering, or science. They conceived of ideas, invented machines, and sometimes founded corporations. By the time a new generation was born in the 1950s and 1960s, both computers and the discipline of computer science were available as a career. However, just as the children of immigrants often strike out on their own and create a new or hybrid culture, many of the most interesting and innovative members of the new generation valued the intense experience of the machine over formal credentials and created a new way of work and life. They proudly took the name "hackers." One of them was Eric Steven Raymond, whose mastery of operating systems and software was matched by his passion and skill as an advocate for open source software.

Raymond was born in about 1958 in Boston but has lived most of his life in Pennsylvania. He studied mathematics and philosophy at the University of Pennsylvania but did not receive a degree. He is a self-taught programmer and computer scientist with a prodigious knowledge of languages and operating systems.

Raymond's résumé begins with work at the Wharton School Computer Center, where from 1977 to 1979 he worked in development and support for software written in APL, LISP, and Pascal. He also wrote his first computer manual (for LISP). Later he would write numerous articles, books, and HOW-TO files on a wide variety of computer topics.

Raymond then worked from 1980 to 1981 at Burroughs Corporation's Federal and Special Systems Group, where he helped develop a variety of artificial intelligence (AI) software, including theorem-proving systems and specialized languages. His next excursion was into the burgeoning world of microcomputing where Raymond, as lead programmer of MicroCorp, developed business software and a program that translated specially structured BASIC statements into code that would run with IBM (Microsoft) BASIC.

In the mid-1980s, Raymond entered the UNIX world, with which he would become most closely associated. Moving between a variety of IBM and DEC machines, Raymond developed software and provided administration and support for both "flavors" (Berkeley BSD and AT&T) of UNIX.

Having acquired this breadth and depth of experience, Raymond in 1985 launched a career as an independent consultant. Besides writing a variety of software for local businesses ranging from a medical practice to a truck-dispatch service, Raymond increasingly became both a student (he often refers to himself as an anthropologist) and an explainer and advocate for "hacker culture."

It is important to note at this point that the term *hacker* as used by Raymond and by many of the most creative innovators in the computer field does not mean people who maliciously and often mindlessly run cracking scripts or insert viruses into computer systems. In his book *The Hacker Dictionary*, Raymond defines the "hacker ethic" as follows:

> 1. The belief that information-sharing is a powerful positive good, and that it is an ethical duty of hackers to share their expertise by writing open-source code and facilitating access to information and to computing resources wherever possible.
> 2. The belief that system-cracking for fun and exploration is ethically OK as long as the cracker commits no theft, vandalism, or breach of confidentiality.

The second part of the definition is, of course, somewhat controversial, and some hackers (in Raymond's sense) refrain from even benign system-cracking. Hackers generally prefer to use the term *cracker* to refer to the malicious hacker.

The Hacker's Dictionary is occasionally republished in revised form, but the freshest copy is available on-line as the Jargon File. It is an invaluable resource for researchers who want to understand a culture that has developed over decades. Like such groups as the Society for Creative Anachronism, neopagans, shooting enthusiasts, and science fiction fans (all of which include Raymond) the hacker culture has not only its special jargon but its traditions, mores, and worldview.

Raymond also made a major contribution to revising the software infrastructure for the Netnews newsgroup system. Although Netnews has been largely eclipsed by the World Wide Web, chat, and conferencing systems, it played an important part in the 1980s and 1990s as a channel for both technical information and the forming of communities of interest.

From 1993 to 1999, Raymond organized and ran a nonprofit project to provide free Internet access to the citizens of Chester County, Pennsylvania. His responsibilities ran the full gamut from technical specifications to publicity and fund-raising. Raymond also served from 1998 to 2002 as a member of the board of directors of VA Linux Systems, a major vendor of the open source operating system.

Raymond is perhaps best known to the general computing community as a tireless and articulate advocate for open source software. The open source movement, begun by RICHARD STALLMAN and his GNU Project, is sometimes mistakenly described as "free software." Actually open source is not about what software costs, but what people are allowed to do with it. According to a statement by Raymond on the Electronic Frontier Foundation website

> The core principles of open source are transparency, responsibility, and autonomy. As open source developers, we expose our source code to constant scrutiny by expert peers. We stand behind our work with frequent releases and continuing inputs of service and intelligence. And we support the rights of developers and artists to make their own choices about the design and disposition of their creative work.

In June 1998, Raymond and Bruce Perens cofounded the Open Source Initiative, a nonprofit foundation that provides education about the open source concept and certifies products as open source. Raymond's 1999 collection of essays, *The Cathedral and the Bazaar*, recounts his study and participation in the hacker culture and describes what he considers to be the key benefits of open source software. He believes that the "bazaar," or freewheeling market created by open source, produces more robust and reliable software than proprietary efforts even by companies with the resources of a Microsoft. To capture it in a slogan, Raymond says that "given enough eyeballs, all bugs are shallow." In other words, with many people communicating and collaborating, it is much easier to fix problems.

Raymond can often be found on panels at science fiction conventions and he continues to enjoy live-action role-playing, in which he, not surprisingly, often portrays a formidable wizard.

Further Reading

Dibona, Chris. *Open Sources: Voices from the Open Source Revolution*. Sebastopol, Calif.: O'Reilly, 1999.

Levy, Steven. *Hackers: Heroes of the Computer Revolution*. Updated ed. New York: Penguin, 2001.

Open Source Initiative. Available on-line. URL: http://www.opensource.org. Downloaded on December 5, 2002.

Raymond, Eric. *The Cathedral and the Bazaar: Musings on Linux and Open Source by an Accidental Revolutionary*. Revised and expanded ed. Sebastapol, Calif.: O'Reilly, 2001.

———. *The New Hacker's Dictionary*. 3rd ed. Cambridge, Mass.: MIT Press, 1996. Also available on-line. URL: http://catb.org/~esr/jargon/html. Updated in September 2002.

Rees, Mina Spiegel

(1902–1997)
American
Mathematician, Computer Scientist

Although the applicability of computers to mathematics seems obvious now, it was not obvious how computers could be made available to the many mathematicians who could benefit from them, or how they should be trained to use them. At the same time, the role of mathematics in technology and thus in society changed dramatically in the course of the 20th century. Mina Rees played an important role in that transformation.

Rees was born on August 2, 1902, in Cleveland, Ohio. When she was only two, her family moved to the Bronx, New York. Her excellent grades (particularly in math) led to her eighth-grade teacher suggesting that she apply to Hunter College High School, an advanced program sponsored by Hunter College.

Rees graduated from high school as class valedictorian, and then enrolled in Hunter College. In 1923 she graduated with an A.B. degree in mathematics with top academic honors (Phi Beta Kappa and summa cum laude); she had also served as student body president and yearbook editor.

While working on her master's degree at Columbia University, Rees taught mathematics at Hunter College High School. In 1926 she received her M.A. degree. She obtained a post as an instructor at Hunter, while continuing her studies at Columbia. However, she was quoted in her obituary notice by the American Mathematical Society (AMS) as recalling:

> When I had taken four of their six-credit graduate courses in mathematics and was beginning to think about a thesis, the word was conveyed to me—no official ever told me this but I learned—that the Columbia mathematics department was really not interested in having women candidates for Ph.D's. This was a very unpleasant shock.

Undeterred, she enrolled at the University of Chicago and earned her doctorate in 1931. She then returned to Hunter as an assistant professor, becoming an associate professor in 1940.

In 1943, during the height of World War II, Rees left Hunter to work as a technical aide and administrative assistant to the chief of the Applied Mathematics Panel of the National Defense Research Committee. There she learned to recruit and manage other mathematicians working on war-related projects, including advances in rocket propulsion and hydrofoil boats as well as advances in computers. After the war, Rees was awarded the President's Certificate of Merit and, from Great Britain, the King's Medal for Service in the Cause of Freedom.

After the war, Rees decided not to return to academia. In 1946, she moved to the Office of Naval Research. She ran its mathematical branch and then from 1949 to 1953 served as director of the Mathematical Sciences Division. Besides helping to shape the navy's technological transformation in the electronic age, she also became increasingly involved with computer-related projects such as the Whirlwind at the Massachusetts Institute of Technology, the first real-time computer simulator system. The Institute of Electrical and Electronics Engineers (IEEE) *Annals of the History of Computing* later recounted that

Mathematician Mina Rees bridged the gap between the world of mathematics and the new technology of computing. A gifted administrator, Rees oversaw important early computer projects such as Whirlwind, the first computer simulator with a graphics display and interactive input. Later, she became the first woman president of the prestigious American Association for the Advancement of Science. *(Courtesy of the Graduate Center, City University of New York)*

> applied mathematics became a respected and growing field . . . due in large part to the influence of Rees, who early on recognized that adequate funding was not being supplied by the United States' computing research and development programs that focused on using the appropriate mathematical methods in connection with automatic digital equipment. Rees helped to initiate and provide funding in those areas.

Built under the overall direction of JAY W. FORRESTER, Whirlwind was not simply a bigger, faster computer, but one that required a new architecture and the development and interfacing of new data storage methods (including magnetic tape) and input/output devices such as Teletypes. As the project continued, Rees became more skeptical and concerned that the engineers had insufficient mathematical experience to deal with the design issues. As the project became bogged down and increasingly over budget, Reese and naval officials considered trying to downsize the project. Whirlwind did become operational in the early 1950s, but only after the air force had joined in the funding.

While working military projects, Rees became involved in the new discipline of operations research, which can be described as the scientific study of organizations and processes. This field would become increasingly important for the complicated logistics requirements of the modern military.

Rees was not primarily a researcher, but she did write and review papers on mathematical subjects such as division algebra, statistics, and in particular, computer applications. She also wrote about the importance of mathematics education and asserted that "it is the duty of teachers and research mathematicians and administrators to aid young people to discover the richness and the variety of mathematics and to seek careers that will provide intellectual as well as other rewards."

In an August 1954 article for *Scientific Monthly*, Rees suggested that researchers prepare to take advantage of a new generation of computers:

> In the immediate future, we may expect smaller and faster machines with bigger appetites—ability to handle more complex problems more quickly. The devices that will make such machines possible—the transistor, the magnetic-core memory, the many types of miniaturized components—are well understood.

Although she worked largely behind the scenes, the mathematical and technical community was well aware of her vital role. In 1953, the American Mathematical Society declared in a resolution that

> Under her guidance, basic research in general, and especially in mathematics, received the most intelligent and wholehearted support. No greater wisdom and foresight could have been displayed and the whole post-war development of mathematical research in the United States owes an immeasurable debt to the pioneer work of the Office of Naval Research and to the alert, vigorous and farsighted policy conducted by Miss Rees.

That same year, Rees finally returned to academia, becoming a full professor at Hunter College and then dean of the faculty. When the City University of New York (CUNY) established its first graduate program in 1961, it hired Rees to serve as its first dean. In 1968, she became provost of the CUNY Graduate Division, and the following year she became president of the Graduate School and University Center. In 1972, the year she retired from CUNY, she became the first woman president of the prestigious American Association for the Advancement of Science (AAAS). The *New York Times* took that occasion to comment that

> the examples set by Marie Curie, Lisa Meitner, Margaret Mead, Mina Rees and many others prove that scientific creativity is not a male monopoly. The gross under-representation of women among scientists implies, therefore, a very substantial loss for American society through its failure fully to utilize all the talented women who might have gone into science.

In 1962, Rees received the first Award for Distinguished Service to Mathematics of the American Mathematical Society. She was also awarded the Public Welfare Medal of the National Academy of Sciences in 1983 and the IEEE Computer Society Pioneer Award (1989). Rees died on October 25, 1997, at the age of 95.

Further Reading

Association for Women in Mathematics. "Biographies of Women Mathematicians." Available on-line. URL: http://www.agnesscott.edu/lriddle/women/women.htm. Updated on August 27, 2002.

Cortada, James W. *Historical Dictionary of Data Processing Biographies*. Westport, Conn.: Greenwood Press, 1987.

Green, Judy, et al. "Mina Spiegel Rees (1902–1997)." *Notices of the AMS*, vol. 45, August 1998.

Available on-line. URL: http://www.ams.org/notices/199807/memorial.rees.pdf.

Lee, J. A. N. *Computer Pioneers*. Los Alamitos, Calif.: IEEE Computer Society Press, 1995.

Rees, Mina. "The Computing Program of the Office of Naval Research, 1946–53." *Annals of the History of Computing*, 4, no. 2 (1982): 102–120.

Rheingold, Howard
(1947–)
American
Writer

On his website, Howard Rheingold says that he "fell into the computer realm from the typewriter dimension, then plugged his computer into his telephone and got sucked into the net." A prolific writer, explorer of the interaction of human consciousness and technology, and chronicler of virtual communities, Rheingold has helped people from students to businesspersons to legislators understand the social significance of the Internet and communications revolution.

Rheingold was born in 1947 in Tucson, Arizona. He was educated at Reed College in Portland, Oregon, but has lived and worked for most of his life in the San Francisco Bay Area. As he recalled later in a speech, Rheingold bought his first personal computer (PC) mainly because he thought word processing would make his work as a writer easier. In 1983, he bought a modem that cost $500 and ran at the barely adequate speed of 1200 baud, or bits per second. He was soon intrigued by the "rich ecology of the thousands of PC bulletin board systems that ran off single telephone lines in people's bedrooms." Interacting with these often tiny cyberspace villages helped Rheingold develop his ideas about the nature and significance of virtual communities.

In 1985, Rheingold joined the WELL (Whole Earth 'Lectronic Link), a unique and remarkably persistent community that began as an unlikely meeting place of Deadheads (Grateful Dead fans) and computer hackers. Compared to most bulletin boards, the Well was more like the virtual equivalent of the cosmopolitan San Francisco Bay Area. Later, Rheingold recalled in a speech on *BBC Online* that

> In the fifteen years since I joined the WELL, I've contributed to dozens of such fund-raising and support activities. I've sat by the bedside of a dying, lonely woman, who would have died alone if it had not been for people she had previously known only as words on a screen. I've danced at four weddings of people who met online. I've attended four funerals, and spoke at two of them.

The sum and evaluation of these experiences can be found in Rheingold's seminal book *The Virtual Community* (1993; revised 2000), which represents both a participant's and an observer's tour through the on-line meeting places that had begun to function as communities. In addition to the WELL, Rheingold also explores MUDs (multi-user dungeons) and other elaborate online fantasy role-playing games; NetNews (also called Usenet) groups; chat rooms; and other forms of on-line interaction. (Toward the end of the 1990s, much conferencing became Web based, and a new phenomenon, the web log, or "blog," allowed people to maintain a sort of interactive diary. Rheingold's use of his website for communicating his thoughts pretty much anticipates the blog.)

Suggesting that the question of whether online communities are "real" communities may be missing the point. Rheingold chronicled the romances, feuds ("flame wars"), and growing pains that made the WELL seem much like a small town or perhaps an artist's colony that just happened to be in cyberspace. Rheingold manages the Brainstorms Community, a private webconferencing community that allows for thoughtful discussions about a variety of topics.

Around 1999, Rheingold started noticing the emergence of a different kind of virtual community—a mobile, highly flexible, and adaptive one. In his 2002 book *Smart Mobs*, Rheingold gives examples of groups of teenagers coordinating their activities by sending each other text messages on their cell phones. In Seattle, Washington, in 1999, anti–World Trade Organization protesters used mobile communications and websites to rapidly shift and "swarm" their objectives, often outflanking police who relied upon traditional communications and chains of command. Rheingold believes that the combination of mobile and network technology may be creating a social revolution as important as that triggered by the PC in the 1980s and the Internet in the 1990s.

Rheingold has written many books and articles. In 1994 he updated the *Whole Earth Catalog*, a remarkable resource book by Stewart Brand that had become a bible for the movement toward a more self-sufficient life in the 1970s. Like LEE FELSENSTEIN and TED NELSON, Rheingold saw the computer (and computer networks in particular) as a powerful tool for creating new forms of community. The original edition of his book *Tools for Thought* (1985) with its description of the potential of computer-mediated communications seems prescient today after a decade of the Web. Rheingold's *Virtual Reality* (1991) introduced the immersive technology that was being pioneered by such researchers as JARON LANIER.

Another track that Rheingold pursued in the 1980s was the exploration of consciousness and cognitive psychology. His books in this area include *Higher Creativity* (written with Willis Harman, 1984), *The Cognitive Connections* (written with Howard Levine, 1986) and *Exploring the World of Lucid Dreaming* (written with Stephen LaBerge, 1990).

Rheingold ventured into publishing in 1994, helping design and edit *HotWired*, which he soon quit "because I wanted something more like a jam session than a magazine."

In 1996, Rheingold launched Electric Minds, an innovative company that tried to offer virtual community-building services while attracting enough revenue from contract work and advertising to become self-sustaining and profitable in about three years. He received financing from the venture capital firm Softbank. After only a few months, the company was out of business. Rheingold believes that what he learned from the experience is that venture capitalists, who want a quick and large return on the investment were "not a healthy way to grow a social enterprise."

Rheingold then started a more modest effort, Rheingold Associates, which "helps commercial, educational, and nonprofit enterprises build on-line social networks and knowledge communities."

According to Rheingold and coauthor Lisa Kimball, some of the benefits of creating such communities include the ability to get essential knowledge to the community in times of emergency, to connect people who might ordinarily be divided by geography or interests, to "amplify innovation" and to "create a community memory" that prevents important ideas from getting lost. Rheingold continues to both create and write about new virtual communities.

Further Reading

Hafner, Katie. *The Well: A Story of Love, Death & Real Life in the Seminal Online Community*. New York: Carroll & Graf, 2001.

"Howard Rheingold." Available on-line. URL: http://www.rheingold.com. Updated on July 31, 2002.

Kimball, Lisa, and Howard Rheingold. "How Online Social Networks Benefit Organizations." Available on-line. URL: http://www.rheingold.com/Associates/onlinenetworks.html. Downloaded on December 2, 2002.

Rheingold, Howard. "Community Development in the Cybersociety of the Future." BBC Online. Available on-line. URL: http://www.partnerships.

org.uk/bol/howard.htm. Downloaded on November 3, 2002.

———. *Tools for Thought: The History and Future of Mind-Expanding Technology.* 2nd rev. ed. Cambridge, Mass.: MIT Press, 2000.

———. *The Virtual Community: Homesteading on the Electronic Frontier.* Rev. ed. Cambridge, Mass.: MIT Press, 2000. (The first edition is also available on-line. URL: http://www.rheingold.com/vc/book.)

Richards, Dan. "The Mind of Howard Rheingold." *Mindjack,* Issue 9/1/1999. Available on-line. URL: http://mindjack.com/interviews/howard1.html. Downloaded on January 28, 2003.

The WELL. Available on-line. URL: http://www.well.com. Downloaded on November 3, 2002.

Ritchie, Dennis

(1941–)
American
Computer Scientist

Although most personal computer (PC) users are familiar with Windows or perhaps the Macintosh, many computers in universities, research laboratories, and even Web servers and Internet service providers run the UNIX operating system. Together with Ken Thompson, Dennis Ritchie developed UNIX and the C programming language—two tools that have had a tremendous impact on the world of computing for three decades.

Ritchie was born on September 9, 1941, in Bronxville, New York. He was exposed to communications technology and electronics from an early age because his father was director of the Switching Systems Engineering Laboratory at Bell Laboratories. (Switching theory is closely akin to computer logic design.) Ritchie attended Harvard University and graduated with a B.S. degree in physics. However, by then his interests had shifted to applied mathematics and in particular, the mathematics of computation, which he later described as "the theory of what machines can possibly do." For his doctoral thesis he wrote about recursive functions—recursion is the ability of a function to invoke itself repeatedly until it reaches a limiting value. This topic was proving to be important for the definition of new computer languages in the 1960s.

In 1967, Ritchie decided that he had had enough of the academic world. Without finishing the requirements for his doctorate, he started work at Bell Labs, his father's employer. Bell Labs is an institution that has made a number of key contributions to communications and information theory.

By the late 1960s, computer operating systems had become increasingly complex and

Together with Kenn Thomson, Dennis Ritchie developed the UNIX operating system and the C programming language, two of the most important developments in the history of computing. *(Photo courtesy of Lucent Technologies' Bell Labs)*

unwieldy. As typified by the commercially successful IBM System/360, each operating system was proprietary, had many hardware-specific functions and tradeoffs in order to support a family of upwardly-compatible computer models, and was designed with a top-down approach. This meant that software written for an IBM computer, for example, would not run on a Burroughs computer without the necessity of changing large amounts of system-specific code.

During his graduate studies, however, Ritchie had encountered a different approach to designing an operating system. A new system called Multics was being designed jointly by Bell Labs, the Massachusetts Institute of Technology (MIT), and General Electric. Multics was quite different from the batch-processing world of mainframes: It was intended to allow many users to share a computer. Ritchie had also done some work with MIT's Project Mac. The MIT computer students, the original "hackers" (in the positive meaning of the term), emphasized a cooperative approach to designing tools for writing programs. This, too, was quite different from IBM's highly structured and centralized approach.

Unfortunately, the Multics project grew increasingly unwieldy. Bell Labs withdrew from the Multics project in 1969, and Ritchie and his colleague Ken Thompson then decided to apply many of the same principles to creating their own operating system. Bell Labs did not want to support another operating system project, but they eventually let Ritchie and Thompson use a DEC PDP-7 minicomputer. Although small and already obsolete, the machine did have a graphics display and a Teletype terminal that made it suitable for the kind of interactive programming they preferred. They decided to call their system UNIX, punning on Multics by suggesting something that was simpler and better integrated.

Instead of designing from the top down, Ritchie and Thompson worked from the bottom up. They designed a way to store data on the machine's disk drive and gradually wrote the necessary utility programs for listing, copying, and otherwise working with the files. Thompson did the bulk of the work on writing the operating system, but Ritchie made key contributions such as the idea that devices (such as the keyboard and printer) would be treated the same way as other files. Later, he reconceived data connections as "streams" that could connect not only files and devices but applications and data being sent using different protocols. The ability to flexibly assign input and output, as well as to direct data from one program to another, would become hallmarks of UNIX.

When Ritchie and Thompson successfully demonstrated UNIX, Bell Labs adopted the system for its internal use. UNIX turned out to be ideal for exploiting the capabilities of the new PDP-11 minicomputer. As Bell licensed UNIX to outside users, a unique community of user-programmers began to contribute their own UNIX utilities. This decentralized but cooperative culture would eventually lead to the advocacy of freely shared "open source" software by activists such as RICHARD STALLMAN and ERIC S. RAYMOND.

In the early 1970s, Ritchie also collaborated with Thompson in creating C, a streamlined version of the earlier BCPL and CPL languages. C was a "small" language that was independent of any one machine but could be linked to many kinds of hardware, thanks to its ability to directly manipulate the contents of memory. C became tremendously successful in the 1980s. Since then, C and its offshoots C++ and Java have become the dominant languages used for most modern programming.

Ritchie also played an important role in defining the standards for C and in promoting good programming practices. He coauthored the seminal book *The C Programming Language* with BRIAN KERNIGHAN.

Ritchie and Thompson still work at Bell Labs' Computing Sciences Research Center. (When AT&T spun off many of its divisions, Bell Labs became part of Lucent Technologies.) Ritchie has developed an experimental operating system called Plan 9 (named for a cult science-fiction movie). Plan 9 attempts to take the UNIX philosophy of decentralization and flexibility even further, and is designed especially for networks where computing resources are distributed.

Ritchie has received numerous awards, often given jointly to Thompson. These include the Association for Computing Machinery Turing Award (1985), the Institute of Electrical and Electronics Engineers Hamming Medal (1990), the Tsutomu Kanai Award (1999), and the National Medal of Technology (also 1999).

Further Reading

"Dennis Ritchie Home Page." Available on-line. URL: http://www.cs.bell-labs.com/who/dmr. Updated in October 2002.

Kernighan, B. W., and Dennis M. Ritchie. *The C Programming Language*. 2nd ed. Upper Saddle River, N.J.: Prentice Hall, 1989.

Lohr, Steve. *Go To*. New York: Basic Books, 2001.

Ritchie, Dennis M., and Ken Thompson. "The Unix Time-Sharing System." *Communications of the ACM* 17, no. 7 (1974): 3365–3375.

Slater, Robert. *Portraits in Silicon*. Cambridge, Mass.: MIT Press, 1987.

Roberts, Lawrence
(1937–)
American
Computer Scientist

The year 1969 saw the culmination of one great technical achievement and the beginnings of another. In July of that year, humans first set foot on the Moon. In October, a group of researchers successfully sent a message between computers using a new kind of networking system. As with the Apollo Project, that system, which eventually became the worldwide Internet, was the product of a number of people. These include LEONARD KLEINROCK, who came up with the use of packet-switching to transmit messages between computers, and VINTON CERF and ROBERT KAHN, who created the actual transmission rules, or protocol, called TCP/IP. But the overall design and management of the project was the responsibility of Lawrence Roberts.

Roberts was born on December 21, 1937, in Connecticut. He attended the Massachusetts Institute of Technology (MIT), earning his bachelor's degree in 1959, master's degree in 1960, and Ph.D. in 1963. After getting his doctorate Roberts joined the Lincoln Laboratory at MIT, where he worked with the transistorized computers TX-O, TX-1, and TX-2. These machines were the forerunners of what would become the minicomputer.

While at Lincoln Lab, Roberts met J. C. R. LICKLIDER, who had developed and publicized a concept for what he rather grandiosely called "the intergalactic network"—a proposal to link computers worldwide using common rules for sending and receiving data. In 1962, Licklider had become director of a new computer research program run by the Defense Department's ARPA (Advanced Research Project Agency). Licklider's work inspired Roberts to begin thinking about how such a network might actually be built. Part of the answer came in 1964, when Leonard Kleinrock wrote his seminal book *Communications Nets* on packet-switching, the system where data is broken into separately addressed portions (packets) and routed from computer to computer over a network.

The following year, Roberts successfully connected his MIT TX-2 computer to a Q-32 computer operated by Thomas Marrill in California. For Roberts, this experiment showed two things. Computers could make a connection in real time and access one other's data and programs. However, the existing telephone system was not reliable or flexible enough. The phone system uses what is called circuit switching: Each call is assigned a specific circuit or route through the system. This means that if circuit noise or some other problem leads to loss of data or interruption of service, the connection itself will be lost. Roberts and Marill reported on the results of their experiment in a paper titled "Toward a Cooperative Network of Time-Shared Computers."

Roberts became even more convinced that Kleinrock's packet-switching, idea would be far superior to circuit-switching. With packet-switching, communication does not depend on the existence (and persistence) of any particular circuit or route. Since data is dispatched in routable packets, the network system itself can manage the flow of packets and, if necessary, reroute them to a different circuit.

Licklider and other researchers at ARPA shared Roberts's ideas, and the head of the ARPA Information Processing Technology Office (IPTO), Robert Taylor, received $1 million in funding for a new computer network. In 1966, he hired Roberts to create the overall design and plan for what would become ARPANET. This plan was published under the title of "Multiple Computer Networks and Intercomputer Communication." Note the words *multiple* and *intercomputer*—this is the basis for the Internet, which is not a network but a "network of networks." The plan specified such matters as how users would be identified and authenticated, the sequence of transmission of characters, the method for checking for errors, and the retransmission of data if errors occurred.

Roberts also wrote a plan for setting up and administering the network. This plan, titled "Resource Sharing Computer Networks," emphasized that the ability of users to log into remote computers and run programs there would greatly increase the resources potentially available to researchers, as well as the efficient utilization of computer systems.

In June 1968, Taylor approved Roberts's plan. Roberts and his colleagues at ARPA worked together with a group at Bolt, Beranek, and Newman, a pioneer computer networking company. They designed and implemented the Interface Message Processor (IMP). This was a minicomputer configured to serve as a bridge and link between computers on the new network. (The IMP was about the size of a refrigerator. Today its function is served by a much smaller machine called a router.)

On October 29, 1969, the first successful full-fledged network messages were sent by Kleinrock and his colleagues between computers at the University of California, Los Angeles and the Stanford Research Institute. This date is often considered to be birthdate of the Internet. By now, Roberts had become head of the ARPA computer program after Taylor's departure.

In 1973, Roberts left the ARPA IPTO to enter the private sector as the chief executive officer (CEO) of Telenet, the first private packet-switched network carrier. Meanwhile, new applications for computer networks were emerging. As Roberts recalled to interviewer Bruce Sullivan, Roberts had originally thought of the network as primarily an extension of existing data services and applications:

> [In 1969 what] I was envisioning was that all the mainframe computers would be able to exchange all their data, and allow anybody to get at any information they wanted at any time they wanted throughout the world, so that we would have what is effectively the Web functionality today.

However, only a few years later, as Roberts recalled:

> By '71, we got the network running and started using it and found that e-mail was a major factor. We realized that it would also be replacing the mail, and so on for communications, and then soon

> after that I realized it would replace it for voice, eventually. In '81, I made a speech where I said it would take 20 more years, but voice would be converted in 2001 to the Internet. And that's basically about what's happening.

Roberts wrote RD, the first program that allowed users to read and reply to email.

In 1979, Roberts left Telenet after it was sold to GTE. In 1982 Roberts served briefly as president and CEO of the express company DHL. He then served for 10 years as chairman and CEO of NetExpress, a company specializing in packetized ATM (asynchronized transfer mode) and facsimile equipment. Through the 1990s, Roberts was involved with other networking companies including ATM Systems (1993–98), and Packetcom.

In the 21st century, Roberts continues to do networking research. As founder, chairman, and CEO of Caspian Networks, he closely monitored the continuing growth of the Internet. He reported that Internet traffic had increased fourfold between April 2000 and April 2001. (However, some critics said that this represented a growth in capacity, not actual usage.) Roberts resigned most of his posts with Caspian Networks in 2002, but retained the role of chief technology officer. Ironically, Roberts reported in 2002 that he still could not get broadband Internet access to his own home.

As one of a handful of Internet pioneers, Roberts has received numerous awards, including the Harry Goode Memorial Award of the American Federation of Information Processing Societies, 1976; the Institute of Electrical and Electronics Engineers (IEEE) Computer Pioneer Award (charter recipient); the Interface Conference Award, the L. M. Ericsson Prize (1982); the IEEE Computer Society Wallace McDowell Award (1990); the Association for Computing Machinery SIGCOMM Award (1998); and the National Academy of Engineers Charles Stark Draper Prize (2001).

Further Reading

"Dr. Lawrence G. Roberts Home Page." Available on-line. URL: http://packet.cc. Downloaded on November 3, 2002.

Hafner, Katie, and Matthew Lyon. *Where Wizards Stay Up Late: The Origins of the Internet*. New York: Touchstone Books, 1998.

Haring, Bruce. "Who Really Invented the Net?" *USA Today Tech Report*, September 2, 1999. Available on-line. URL: http://www.usatoday.com/life/cyber/tech/ctg000.htm.

Sullivan, Bruce. "Internet Founder Working on Secret Optical Brew." *ISP Business News* 7, January 8, 2001, n.p.

S

⊠ Sammet, Jean E.
(1928–)
American
Computer Scientist

Among the pioneering generation of programmers and computer scientists, there were a few remarkable figures who helped bridge the gap between the craft of programming and the science of data processing. Jean Sammet was one such bridge-builder, as well as becoming one of the first historians of computing.

Sammet was born on March 23, 1928, in New York City. She proved to be an exceptional math student from an early age. She attended Mount Holyoke College and received a B.A. degree in mathematics in 1948. She then entered the University of Illinois and earned her master's degree in only one year. At the university, she taught mathematics for three years before returning to New York to become a teaching assistant at Barnard College.

In 1953, she went to work for Sperry, the company that had taken over the development and marketing of Univac, the first commercial electronic digital computer. From 1955 to 1958, she was the leader of a programming group. During this time, programming was in transition, growing out of the ad hoc efforts involving the earliest computers into a systematic discipline that could be taught and practiced consistently.

In pursuit of this goal, Sammet lectured on programming principles at Adelphi College (1956–57); in 1958, she taught one of the first courses in programming in FORTRAN. FORTRAN was the first widely used "modern" programming language in that it used names, not numeric addresses, to stand for variables, and made it relatively easy for mathematicians, scientists, and engineers to write computer code that corresponds to mathematical formulas (hence the name FORmula TRANslator).

From 1958 to 1961, Sammet worked at Sylvania Electric Products as a programmer. Sammet also headed the development of software tools (such as a compiler) for a military program called MOBIDIC, which was intended to build a series of "mobile digital computers" that could be mounted in trucks for field use. This work provided Sammet with detailed hands-on experience in program language design, structure, and implementation.

During that time, Sammet also became involved with the design of perhaps the most successful programming language in history, COBOL (Common Business-Oriented Language). From June to December 1959, she served on the short-range committee that developed the initial specifications for the language. (It should be noted that although the earlier work of GRACE MURRAY HOPPER had considerable influence on the design of COBOL, Hopper did not actually serve on the committee that developed the language.)

In 1961, Sammet moved to IBM, where she organized and managed the Boston Programming Center in the IBM Data Systems Division. She developed the concept for FORMAC, the first generalized language for manipulating algebraic expressions, and then managed its development. For this work she received an IBM Outstanding Contribution Award in 1965. Sammet continued working in various posts at IBM, functioning as both a lecturer and a consultant.

Sammet was very active in many of the activities of the Association for Computing Machinery (ACM), a preeminent computing organization. She served as the organization's vice president from 1972 to 1974; then, from 1974 to 1976, Sammet served as its first woman president. In 1977, she was made a member of the National Academy of Engineering.

In 1978, Sammet became the IBM division manager for Ada, the new structured programming language that was being promoted by the federal government. Besides making plans for implementing and using the language at IBM, she also served as an IBM representative for Ada standardization activities. She continued these activities until her retirement in 1988.

Besides being an expert in programming languages, Sammet also became one of the first historians of the field. Her 1969 book *Programming Languages: History and Fundamentals* combined both aspects of her work. During the 1970s, as participants in the field began to think more about the need to preserve and celebrate its early history, Sammet attended a number of conferences and symposia. Perhaps her most significant achievement in this respect was when she conceived and planned the first ACM SIGPLAN History of Programming Languages (HOPL) conference in 1978, and chaired the second conference (HOPL-2) in 1983. Sammet was also active in the American Federation of Information Processing Societies (AFIPS), serving from 1977 to 1979 as chairman of its History of Computing Committee and helping create the Institute of Electrical and Electronic Engineers (IEEE) journal *Annals of the History of Computing*.

Sammet also served from 1983 to 1998 on the board of directors of the Computer Museum, and (from 1991) the executive board of the Software Patent Institute. She has maintained extensive archives of her work that offer scholars an unparalleled view at how the institutions of computing functioned in their formative and mature years.

Awards received by Sammet include the ACM Distinguished Service Award (1985) "For dedicated, tireless and dynamic leadership in service to ACM and the computing community; for advancing the art and science of computer programming languages and recording its history." She was also part of the charter group of ACM Fellows (1994) and received the Lovelace Award of the Association for Women in Computing (1989).

Further Reading

Association for Women in Computing. Available online. URL: http://www.awc-hq.org. Downloaded on November 3, 2002.

Lee, J. A. N. *Computer Pioneers*. Los Alamitos, Calif.: IEEE Computer Society Press, 1995.

Sammet, Jean E. "The Early History of COBOL," in *History of Programming Languages*, Richard L. Wexelblat, editor. New York: Academic Press, 1981.

———. *Programming Languages: History and Fundamentals*. Upper Saddle River, N.J.: Prentice-Hall, 1969.

Samuel, Arthur Lee
(1901–1990)
American
Computer Scientist

When researchers first accepted the challenge of creating artificial intelligence (AI) they naturally gravitated toward programming an artificial game player. Board games in particular offered several advantages as an arena for AI experiments. The rules were clear, the domain was restricted, and the experience of expert human players was available in the form of recorded games. As J. A. N. Lee notes, "Programs for playing games fill the role in AI research that the fruit fly (*Drosophila*) plays in genetics. *Drosophilae* are convenient for genetics because they breed quickly, and games are convenient for AI because it is easy to compare computer performance with that of people."

CLAUDE E. SHANNON and ALAN TURING laid much of the groundwork at the beginning of the 1950s and began to devise algorithms that would allow a computer chess player to choose reasonable moves. By the end of the decade, ALAN NEWELL and HERBERT A. SIMON had made considerable progress.

The first truly impressive computer player was devised by Arthur Samuel for checkers. Although simpler than chess, checkers (draughts) offered plenty of challenge. Samuel would use his increasingly powerful checkers programs as a platform for developing and testing basic learning strategies that would become part of modern AI research.

Samuel was born in 1901 in Emporia, Kansas. After getting his B.S. degree at the College of Emporia in 1923, he studied at the Massachusetts Institute of Technology (MIT) for his master's degree in electrical engineering (awarded in 1926) while working intermittently at the General Electric facility in Schenectady, New York. He then taught at MIT as an instructor in electrical engineering for two years.

In 1928, Samuel moved to Bell Telephone Laboratories, one of the nation's foremost research facilities in the emerging discipline of electronics. Samuel's research focused on the behavior of space charges between parallel electrodes in vacuum tubes. Later, during World War II, he worked on developing the TR-box, a device that automatically disconnects a radar unit's receiver while the transmitter is running, thus preventing the transmitter from burning out the receiver.

After the war, Samuel became a professor of electrical engineering at the University of Illinois. The successful demonstration of ENIAC, the first large-scale electronic digital computer, was inspiring many institutions to build their own computing machines. Samuel became involved in the Illinois effort. During this time, he first became interested in the idea of programming a computer to play checkers. However, he did not finish the program, probably because delays in the computer project meant there was no machine to test it on.

In 1949, Samuel joined IBM's Poughkeepsie laboratory, where the preeminent maker of office equipment was trying to enter the nascent computer market. Samuel worked on the development of IBM's first major computer, the 701. This machine, like ENIAC, used vacuum tubes for its logic functions. For memory, it used a Williams tube. This was a cathode-ray tube similar to that used in a television. However, instead of showing a picture, the tube's electron gun deposited charges on the inner surface of the tube. These charges represented the ones and zeroes of binary computer data. Samuel was able to improve this technology considerably, increasing the tube's data storage capacity from 512 bits to 2,048, as well as improving the tube so that on the average it only failed once every half hour.

While working on hardware, Samuel did not forget about software. He revived his checkers program and wrote a working version for the 701. Just before the Samuels Checkers-Playing

Program was demonstrated, IBM's founder and president, THOMAS J. WATSON SR., predicted that a successful demonstration would raise the price of IBM stock by 15 points.

The demonstration was successful and the prediction correct. Samuel's program did not merely apply a few simple principles and check for the best move in the current position. Instead, he "trained" the program by giving it the records of games played by expert checker players, with the good and bad moves indicated. The program then played through the games and tried to choose the move in each position that had been marked as good. By adjusting its criteria for choosing moves, it was able to choose the correct move more often. In other words, the program was learning to play checkers like an expert.

Samuel continued to work on his program. In 1961, when asked to contribute the best game it had played as an appendix to an important collection of AI articles called *Computers and Thought*, Samuel instead decided to challenge the Connecticut state checkers champion, the nation's fourth-ranked player, to a game. Samuel's program won the game, and the champion graciously provided annotation and commentary to the game, which was included in the book.

Besides demonstrating the growing ability of computers to "think" as well as compute, Samuel's success helped shape the ongoing development of computers. Computer designers began to think more in terms of computers being logical and symbolic information processors rather than just very fast calculators. The ability of computers to process logical instructions was thus expanded and improved.

The techniques developed by the checkers program became standard approaches to games and other AI applications. One relatively powerful technique is alpha-beta pruning. In an article on machine learning, Samuel described it as follows: "a technique for not exploring those branches of a search tree that analysis indicates not to be of further interest either to the player making the analysis (this is obvious) or to his opponent (and this is frequently overlooked)." This ability to quickly winnow out bad moves and focus on the more promising sequences would become even more important in chess, where there are far more possible moves to be considered.

In 1966, Samuel retired from IBM and went to Stanford University, where he continued his research. He did some work on speech recognition technology and helped develop the operating system for SAIL (the Stanford Artificial Intelligence Laboratory). He also mentored a number of researchers who would carry on the next generation of AI work. Many of his students later fondly recalled his kindness and helpfulness. Samuel continued to work on software well into his eighties. His last major project involved improving a program for printing text in multiple fonts. He also wrote software documentation and a partial autobiography.

Samuel died on July 29, 1990, following a struggle with Parkinson's disease. In his long career, he had been honored by being elected a fellow of the Institute of Electrical and Electronic Engineers (IEEE), the American Physical Society, the Institute of Radio Engineers, and the American Institute of Electrical Engineers. He was also a member of the Association for Computing Machinery and the American Association for the Advancement of Science. He also received the IEEE Computer Pioneer Award (1987).

Further Reading

Hsu, Feng-Hsiung. *Behind Deep Blue: Building the Computer that Defeated the World Chess Champion*. Princeton, N.J.: Princeton University Press, 2002.

Lee, J. A. N. *Computer Pioneers*. Los Alamitos, Calif.: IEEE Computer Society Press, 1995.

Samuel, Arthur L. "Some Studies in Machine Learning Using the Game of Checkers." In *Computers and Thought*, Edward A. Feigenbaum and Julian Feldman, editors. New York: McGraw-Hill, 1983, pp. 71–105.

Shannon, Claude E.

(1916–2001)
American
Mathematician, Computer Scientist

In the modern world, information and communication are inextricably bound. Data travels over a complex communications network that includes everything from decades-old phone lines to optical fiber, wireless, and satellite links. The information age would not have been possible without a fundamental understanding of how information could be encoded and transmitted electronically. Claude Shannon developed the theoretical underpinnings for modern information and communications technology and then went on to make important contributions to the young discipline of artificial intelligence (AI).

Shannon was born in Gaylord, Michigan, on April 30, 1916. His father was an attorney and probate judge, his mother a language teacher and a high school principal. Shannon's extended family included a grandfather who had patented several inventions and a distant cousin who was a rather more famous inventor—Thomas Edison. (Shannon's sister Catherine would become a professor of mathematics.)

As a boy, Shannon loved to tinker with both mechanical and electrical equipment. Moving well beyond the prepackaged possibilities of Erector sets, he not only endlessly tinkered with radios but even built a working telegraph system linking his house with a friend about half a mile away. He persuaded the local phone company to give him some surplus equipment, and then "upgraded" the service from telegraph to telephone.

Shannon enrolled in the University of Michigan both prepared for and inclined toward a career in electrical engineering. However, as he began to take mathematics courses he decided that he was interested in that field as well, particularly in symbolic logic, where the logical algebra of GEORGE BOOLE would prove increasingly important for telephone switching systems and especially the development of electronic computers. Thus, when Shannon graduated in 1936 he had earned bachelor's degrees in both mathematics and electrical engineering.

One day Shannon saw a notice on a bulletin board looking for someone to run a machine called a differential analyzer at the Massachusetts Institute of Technology (MIT). He took the job and also enrolled for graduate study, receiving a master's degree in electrical engineering and a Ph.D. in mathematics, both in 1940.

The differential analyzer had been built by VANNEVAR BUSH. It was an analog computer. Unlike today's digital computers, an analog computer uses the interaction of continually varying physical quantities, translated into varying mechanical force or voltage levels. To solve a differential equation with the differential analyzer, Shannon had to translate the various variables and constants into a variety of physical settings

Claude Shannon developed the fundamental theory underlying modern data communications, as well as making contributions to the development of artificial intelligence, such as algorithms for computer chess-playing and a mechanical "mouse" that could learn its way through a maze. *(Photo courtesy of Lucent Technologies' Bell Labs)*

and arrangements of the machine's intricate electromechanical parts.

Unlike earlier purely mechanical analog computers, the differential analyzer was driven by electrical relay and switching circuits. Shannon became interested in the underlying mathematics of these control circuits. He then realized that their fundamental operations corresponded to the Boolean algebra he had studied in undergraduate mathematics classes. It turned out that the seemingly abstract Boolean AND, OR, and NOT operations had a practical engineering use. Any circuit could, in principle, be described as a combination of such logical operations.

Shannon used the results of his research in his 1938 M.S. thesis, titled "A Symbolic Analysis of Relay and Switching Circuits." This work, often considered to be one of the most important master's theses ever written, was honored with the Alfred Nobel prize of the combined engineering societies (this is not the same as the more famous Nobel Prize.)

Along with the work of ALAN TURING and JOHN VON NEUMANN, Shannon's logical analysis of switching circuits would become essential to the inventors who would build the first digital computers in just a few years. (Demonstrating the breadth of his interests, Shannon's Ph.D. thesis would be in an entirely different application—the algebraic analysis of problems in genetics.)

In 1941, Shannon joined Bell Laboratories, perhaps America's foremost industrial research organization. The world's largest phone company had become increasingly concerned with how to "scale up" the burgeoning telephone system and still ensure reliability. The coming of war also highlighted the importance of cryptography—securing one's own transmissions while finding ways to break opponent's codes. Shannon's existing interests in both data transmission and cryptography neatly dovetailed with these needs.

For security reasons, Shannon's paper titled "A Mathematical Theory of Cryptography" would not be published until after the war. But Shannon's most lasting contribution would be to the fundamental theory of communication. His formulation explained what happens when information is transmitted from a sender to a receiver—in particular, how the reliability of such transmission could be analyzed.

Shannon's 1948 paper "A Mathematical Theory of Communication" was published in *The Bell System Technical Journal*. Shannon identified the fundamental unit of information: the binary digit, or "bit," which would become familiar to computer users. He showed how to measure the redundancy (duplication) within a stream of data in relation to the transmitting channel's capacity, or bandwidth. Finally, he showed methods that could be used to automatically find and fix errors in the transmission. In essence, Shannon founded modern information theory, which would become vital for technologies as diverse as computer networks, broadcasting, data compression, and data storage on media such as disks and CDs.

One of the unique strengths of Bell Labs is that it did not limit its researchers to topics that were directly related to telephone systems or even data transmission in general. Like Alan Turing, Shannon became interested after the war in the question of whether computers could be taught to perform tasks that are believed to require true intelligence. He developed algorithms to enable a computer to play chess and published an article on computer chess in *Scientific American* in 1950. He also became interested in other aspects of machine learning, and in 1952 he demonstrated a mechanical "mouse" that could solve mazes with the aid of a circuit of electrical relays.

The mid-1950s proved to be a very fertile intellectual period for AI research. In 1956, Shannon and AI pioneer JOHN MCCARTHY put out a collection of papers titled "Automata Studies." The volume included contributions by two other seminal thinkers, JOHN VON NEUMANN and MARVIN MINSKY.

Although he continued to do research, by the late 1950s Shannon had changed his emphasis to teaching. As Donner Professor of Science at MIT (1958–78) Shannon delivered lectures that inspired a new generation of AI researchers. During the same period, Shannon also explored the social impact of automation and computer technology as a Fellow at the Center for the Study of Behavioral Sciences in Palo Alto, California.

Even in retirement Shannon was an energetic and stimulating presence. A *Scientific American* writer recorded a visit to Shannon's home:

> Without waiting for an answer, and over the mild protests of his wife, Betty, he leaps from his chair and disappears into the other room. When I catch up with him, he proudly shows me his seven chess-playing machines, gasoline-powered pogostick, hundred-bladed jackknife, two-seated unicycle and countless other marvels.

Shannon would receive numerous prestigious awards, including the Institute of Electrical and Electronics Engineers Medal of Honor and the National Medal of Technology (both in 1966). Shannon died on February 26, 2001, in Murray Hill, New Jersey.

Further Reading

Horgan, John. "Claude E. Shannon: Unicyclist, Juggler and Father of Information Theory." *Scientific American*, January 1990, pp. 22ff.

Shannon, Claude Elwood. "A Chess-Playing Machine." *Scientific American*, February 1950, pp. 48–51.

———. "A Mathematical Theory of Communication." *Bell System Technical Journal* 27 (July and October 1948): 379–423, 623–656. Also available on-line. URL: http://cm.bell-labs.com/cm/ms/what/shannonday/paper.html. Downloaded on December 5, 2002.

Waldrop, M. Michael. "Claude Shannon: Reluctant Father of the Digital Age." *Technology Review*. July/Aug. 2001, n.p. Also available on-line. URL: http://www.techreview.com/articles/waldrop0701.asp. Downloaded on December 5, 2002.

Simon, Herbert A.

(1916–2001)
American
Computer Scientist, Scientist

There is science, and then there is the science of science itself. Herbert Simon used innovative computer simulations to develop a new understanding of how scientists and mathematicians solve problems and how people make decisions.

Simon was born on June 15, 1916, in Milwaukee, Wisconsin. His father was an electrical engineer and inventor; his mother was an accomplished pianist. Young Simon was a bright student who skipped three semesters in high school. He would later describe himself as "introspective, bookish, and sometimes lonely" in school—yet paradoxically, he was effective socially, becoming president of most of the clubs he joined.

Simon entered the University of Chicago when he was only 17. While studying for his B.A. degree in political science (awarded in 1936) Simon studied the operation of the Milwaukee Recreation Department. This study in public administration inspired what would be the core concern of Simon's research career—the process of decision making, whether by people or computers.

After graduation, Simon worked for several years for the International City Manager's Association. As an assistant to Clarence Ridley (who had been one of his teachers), Simon helped devise mathematical methods for evaluating the effectiveness or efficiency of municipal services. While doing this work, Simon was introduced to automated information processing in the form of IBM punch-card tabulation equipment. This made him aware of the potential value of the new computing technology that would emerge in the 1940s.

In 1939, Simon moved to the University of California, Berkeley, to head a Rockefeller Foundation–funded study of local government.

During this time, he also completed the work for his University of Chicago Ph.D. (awarded in 1943). His doctoral dissertation, which would later be published in book form in 1947 as *Administrative Behavior*, was an analysis of decision-making in the hierarchies of organizations. Later, Simon would explain to Constance Holden of *Psychology Today* that during this time behaviorism (stimulus-response and conditioning) was king and the analysis of cognitive behavior was out of fashion. "You couldn't use a word like 'mind' in a psychology journal—you'd get your mouth washed out with soap."

Meanwhile, Simon had joined the political science faculty at the Illinois Institute of Technology; in 1946, he became chair of the department. In 1949, he joined the new business school at Carnegie Institute of Technology in Pittsburgh, which later became Carnegie Mellon University (CMU).

Simon would spend the rest of his career at CMU and it was there that his interests in decision making and information processing would be joined in the emerging discipline of artificial intelligence (AI). In 1952 Simon met ALLEN NEWELL, who would became his closest collaborator. Simon noted that "We both viewed man's mind as a symbol-manipulating or information-processing system, but we lacked the language and the technology that were needed."

In 1955–56 Simon, Newell, and Clifford Shaw began to create computer models to simulate the way people solved problems. They had people work through logic problems while writing down their reasoning process step by step—premises, assertion, deductions, and so on. The scientists then wrote a computer program that could work with a set of premises in the same way. This program, called Logic Theorist (and a later, more elaborate version, called General Problem Solver, or GPS) coincided with the seminal Dartmouth Summer Conference that both publicized the newly named field of artificial intelligence and set its basic agenda for decades to come.

Traditionally, logicians and analysts had tended to operate from the assumption of complete knowledge—that is, that decision makers or problem solvers had, or at least had access to, all the knowledge they needed to solve a problem. Simon and J. R. Hayes, however, had explored the process of solving poorly structured or partially understood problems, and wrote a program called Understand, which built up a series of increasingly complete problem definitions and then proceeded to solve the problem.

Simon then brought a similar insight to economics. Economists, too, tended to write about a market in which all the participants had perfect or complete knowledge with which they could act in such a way as to maximize profits (or minimize losses). Simon pointed out that in actual business decisions, information is incomplete and thus decision makers had to take uncertainty into consideration and arrive at a compromise. He called this behavior or strategy "satisficing." "Bounded rationality," Simon's new approach to understanding economic decision making, would gain in influence through the 1960s and 1970s and would earn him the Nobel Prize in economics in 1978.

During the 1980s, Simon continued his research and writing, moving seamlessly between economics, psychology, and computer science and helping foster connections between the disciplines in the curriculum at CMU. In addition to completing the second volume of a book called *Models of Thought*, Simon also published an autobiography, *Models of My Life*. In his introduction to the latter book, he tried to explain how he had approached his multifaceted work:

> I have been a scientist, but in many sciences. I have explored mazes, but they do not connect into a single maze. My aspirations do not extend to achieving a single consistency in my life. It will be enough if I can play each of my roles creditably, borrowing sometimes from

one for another, but striving to represent fairly each character when he has his turn on stage.

In addition to the 1978 Nobel Prize in economics, Simon has received many other awards and positions. These include his election to the National Academy of Sciences, becoming chairman of the National Research Council Division of Behavioral Sciences (1967), and his appointment to the President's Science Advisory Committee (1968). He received the American Psychological Association Distinguished Scientific Contribution Award (1969), the Association for Computing Machinery Turing Award (1975; shared with Alan Newell), and the National Medal of Science (1986).

Further Reading

Holden, Constance. "The Rational Optimist: Will Computers Ever Think Like People? This Expert in Artificial Intelligence Asks, Why Not?" *Psychology Today*, October 1986, pp. 54ff.

McCorduck, Pamela. *Machines Who Think*. New York: W. H. Freeman, 1979.

Simon, Herbert. *Administrative Behavior*. 3rd ed. New York: Macmillan, 1976.

———. *Models of My Life*. New York: Basic Books, 1991.

———. *Models of Thought*. New Haven, Conn.: Yale University Press, 1979.

Stallman, Richard

(1953–)
American
Computer Scientist, Writer

Most of the software familiar to today's computer users comes in a box with the name of a corporation such as Microsoft on it. The source code, or the instructions that make the software run, is a proprietary secret. Thus the program can be fixed or improved only by the company that sells it. Users are consumers with a generally passive relationship to the product.

Richard Matthew Stallman had a different idea about how software should be developed and distributed. Stallman, who himself created some of the finest software development tools, believed that the source code should be made freely available so that a community of programmers and users could continually improve it. He thus became the highly visible leader of a loosely organized but influential movement called "open source."

Stallman was born on March 16, 1953, in New York. He showed aptitude for mathematics from a young age, starting to explore calculus when he was eight. Later, while in a summer camp, he found that one of the counselors had a manual for the IBM 7094 mainframe computer. The boy read through it and began to write simple programs. Even though he had no access to an actual computer, Stallman was hooked. As he recalled to Steven Lohr: "I just wanted to write programs. I was fascinated," he recalled. In high school, Stallman signed up for an IBM program that gave promising students access to time on a computer. He quickly became such a proficient programmer that IBM hired him to work for them during his last summer before college.

Stallman then enrolled in Harvard University, where he received a B.A. degree in physics in 1974. He soon was living what writer Steven Levy would later describe as the "hacker ethic" of untrammeled access to technology. The time-sharing computer at Harvard normally deleted a student's work when he or she logged off the system, and there was no disk or tape storage. However, Stallman discovered that if he logged onto two terminals instead of one, the system would not remove his data. As Stallman explained to Lohr "I did not like rules. I still don't. It was an intellectual exercise expressing what I thought of rules." Certainly, it would not be the last time Stallman challenged rules that he thought were pointless and limiting.

In 1971, while still a Harvard undergraduate, Stallman began to make a name for himself at the Artificial Intelligence Laboratory at the nearby Massachusetts Institute of Technology (MIT). He helped improve the MIT operating system called ITS (Incompatible Time Sharing, a joking reference to the CTS, or Compatible Time Sharing, system that MIT hackers considered to be pathetic). At a time when operating systems were normally controlled by computer makers, Stallman said to Lohr that at MIT "anybody with ability was welcome to come in and try to improve the operating system. We built on each other's work."

Stallman essentially lived a double life. By day, he was a hardworking graduate student; at night, he was a member of the crew of MIT hackers who would be made famous by Steven Levy. Levy observed, "When the people in the lab discovered after the fact that he was simultaneously earning a magna cum laude degree in physics at Harvard, even those master hackers were astonished."

Stallman also gained a reputation for taking the initiative in the endless war between MIT officials and hackers over computing resources. As he recalled to Levy:

> The terminals were thought of [as] belonging to everyone, and professors locked them up in their offices on pain of finding their doors broken down. . . . Many times I would climb over ceilings or underneath floors to unlock rooms that had machines in them that people needed to use, and I would usually leave behind a note explaining to the people that they shouldn't be so selfish as to lock the door.

When administrators began to require that users have passwords, Stallman encouraged his fellow hackers to use just a simple carriage return.

However, Stallman was not merely a gadfly. He did serious, high-quality work in developing software tools. In particular he developed Emacs, an editor for writing programs. It is perhaps his finest achievement as a software engineer. Far more than a text editor, Emacs was itself programmable and extendable using a language similar to LISP. Since LISP was the language of choice for artificial intelligence programming, researchers happily took to the powerful Emacs for help in writing their programs. Stallman and a variety of collaborators would continue to extend Emacs through the 1980s, such as by creating a multiwindow version for use with X-Windows, a popular graphical user interface for UNIX.

By the mid-1970s, however, commercialism was invading the UNIX world. Stallman's supervisor Russ Noftsker left MIT to start his own LISP development company, Symbolics. He recruited many of the top MIT hackers to join him. This angered Stallman, who believed that this exodus to the private sector had destroyed the unique cooperative environment at MIT. Stallman formed a rival company, LMI, which developed cheaper alternatives to the software Symbolics was trying to sell to MIT and others.

In 1983, Stallman began the GNU Project. GNU is a recursive acronym that stands for "GNU's not UNIX." At the time, anyone who wanted to use UNIX had to buy an expensive license because the operating system was the property of AT&T, whose Bell Laboratories had developed it in the early 1970s. However, it was legal to create a version of UNIX that was functionally compatible (that is, responded to commands in exactly the same way), provided that the copyright-protected AT&T source code was not used. Stallman decided to do just that.

Over the next two decades, Stallman and a large number of collaborators around the world worked to create GNU as a free and compatible alternative to UNIX. One of Stallman's most important contributions to the effort was the GNU C compiler, a "portable" compiler that works with dozens of different operating systems. Stallman also developed the GNU symbolic debugger (GDB). Using these tools, Stallman and his collaborators created free, compatible GNU versions of most of the standard UNIX utilities.

In 1985, Stallman left MIT to found the Free Software Foundation. Besides promoting the GNU project, the FSU also promotes the concept of open source software. The essence of open source was stated in Stallman's original "GNU Manifesto" as follows: "Everyone will be permitted to modify and redistribute GNU, but no distributor will be allowed to restrict its further redistribution." In order to facilitate this idea, Stallman created the General Public License (GPL), which requires not only that source code be included with any released version of a program (the open source concept), but additionally, that any improved or expanded version be free.

Stallman is often acerbic when asked about "open source" software that actually comes with restrictions or that is part of a corporate marketing strategy (as with Netscape). He complains that "the important ideas have been sanitized away. Open source [has become] corporate-friendly." Another sore spot for Stallman subject is Linux, the open source UNIX variant created by LINUS TORVALDS. Although he acknowledges Torvalds's achievement in creating the kernel (the "core" of the operating system's functions), he is unhappy that many people do not realize that most of the utilities used by Linux users were created by the GNU project. Stallman urges people to refer to the new operating system as "GNU/Linux."

Continuing to hack interesting projects and promote his pure vision of free software, Stallman pursues a number of avocations. He has written in *The Hacker's Dictionary* that his "hobbies include affection, international folk dance, flying, cooking, physics, recorder, puns, science fiction, fandom and programming: I magically get paid for doing the last one."

Stallman received a MacArthur Foundation fellowship in 1990, the Association for Computing Machinery Grace Hopper Award (for achievement by a computer scientist under 30) in 1991, and the Institute of Electrical and Electronics Engineers Computer Pioneer Award of 1998 (shared with Linus Torvalds).

Further Reading

DiBona, Chris, Sam Ockman, and Mark Stone, editors. *Open Sources: Voices from the Open Source Revolution*. Sebastopol, Calif.: O'Reilly, 1999.

Free Software Foundation. Available on-line. URL: http://www.fsf.org/fsf. Updated on June 12, 2002.

Levy, Steven. *Hackers: Heroes of the Computer Revolution*. Updated ed. New York: Penguin, 2001.

Lohr, Steven. *Go To*. New York: Basic Books, 2001.

Stallman, Richard M. "The GNU Manifesto." (1985, 1993). Available on-line. URL: http://www.gnu.org/gnu/manifesto.html. Downloaded on December 6, 2002.

———. *Gnu Emacs Manual*. Lincoln, Nebr.: iUniverse.com, 2000.

———. "The Right to Read." *Communications of the ACM* 40 (February 1997): 85–87.

Stibitz, George

(1904–1995)
American
Mathematician, Inventor

Between the hand-cranked calculator and the fully electronic digital computer came a transitional period in which many of the ideas behind the modern computer were first implemented using electromechanical technology. George Stibitz was a pioneer in using the binary logic of GEORGE BOOLE to build circuits that could do arithmetic.

Stibitz was born on April 20, 1904, in York, Pennsylvania. His mother had been a mathematics teacher before marriage, and his father was a professor of theology. The family moved to Dayton, Ohio, shortly after Stibitz's birth.

Young Stibitz showed considerable academic aptitude and was enrolled in a special high school.

Stibitz then enrolled in Denison University in Granville, Ohio, receiving a bachelor's degree in mathematics in 1926. He went on to earn a master's degree in physics at Union College in Schenectady, New York, the following year. After working for a year at the General Electric facility in Schenectady, Stibitz enrolled in Cornell University, where he received his Ph.D. in mathematical physics in 1930.

By 1931, jobs were hard to come by, but despite the Great Depression Stibitz found not just a job but probably the ideal position—he became a researcher at the Bell Telephone Laboratory. The telephone giant's farsighted executives were concerned about the growing complexity and extent of the telephone system and the need to further automate its operations. Calls that used to be connected by operators were increasingly being made automatically by sequences of relays—electromagnetically activated switches. Telephone engineers needed a better overall theory that could guide them in building complicated switch sequences.

Stibitz was hired to study this matter. At the time, he was not very familiar with how relays worked, but as he studied them he realized that they were essentially binary (two-valued, on and off) devices. This meant that the algebraic logic that English mathematician George Boole had defined almost a century earlier was directly applicable to the behavior of the relays. For example, two relays in a series corresponded to the AND operation, while two in parallel worked like an OR. More complicated arrangements could be constructed by combining these and a few other basic operations.

By 1937, Stibitz had moved beyond understanding relays to using them for a new purpose that had nothing to do with telephones. One of the consequences of Boolean algebra is that arithmetic operations such as addition can be defined in terms of a sequence of logical operations. As a demonstration, Stibitz assembled two relays, connected them to flashlight bulbs and a battery, and came up with a two-digit binary adder. With 1 representing a closed switch and 0 an open one, if the adder were set to 01 and another 1 pulse was fed to it, the result would be 10—binary for 1 + 1 = 2. This first demonstration device was dubbed the "model K" because Stibitz had built it in his kitchen. Today, a replica is on display at the Smithsonian Institution.

After he demonstrated the device to his colleagues at Bell Labs, Stibitz built a multidigit version and started to design a calculator based on binary relay circuits. However, the binary number system was quite unfamiliar to most people at the time. Stibitz therefore came up with a way to convert the output of the binary circuits to ordinary decimal numbers. He did this through the clever method of adding three to each of the pair of binary numbers being fed into the adder. This total addition of six meant that a total of decimal nine would become 15, which is 1111 in binary. Therefore, the next number would create an overflow (or carry) in both decimal and binary form. Further, the binary complement of a sum in this "excess-3" format was the same as the binary form of the decimal complement. This created a sort of "bridge" between the binary and decimal representations.

At first, the Bell executives were not terribly impressed with these developments. After all, electromechanical calculators were readily available. But gradually they realized that they were going to have to tackle problems that required too much calculation to be handled by mechanical calculators. They therefore funded a full-scale model of a calculator based on Stibitz's logical arithmetic.

In January 1940, the new device was ready. It was called the Complex Number Calculator (CNC) Model 1, because it was designed to work with complex numbers (numbers consisting of both real and imaginary parts). Such numbers turn up regularly in physics and electronics work. Stibitz drew a schematic for a machine that could handle eight-digit complex numbers. He originally suggested using existing rotary telephone switches, but a colleague, S. B. Williams had the inspired idea of using newly developed crossbar switches to store the numbers, and relays to transmit the numbers through the machine.

The final version of the machine had 10 crossbar switches and 450 relays. It could multiply two eight-digit complex numbers after about 30 seconds of furious clicking and clacking. While very slow by modern computer standards, it was considerably faster than having human operators perform the equivalent sequence of operations on ordinary desk calculators—and assuming the numbers were input correctly, there was no need to worry about operator error.

Another ingenious aspect of the Model 1 is that it used Teletypes as input and output devices. This not only made for much easier data entry and printouts, it allowed for something that astonished the audience at the annual meeting of the American Mathematical Society at Dartmouth College in New Hampshire in September 1940. The audience was provided with a Teletype and invited to type numbers. The numbers were converted to codes and sent along a telegraph line to the Model 1, which was in New York. The machine performed the calculation and sent the results back along the line to Dartmouth, where they were printed out for the audience to see. This was the first-ever demonstration of remote computing, and showed how data could be converted to codes and transmitted using an existing communications system. In a sense, today's Internet is a remote descendant of what happened that day.

When World War II arrived, so did the military's voracious appetite for calculations for artillery tables. Bell Labs gave Stibitz leave to join the National Defense Research Council, where he worked on improved versions of the CNC. Model 2 allowed sequences of instructions—a program—to be loaded from a punched tape, allowing the same set of calculations to be performed on different sets of data. He even incorporated error correction codes. Model 5, which came into service in 1945, added the use of the more flexible floating point decimal representation. It also provided a form of temporary memory by saving intermediate results on punched tape and feeding them back into later stages of the calculation. The program could even use instructions to "jump" to a different tape based on the result of a test—the equivalent of an IF . . . THEN statement in modern programming.

After the war, Stibitz did not return to Bell Labs. Instead, he moved to Vermont, where he became an independent consultant in applied mathematics and worked on projects for a number of government agencies and corporations. In 1946, he joined the Department of Physiology at Dartmouth College, where he worked with medical researchers to use computers to explore the diffusion of gases and drugs through the body. He retired in 1972 but continued to have a lively interest in computing.

In the end, relay computers like Stibitz's and one (unknown to Stibitz) also designed by KONRAD ZUSE in Germany proved to be too slow. The path to the modern computer lay instead through electronics, as demonstrated by ENIAC, built by J. PRESPER ECKERT and JOHN MAUCHLY. However, Stibitz had laid a solid foundation in logic circuit design, numeric representation, input, output, and even data communications.

Stibitz died on January 31, 1995. He and Konrad Zuse shared the Harry Goode Award of the American Federation of Information Processing Societies (1965). Stibiz also received the Institute of Electrical and Electronics Engineers Computer Pioneer Award (1982).

Further Reading

Bernstein, Jeremy. *Three Degrees Above Zero: Bell Labs in the Information Age*. New York: Charles Scribner's Sons, 1984.

Goldstine, H. H. *The Computer from Pascal to von Neumann*. Princeton, N.J.: Princeton University Press, 1993.

Stibitz, George R., and Jules A. Larrivee. *Mathematics and Computers*. New York: McGraw-Hill, 1957.

Stoll, Clifford

(1950–)
American
Writer, Programmer, Scientist

Clifford Stoll did not start out having much interest in computers. He was an astronomer, and found that field to be quite absorbing enough. However, one day when he was working at Lawrence Berkeley Laboratory, he was asked to help the computer staff find out why two accounts on the lab's computers differed by 75 cents. The resulting odyssey put Stoll on the trail of a hacker-spy and on an obstacle course created by reluctant administrators and uninterested Federal Bureau of Investigation (FBI) agents. His remarkable experience led him to write a best-selling book called *The Cuckoo's Egg* and launched him on a new career as a critic (some would say gadfly) devoted to challenging what he sees as excessive hype about the capabilities of computers and warning of society's excessive dependence on the machines.

Stoll was born on June 4, 1950, in Buffalo, New York. He attended the University of Buffalo, receiving his B.A. degree in 1973, and then earned his doctorate in astronomy from the University of Arizona in 1980. During the 1980s, he worked as an astronomer and astrophysicist at a number of institutions, including the Space Telescope Institute in Baltimore, the Keck Observatory in Hawaii, and the Harvard-Smithsonian Center for Astrophysics in Massachusetts. It was while working at the Lawrence Berkeley Laboratory in 1988 that he encountered the now-famous 75-cent error. Next, as Stoll recounted to an interviewer: "I . . . look at the accounting system, and lo! The program's working right. I go and look at my files and say, "Oh, jeez. Here's somebody breaking into my computer."

Stoll became concerned because the laboratory worked on a number of government projects, and if someone was embezzling money or perhaps stealing classified information, the government funders—and ultimately, Congress—might become quite unhappy, leading to serious repercussions for the lab's work. But also, as a scientist, Stoll became intrigued with the breakin as a phenomenon. How was the thief getting in? To what extent were the lab's computer systems compromised?

When he found the files that the unknown hacker had retrieved and squirreled away in an account on the lab computer, Stoll became considerably more concerned. Now he was thinking, as he recalled in *The Cuckoo's Egg*

> "Why is somebody breaking into a military computer, stealing information, [and] . . . copying it across networks into some machine in Europe? Why is it that this person is obsessed with things like Strategic Defense Initiative [and] . . . North American Air Defense?" Immediately, it comes up in my mind, "Oh. We're looking at a spy."

Stoll realized that he could simply erase the unknown account and block it, and the hacker would likely go elsewhere. But if this was just the tip of an espionage iceberg, did he have an obligation to let the hacking continue while he tried to trace its origin? To his dismay, however, Stoll found that the lab administrators just wanted to avoid liability. Further, when he contacted the FBI, they essentially said they were not interested in some mysterious petty computer fraud. However, Stoll had better luck with the Central Intelligence Agency (CIA), which began to work with him as he started to trail the mysterious hacker.

Not being a computer expert, Stoll had to teach himself about computer security as he worked. He gradually followed the hacker's files back to their sources, which included missile test ranges in Alabama and White Sands, a defense contractor in McLean, Virginia, and finally to a house in Hanover, Germany and a freelance spy, a 25-year-old German hacker named Markus Hess who was selling U.S. government secrets to the KGB in exchange for money and cocaine.

As a result of Stoll's year-long effort, which he chronicled in *The Cuckoo's Nest*, many Americans, including intelligence officials, became aware for the first time how vulnerable the growing computer networks were to espionage, and that the supposed firewall between classified and public information was porous. (A Public Broadcasting Service special, *The KGB, The Computer, and Me* was based on the book.)

Stoll's adventures in cyberspace led through the 1990s to explorations into other aspects of computer use. Stoll became convinced that the Internet and the "information highway" that was being touted so exuberantly in the mid-1990s was too often a road to nowhere. In his 1995 book *Silicon Snake Oil,* he suggests that too many people are becoming passive "Internet addicts," just as an earlier generation had been mesmerized by TV. In his many public appearances, Stoll has a number of one-liners at hand. For example, he asks, "Why are drug addicts and computer aficionados both called 'users'?"

One of Stoll's biggest concerns is the overemphasis on bringing computers into the schools. In *Silicon Snake Oil,* he argues that

> Our schools face serious problems, including overcrowded classrooms, teacher incompetence, and lack of security. Local education budgets barely cover salaries, books, and paper. Computers address none of these problems. They're expensive, quickly become obsolete, and drain scarce capital budgets.

On a more philosophical level, Stoll has argued that computers are likely to be used as a substitute for the intimate relationship between teacher and student that he believes is essential for real education. He warns that what is seen on the screen, however intricate and dazzling, is no substitute for the experience of human relationships and the natural world. Stoll's next book, *High Tech Heretic*, continued those themes.

Of course, Stoll has attracted his own share of critics. One critic, Etelka Lehoczky, writing in *Salon*, suggests that "for all his familiarity with computers, Stoll's view of their place in the world has all the good-vs.-evil simplicity of a born-again Christian's—or a recovering alcoholic's." Others complain that Stoll offers little practical advice on how best to use computer technology in education and elsewhere. Stoll in turn denies that he is simply negative about computers and their role: As he insists to interviewer John Gerstner of *Communication World*

> Nor am I a curmudgeon. I am not cynical. I feel it is important to be skeptical, but not cynical. I have hope for the future. I look forward with optimism, but I am cynical of claims made for the future. I'm an astronomer and a physicist and as such I am paid to be skeptical.

Further Reading

"Cliff Stoll Resources." Available on-line. URL: http://www.badel.net/resources/stoll.htm. Posted on February 26, 2000.

"Clifford Stoll." Available on-line. URL: http://www.crpc.rice.edu/CRPC/GT/msirois/Bios/stoll.html. Updated on September 3, 2000.

Gerstner, John. "Cyber-Skeptic: Cliff Stoll." *Communication World*, June–July 1996, n.p.

Stoll, Clifford. *The Cuckoo's Egg: Tracking a Spy Through the Maze of Computer Espionage*. New York: Pocket Books, 2000.

———. *High Tech Heretic: Reflections of a Computer Contrarian*. New York: Anchor Books, 2000.
———. *Silicon Snake Oil: Second Thoughts on the Information Superhighway*. New York: Doubleday, 1995.

Stroustrup, Bjarne
(1950–)
Danish
Computer Scientist

By the late 1970s, researchers such as KRISTEN NYGAARD and ALAN C. KAY had laid the foundations for a new way of thinking about computer programs. This paradigm, called object-oriented programming, provided a way for programs to model the real-world behavior of different kinds of objects and their properties and functions. However, languages such as Simula and Smalltalk were confined mainly to research laboratories. The language used for most non-mainframe programming was C, which was serviceable but not object oriented. Bjarne Stroustrup had the idea of taking a popular language and recasting it as an object-oriented one, making it suitable for the complex software architecture of the late 20th century.

In the 1980s, Bjarne Stroustrup created the object-oriented C++ language. By letting programmers organize complex software better while building on their existing skills, C++ became the most popular language for general applications programming. *(Photo courtesy of Bjarne Stroustrup)*

Stroustrup was born on December 30, 1950, in Århus, Denmark. Growing up, Stroustrup seemed more interested in playing soccer than engaging in academic pursuits. However, one teacher who saw his potential gradually awakened the boy's interest in mathematics.

As a student at the University of Århus, his interests were not limited to computing (indeed, he found programming classes to be rather dull). However, unlike literature and philosophy, programming did offer a practical job skill, and Stroustrup began to do contract programming for Burroughs, an American mainframe computer company. To do this work, Stroustrup had to pay attention to both the needs of application users and the limitations of the machine, on which programs had to be written in assembly language to take optimal advantage of the memory available.

By the time Stroustrup received his master's degree in computer science from the University of Århus, he was an experienced programmer, but he soon turned toward the frontiers of computer science. He became interested in distributed computing (writing programs that run on multiple computers at the same time) and developed such programs at the Computing Laboratory at Cambridge University in England, where he earned his Ph.D. in 1979.

The 1970s was an important decade in computing. It saw the rise of a more methodical approach to programming and programming languages. It also saw the development of a powerful and versatile new computing environment: the UNIX operating system and C programming language developed by DENNIS RITCHIE and KENNETH THOMPSON and Bell Laboratories. Soon after getting his doctorate, Stroustrup was invited to join the famous Bell Labs. That institution had become notable for giving researchers great freedom to pursue what interested them. Stroustrup was told to spend a year getting to know what was going on at the lab, and then they would ask him what he would like to pursue.

Stroustrup looked for ways to continue his work on distributed computing. However, he was running into a problem: traditional programming languages such as FORTRAN, Algol, or C were designed to have the computer perform just one task, step by step. While a program was likely to divide the processing into subroutines to make it more manageable, the overall flow of execution was linear, detouring into and returning from subroutines (also called procedures or functions) as necessary.

Stroustrup decided that he needed a language that was better than C at working with the various modules running on the different computers. He studied an early object-oriented language called Simula. It had a number of key concepts, including the organization of a program into classes, entities that combined data structures and associated capabilities (methods). Classes and the objects created from them offered a better way to organize large programs. For example, instead of calling on procedures to draw graphics objects such as windows on the screen, a window object could be created that included all the window's characteristics and the ability to draw, move, resize, or delete the window. This approach also meant that the internal data needed to keep track of the window was "encapsulated" in the window object where it could not be inadvertently changed by some other part of the program. The rest of the program communicated with the window object by using its defined methods as an interface.

Stroustrup found that object-oriented programming was particularly suited for his projects in distributed computing and parallel programming. After all, in these systems there were by definition many separate entities running at the same time. It was much more natural to treat each entity as an object.

However, Simula was fairly obscure and produced relatively slow and inefficient code. It was also unlikely that the large community of systems programmers who were using C would switch to a completely different language. Instead, starting in the early 1980s Stroustrup decided to add object-oriented features (such as classes with member functions, user-defined operators, and inheritance) to C. At first he gave the language the rather unwieldy name of "C with Classes" However, in 1985 he changed the name to C++. (The ++ is a reference to an operator in C that adds one to its operand, thus C++ is "C with added features.")

At first, some critics criticized C++ for retaining most of the non-object-oriented features of C (unlike pure object languages such as Smalltalk), while others complained that the overhead required in processing classes made C++ slower than C. However, Stroustrup cleverly designed his C++ compiler to produce not machine-specific code but C source code. In effect, this meant that anyone who already had a C compiler could use C++ and could optimize the C code later if necessary. (Soon the development of more efficient compilers made this generally unnecessary.) The fact that C++ was built upon the familiar syntax of C made the learning curve relatively shallow, and Stroustrup aided it by writing a clear, thorough textbook called *The C++ Programming Language*.

In 1990, *Fortune* magazine included Stroustrup in a cover feature titled "America's Hot Young Scientists." By the early 1990s, there were an estimated 500,000 C++ programmers. The applications programming interfaces that were needed to allow programmers to use the many services of complex operating systems such as Microsoft Windows were reorganized as C++ classes, making learning them considerably easier.

In the mid-1990s, a language called Java was designed by JAMES GOSLING. It can be thought of as a somewhat streamlined C++ designed to run on Web servers and browsers. Although Java has become popular for such applications, it has not displaced C++ for the bulk of applications and systems programming.

With the popularity of C++ came the need to establish official standards. Stroustrup was not pleased with his new responsibilities. He noted in 1994 in *PC Week* that "writing code is fun, planning a standards meeting is not fun. I don't get my hands dirty enough."

Stroustrup has continued to work at AT&T Labs (one of the two successors to the original Bell Labs). He is head of the Large-Scale Program Research Department. Stroustrup received the 1993 Association for Computing Machinery Grace Hopper Award for his work on C++ and is an AT&T Fellow.

Further Reading

"Bjarne Stroustrup's Homepage." Available on-line. URL: http://www.research.att.com/~bs/homepage.html. Downloaded on November 3, 2002.

Gribble, Cheryl. "History of C++." Available on-line. URL: http://www.hitmill.com/programming/cpp/cppHistory.asp. Updated on November 3, 2002.

Stroustrup, Bjarne. *The C++ Programming Language*. 3rd ed. Reading, Mass.: Addison-Wesley, 2000.

———. *The Design and Evolution of C++*. Reading, Mass.: Addison-Wesley, 1995.

Sutherland, Ivan

(1938–)
American
Computer Scientist, Inventor

Today it is hard to think about computers without interactive graphics displays. Whether one is flying a simulated 747 jet, retouching a photo, or just moving files from one folder to another, everything is shown on the screen in graphical form.

For the first two decades of the computer's history, however, computers lived in a text-only world except for a few experimental military systems. During the 1960s and 1970s, Ivan Edward Sutherland would almost single-handedly create the framework for modern computer graphics while designing Sketchpad, the first computer drawing program.

Sutherland was born on May 16, 1938, in Hastings, Nebraska; the family later moved to Scarsdale, New York. His father was a civil engineer, and as a young boy Sutherland was fascinated by the drawing and surveying instruments his father used. This interest in drawing and geometry would be expressed later in his development of computer graphics.

When he was about 12, Sutherland and his brother got a job working for a pioneer computer scientist named Edmund Berkeley. Berkeley would play a key role in the direction of Sutherland's career as a computer scientist. As Sutherland described in a 1989 interview, for the Association for Computing Machinery (ACM), Berkeley

> had a wonderful "personal computer" called Simon. Simon was a relay machine about the size of a suitcase; it weighed about 50 pounds. It had six words of memory of two bits each—so a total of 12 bits of memory. And it could do 2-bit arithmetic, which meant that using double precision arithmetic you could add numbers up to 15 to get sums up to 30.

Berkeley encouraged young Sutherland to play with Simon. Simon could only add, but Sutherland decided to see if he could get it to divide. Eventually, he wrote a program that let the machine divide numbers up to 15 by numbers up to three (remember, the machine only held 12 bits). However, in order to do so, he had to actually rewire the machine so that it could do a conditional test to determine when it should stop.

Berkeley also introduced Sutherland to CLAUDE E. SHANNON, at Bell Labs. Shannon was the creator of information theory and one of the world's foremost computer scientists. Years later, Sutherland would become reacquainted with Shannon at the Massachusetts Institute of Technology (MIT), when Shannon became his dissertation adviser.

Sutherland first attended Carnegie Mellon University, where he received a B.S. degree in electrical engineering in 1959. The following year, he earned an M.A. degree from the California Institute of Technology (Caltech). He then went to MIT to do his doctoral work under Shannon at the Lincoln Laboratory.

At MIT, Sutherland was able to work with the TX-2, an advanced (and very large) transistorized computer that was a harbinger of the minicomputers that would become prevalent later in the decade. Unlike the older mainframes, the TX-2 had a graphics display and could accept input from a light pen as well as switches that served a function similar to today's mouse buttons. The machine also had 70,000 36-bit words of memory, an amount that would not be achieved by personal computers (PCs) until the 1980s. Having this much memory made it possible to store the pixel information for detailed graphics objects.

Having access to this interactive machine gave Sutherland the idea for his doctoral dissertation (submitted in 1963). He developed a program called Sketchpad. Doing so required that he develop a whole new approach to computer graphics. MIT hackers had succeeded in drawing little blobs to represent spaceships in their early-1960s game Spacewar, but drawing realistic objects required developing algorithms for plotting pixels and polygons as well as scaling objects in relation to the viewer's position. Sutherland's Sketchpad could even automatically "snap" lines into place as the user drew on the screen with the light pen. Besides drawing, Sketchpad demonstrated the beginnings of the graphical user interface that would be further developed by such researchers as ALAN C. KAY and DOUGLAS ENGLEBART at the Xerox Palo Alto Research Center in the 1970s and would reach the consumer in the 1980s.

After demonstrating Sktechpad in 1963 and receiving his Ph.D. from MIT, Sutherland took on a quite different task. He became the directory of the Information Processing Techniques Office (IPTO) of the Defense Department's Advanced Research Projects Agency (ARPA). Taking over from J. C. R. LICKLIDER, Sutherland managed a $15 million budget. While continuing his research on graphics, Sutherland also oversaw the work on computer time-sharing and the networking research that would eventually lead to ARPANET and the Internet. In 1966 Sutherland went to Harvard, where he developed the first computerized head-mounted display (HMD)—a device now used in many applications, including aviation and the military.

In 1968, Sutherland and David Evans went to the University of Utah, where they established an IPTO-funded computer graphics research program. Sutherland's group brought computer graphics to a new level of realism. For example, they developed the ability to place objects in front of other objects, which required intensive calculations to determine what was obscured. They also developed an idea, suggested by Evans, called incremental computing. Instead of drawing each pixel in isolation, they used information from previously drawn pixels to calculate new ones, considerably speeding up the rendering of graphics. The results began to approach the realism of a photograph. (The two researchers also founded a commercial enterprise, Evans and Sutherland, to exploit their graphics ideas. It became one of the leaders in the field.)

In 1976, Sutherland left the University of Utah to serve as the chairman of the computer science department at Caltech. Working with a colleague, Carver Mead, Sutherland developed a systematic concept and curriculum for integrated circuit design, which became the specialty of the department. He would later point out that it was the important role that geometry played in laying out components and wires that had intrigued him the most.

An underlying theme of Sutherland's approach to programming is that developing better ways to visualize what one is trying to do is essential. He noted in the ACM interview that "the hard parts of programming are generally hard because it's understanding what the problem is that's hard."

Sutherland sometimes got involved in fun, slightly bizarre side projects. One was a machine dubbed the Trojan Cockroach. In 1979, a colleague, Mark Raibart, had suggested building a robot that could hop on one leg. The half-ton, 18-horsepower, gas-powered, hydraulic-actuated machine could locomote, sort of, on a good day. However, as he told the ACM interviewer it "would break down and then we'd have this half ton of inert iron in the parking lot and we'd have to figure out how to get back home again."

Sutherland left Caltech in 1980 and started a consulting and venture capital firm with Bob Sproull, whom he had met years earlier at Harvard. In 1990, Sun Microsystems bought the company and made it the core of Sun Labs, where Sutherland continues to work as a Sun Microsystems Fellow. Sutherland received the prestigious Turing Award from the Association for Computing Machinery in 1988.

Further Reading

Frenkel, Karen A. "Ivan E. Sutherland 1988 A. M. Turing Award Recipient." *Communications of the ACM* 32 (June 1989): 711ff.

Raibert, M. H., and I. E. Sutherland. "Machines that Walk." *Scientific American*, January 1983, pp. 44–53.

Sutherland, Ivan E. "Sketchpad—A Man-Machine Graphical Communication System." *Proceedings of the Spring Joint Computer Conference*, Detroit, Mich., May 1963.

Sutherland, Ivan E., and C. A. Mead. "Microelectronics and Computer Science." *Scientific American*, September 1977, pp. 210–228.

T

⊠ Thompson, Kenneth
(1943–)
American
Computer Scientist

As primary developer of the UNIX operating system, Kenneth Lane Thompson created much of the programming and operating environment that has been used by universities, research laboratories, and an increasing number of other users since the 1970s.

Thompson was born on February 4, 1943, in New Orleans. His father was a navy fighter pilot, and Thompson's family moved frequently. By the time he was two, the family had lived in San Diego, Seattle, San Francisco, Indianapolis, and Kingsville, Texas, where the family lived during much of Thompson's boyhood.

The boy often hung around a Kingsville shop that serviced the shortwave radios that the isolated oil town used instead of telephones. He learned about electronics by tinkering with the radios in the shop. Thompson started saving his allowance so he could buy a wondrous new component, a transistor, which cost $10. However, one day he found that the shop was selling them for only $1.50, so he was able to buy several. Only later did he suspect that his father had made a quiet arrangement with the shop to provide his son with a special discount. Thompson was also allowed to join radio technicians who serviced radios on oil rigs, climbing up to reach the equipment.

In 1960, Thompson journeyed to a quite different world when he enrolled in the University of California, Berkeley to study electrical engineering. As an undergraduate, he worked in the computer center, writing utility programs and helping other users debug their programs. He also participated in a work-study program at General Dynamics (a defense contractor) in San Diego.

Thompson received his bachelor's degree in electrical engineering from Berkeley in 1965 and earned a master's degree the following year. However, by then his interests had shifted almost completely from electrical engineering to software design. He later recalled, to Robert Slater that "I used to be an avid hacker in an electrical sense, building things. And ever since computers, I find it similar. Computing is an addiction. Electronics is a similar addiction but not as clean. Much dirtier. Things burn out."

In 1966, Thompson went to work at Bell Laboratories, a hotbed of innovation in computer science that had hosted such researchers as CLAUDE E. SHANNON, father of information theory. Thompson fit well into the unconventional "hacker" culture at Bell Labs, sometimes working for 30 hours at a stretch.

At that time, Bell Labs was heavily invested (along with the Massachusetts Institute of Technology (MIT) and General Electric) in a project called Multics. Multics was intended to be a new operating system built from the ground up to accommodate hundreds of users running programs at the same time. However, the ambitious project had bogged down, and in 1969 Bell Labs withdrew from it.

Thompson believed that a multiuser, multitasking operating system could still be built using a simpler approach. Thompson wrote a proposal for such a system, but Bell Labs officials, still unhappy in the wake of the Multics debacle, turned down Thompson's proposal. But Thompson was not willing to give up. He found an unused PDP-7 minicomputer in a storage room. Although it was obsolete, the machine had been used as an interactive terminal for a mainframe. It had a fast hard drive, a graphics display, and a Teletype for entering text—just the right hardware for designing an operating system that would be used interactively.

During the Multics project, Thompson had worked with another young researcher, DENNIS RITCHIE, and the two had found that they worked well together. Ritchie agreed to join Thompson in designing the new operating system. They started by designing a file system—a set of facilities for storing and retrieving data on the disk drive. As Ritchie later recalled to Slater,

> In the process it became evident that you needed various commands and software to test out the file system. You can't just sit there, you've got to make files and copy them. And so we wrote a small command interpreter that would [process] things that you typed to the keyboard . . . that is the essence of an operating system: something to read the commands, something to hold the data.

The process of building the new operating system was very tedious. The PDP-7 did not have a compiler for turning programs into runnable machine code, so Thompson and Ritchie had to write the operating system code on a GE mainframe. A program called a cross-assembler then translated the instructions to ones that could run on the PDP-7. The instructions would be punched from the GE machine onto paper tape that could then be loaded into the PDP-7. Eventually Thompson was able to get a program editor and assembler running on the smaller machine, and life became a little easier.

Finally Thompson, with Ritchie's assistance, had a primitive but workable operating system. The name they chose for it was UNIX (originally spelled Unics), which was a pun meant to suggest that the new system was simpler and better integrated than the behemoth Multics. However, there remained the problem of how to get people to use the new system on contemporary hardware. Given Bell Labs's resistance to their project, they decided to create a sort of software Trojan horse. They told lab officials that they wanted to develop an office text-processing system (what today is called a word processor). After some resistance, the lab agreed to give them a new PDP-11 minicomputer.

It turned out that the new machine's disk drive had not arrived, causing months of delay. However, by that time Thompson had managed to finagle the purchase of a more powerful computer. Thompson and Ritchie were able to set up a demonstration UNIX-based text processing system with three users at Teletypes. UNIX worked surprisingly well, and soon other departments were clamoring for time on the new machine so they could keep up with the growing flood of paperwork around the lab.

Thompson had also had a hand in the improvement of computer languages. At the start of the 1970s, there was much interest in creating a streamlined, well-structured language for systems and general applications programming.

In 1970, Thompson created an experimental language called B. Ritchie in turn came up with C, which would become the dominant computer language by the end of the decade. Thompson then rewrote UNIX in C so that the operating system could be implemented on virtually any machine that had a C compiler. This is what made UNIX into a "portable" operating system that, unlike the dominant personal computer operating systems of the 1980s and 1990s, was not tied to a particular kind of hardware.

Thompson also worked at the University of California, Berkeley, with BILL JOY and Ozalp Babaoglu to create a new version of UNIX that included paging (a way of flexibly managing and extending memory) and other features for the new generation of DEC VAX minicomputers. This became the well-regarded BSD version of UNIX, and a later version (SunOS) would be used by Bill Joy at Sun Microsystems.

By the 1980s, however, Thompson had largely moved on from UNIX to other interests. One of these interests, chess, dated back to his childhood. In 1980, Thompson and Joe H. Conden developed a computer chess program called BELLE. This program won the 1980 World Computer Chess Championship, and by 1983 it was strong enough to earn the official U.S. Chess Federation rank of Master.

Besides lecturing and teaching in various places during the 1980s and 1990s, Thompson also worked on a new operating system called Plan 9. In 1983, Thompson became a Bell Labs Fellow and also shared the Association for Computing Machinery Turing Award with Dennis Ritchie. The citation for the award reads as follows:

> The success of the UNIX system stems from its tasteful selection of a few key ideas and their elegant implementation. The model of the UNIX system has led a generation of software designers to new ways of thinking about programming. The genius of the UNIX system is its framework, which enables programmers to stand on the work of others.

Thompson was also elected to the U.S. Academy of Science and the U.S. Academy of Engineering in 1980, and in 1998 Thompson and Ritchie were awarded the National Medal of Technology. Thompson retired from Bell Labs in 2000.

Further Reading

"The Creation of the UNIX Operating System." 2002. Available on-line. URL: http://www.bell-labs.com/history/unix. Downloaded on November 3, 2002.

Slater, Robert. *Portraits in Silicon*. Cambridge, Mass.: MIT Press, 1987.

Thompson, Kenneth, and Dennis M. Ritchie. *Unix Programmer's Manual*. 6th ed. Santa Barbara, Calif.: Computer Center, University of California, 1979.

———. "The Unix Time-Sharing System: Unix Implementation." *Bell System Technical Journal* 57, no. 6 (1978): 1931–1946.

Tomlinson, Ray

(1941–)
American
Computer Scientist

Every day, millions of e-mail messages zip around the world. The addresses that direct these missives all contain a little symbol, @, between the recipient's name and the address of the mail service. Few e-mail users know that it was Ray Tomlinson who first developed the system of sending messages between computers and incidentally, rescued the @ symbol from the obscure reaches of the keyboard.

Tomlinson was born in 1941 in the tiny town of Vale Mills, New York. As a boy, he incessantly took things apart to see how they worked. When he was 12, his mother, despairing of being able to keep any household appliances in working order, bought him a radio kit, which sparked in Tomlinson a passion for electronics.

Tomlinson attended Rensselaer Polytechnic Institute in New York, earning a bachelor's degree in electrical engineering in 1963. He then went to the Massachusetts Institute of Technology (MIT) and received his master's degree in 1965. Meanwhile, he had become acquainted with computers as an intern at IBM in Poughkeepsie, New York, in 1960. He read a manual and wrote a program without knowing whether it would work. "I didn't know there were assemblers and compilers," he recalled, "but I understood that if I put holes in cards, things would happen."

While continuing at MIT for his doctorate, Tomlinson was supposed to be working for his dissertation on a voice recognition system, mainly the tricky matter of interfacing analog circuits to a digital computer system. But Tomlinson's interest kept wandering from hardware to software. Finally he left MIT without finishing his doctorate and got a job with the pioneering computer networking firm Bolt, Beranek, and Newman (BBN) in Cambridge, Massachusetts, where he still can be found today.

BBN was probably the single most important company for building the hardware infrastructure for the networks that would eventually become ARPANET and then the Internet. According to his former colleague Harry Forsdick, Tomlinson fit perfectly into the effort at BBN, who called him "a silent warrior . . . He is so quiet and self-deprecating but he's been the role model for young programmers at BBN for years. He's the best coder I've ever seen, bar none."

By the early 1970s, the first networks were starting to link computer centers at universities and laboratories. However, these networks were not used by people to send messages, but to transfer files and run programs. The only message facility was something called SNDMSG, which basically let users on a computer leave messages for each other in a file. SNDMSG, however, only allowed a user to communicate with another user with an account on the same computer.

However, in 1971 Tomlinson had also been working on a file-transfer program called CPYNET, which allowed files to be copied over the network from one computer to another. The idea that then came to him, seems simple in retrospect, like most great ideas: Why not combine SNDMSG and CPYNET? With the programs connected, a user at one computer could post a message to a file on his or her computer. CPYNET could then send the file containing the message to the computer where the recipient had an account. That user could then use that computer's SNDMSG program to read the message. Of course, quite a bit of software had to be written to tie together the whole system. For example, there had to be a way to read the address in a message file and figure out which computer it referred to so the proper CPYNET command could be constructed.

After a certain amount of tinkering, Tomlinson got the program to work. Later he would be asked what was in the first e-mail message ever sent between computers. A bit ruefully he remarked that unlike the famous "What hath God wrought?" of Morse or Bell's "Watson, come here, I need you!" the first e-mail message was probably something like QWERTYUIOP—the top line of alphabetical characters on a keyboard.

As with the telegraph and telephone, it took a while for people to figure out what to do with this new capability. Jerry Burchfiel, Tomlinson's colleague at BBN, remembered to interviewer Judith Newman that "when Ray came up with this electronic host-to-host e-mail demo over a weekend, my first reaction was maybe we shouldn't tell anybody. [E-mail] didn't fit under the contract description for the work we were doing." Burchfiel went on to say that "he was like everyone else. I had no idea it would take over the world." But LAWRENCE ROBERTS, head of computer communications research for ARPA (the Advanced Research Projects Agency of the Defense Department), liked the idea of e-mail and decided that it would be built into all the systems BBN designed for its clients.

To set up a standard protocol for e-mail, it was necessary to specify rules for how machine addresses would be constructed. In particular, Tomlinson needed a symbol to separate the user name from the name of the user's host computer. Later he would say to Newman that the choice of @ was pretty obvious: "You couldn't use a single [alphabetical] letter or number, because that would be confusing. It had to be something brief, because terseness was important. As it turns out, @ is the only preposition on the keyboard. I just looked at it, and it was there. I didn't even try any others."

As the volume of e-mail increased, users complained that it was hard to read because it arrived with all messages run together in a single file. Lawrence Roberts, a manager at ARPA, then wrote a program called RD. It allowed users to read, reply to, save or delete messages one at a time.

Although Tomlinson would become known in the computer commmunity as the "guy who invented e-mail," he accomplished many more formidable achievements behind the scenes. For example, he played an important role in developing Telnet, the facility that allows users to log into a remote computer and run programs there. He also contributed to the implementation of TCP/IP, the basic protocol for Internet data transmission.

In 1980, Tomlinson designed a microcomputer called Jericho from selecting chips all the way to the system software. Forsdick describes it as "an amazing tour de force of software and hardware. He has always been the ultimate troubleshooter, whether it was fixing the dishwasher in your kitchen or solving some esoteric problem in microcode."

As for what Tomlinson thinks of today's proliferation of e-mail, he muses:

> It's kind of interesting to contemplate what fraction of a cent I would have to get on the use of every @ sign to exceed Bill Gates's fortune. It's a very tiny amount—like 0.000001 cent or something. Especially with the amount of spam out there. Of course, then I'd be praying for spam, instead of condemning it.

Tomlinson has received several awards, including the 2000 George Stibitz Computer Pioneer Award of the American Computer Museum, the 2001 "Webby" award of the International Academy of Digital Arts and Sciences, and the 2002 *Discover* magazine Award for Innovation in Science and Technology.

Further Reading

Campbell, Tod. "The First E-mail Message." *Pre Text Magazine*. March 1998. Available on-line. URL: http://www.pretext.com/mar98/features/story2.htm. Downloaded on December 6, 2002.

Festa, Paul. "Present at the 'E'-Creation." *CNET News.com*. October 10, 2001. Available on-line. URL: http://news.com.com/2008-1082-274161.html?legacy=cnet. Downloaded on December 6, 2002.

Hafner, Katie. "Billions Served Daily, and Counting." *New York Times*, December 6, 2001, p. G1.

Tomlinson, Ray. "In the Beginning, a Note to Himself." *New York Times*, December 6, 2001, p. G9.

Torres Quevedo, Leonardo

(1852–1936)
Spanish
Engineer, Inventor

In the 18th century, a tradition of building automata, or very complex automatic mechanical devices, flourished. Although a few automata, such as the famous chess-playing "Turk," had a concealed human operator, most relied upon very carefully designed sequences of motion controlled by studs and cams.

The 19th century saw the first complete design for a mechanical digital computer by CHARLES BABBAGE as well as the beginnings of analog computing and automatic controls. The little-known but interesting career of the Spanish inventor Leonardo Torres Quevedo covers roughly the span between Babbage and the first modern digital computers.

Torres Quevedo was born on December 28, 1852, in Santa Cruz de Iguna in the province of Santander, northern Spain. He grew up in Bilbao and spent two years in a Catholic school in France. After completing his secondary education, Torres Quevedo enrolled in the Escuela de Caminos in Madrid, graduating in 1876 with a degree in civil engineering.

However, Torres Quevedo was not particularly interested in civil engineering. A generous inheritance from a distant relative gave him the financial freedom to pursue his many ideas and inventions (including an airship design that would be used in World War I).

His inventions in the computing field began in the 1890s when Torres Quevedo became interested in representing and solving mathematical equations mechanically. In 1893, he demonstrated a machine that used an ingenious combination of logarithmic scales (as in a slide rule) arranged geometrically to solve a wide variety of algebraic equations. The device was demonstrated in 1895 to a meeting of the French Association for the Advancement of Science.

Early in the 20th century, Torres Quevedo became increasingly interested in using electrical controls to operate machines. He also experimented with radio control (which he called *telekino*), building a small boat that could be steered by remote control from several kilometers away, as well as designs for radio-controlled torpedoes.

By 1906, Spanish scientists were so impressed by Torres Quevedo's ideas that they successfully lobbied for government support to establish a national Institute for Applied Mechanics, of which Torres Quevedo was naturally appointed director. (The organization's name was later changed to Institute of Automatics.)

However, Torres Quevedo was much more than an inspired tinkerer. He continually worked on developing new concepts and notations to describe machines on an abstract level, much in the way that ALAN TURING would do about 30 years later.

In 1911, Torres Quevedo built an automated chess machine that unlike the Turk really was automatic, although it could only perform rook and king vs. king endgames. The machine included electrical sensors to pick up the piece locations and a mechanical arm to move pieces.

Torres Quevedo's 1913 paper "Essais sur l'Automatique" (Essays on Automatics) built his ideas upon the earlier work of CHARLES BABBAGE which had been largely forgotten. In his paper, Torres Quevedo described such modern computer ideas as number storage, look-up tables, and even conditional branching (as in IF . . . THEN statements). In 1920, he built a less ambitious device that demonstrated the practicality of a keyboard-driven electromechanical calculator. It was connected to a typewriter that was used as the input/output device. However, he did not attempt to create a commercially viable version of the device.

Torres Quevedo died on December 18, 1936, in Madrid in the midst of the Spanish Civil War.

Further Reading

Cortada, James W. *Historical Dictionary of Data Processing Biographies*. New York: Greenwood Press, 1987.

Ralston, Anthony, Edwin D. Reilly, and David Hemmendinger, editors. *Encyclopedia of Computer Science*. 4th ed. London: Nature Publishing Group, 2000.

Torvalds, Linus
(1969–)
Finnish
Computer Scientist

Finnish programmer Linus Torvalds created Linux, a free version of the UNIX operating system. Its reliability and utility enhanced by a community of volunteer developers, Linux today powers many Web servers as well as providing an alternative to expensive proprietary operating systems for workstations and the desktop. *(© James A. Sugar/CORBIS)*

Linus Torvalds developed Linux, a free version of the UNIX operating system that has become the most popular alternative to proprietary operating systems. Today, reliable Linux systems can be found running countless Web servers and workstations.

Torvalds was born on December 28, 1969, in Helsinki, Finland. His childhood coincided with the microprocessor revolution and the beginnings of personal computing. At age 10, he received a Commodore personal computer (PC) from his grandfather, a mathematician. He learned to write his own software to make the most out of the relatively primitive machine.

In 1988, Torvalds enrolled in the University of Helsinki to study computer science. There he encountered UNIX, a powerful and flexible operating system that was a delight for programmers who liked to tinker with their computing environment. Having experienced UNIX, Torvalds could no longer be satisfied with the operating systems that ran on most PCs, such as MS-DOS, which lacked the powerful command shell and hundreds of utilities that UNIX users took for granted.

Torvalds's problem was that the UNIX copyright was owned by AT&T, which charged $5,000 for a license to run UNIX. To make matters worse, most PCs were not powerful enough to run UNIX, so he would probably need a workstation costing about $10,000.

At the time, the GNU project was underway, led by RICHARD STALLMAN, who had created the concept of open source software. The Free Software Foundation (FSF) was attempting to replicate all the functions of UNIX without using any of AT&T's proprietary code. This would mean that the AT&T copyright would not apply, and the functional equivalent of UNIX could be given away for free. Stallman and the FSF had already provided key tools such as the C compiler (Gnu C) and the Emacs program editor. However, they had not yet created the core of the operating system, known as the kernel. The kernel contains the essential functions needed for the operating system to control the computer's hardware, such as creating and managing files on the hard drive.

After looking at Minix, another effort at a free UNIX for PC, Torvalds decided in 1991 to write his own kernel and put it together with the various GNU utilities to create a working UNIX clone. Originally Torvalds wanted to call the new system Freax (for "free Unix" or "freaks" or "phreaks," a term sometimes used by hackers). However, the manager of the website through which Torvalds wanted to provide his system disliked that name. Therefore, Torvalds agreed to use the name Linux (pronounced LIH-nucks), combining his first name, Linus, and the "nix" from UNIX. In October 1991, the first version, modestly labeled .02, was made available for ftp download over the Internet.

Torvalds adopted the General Public License (GPL) developed by Stallman and the FSU, allowing Linux to be distributed freely by anyone who agreed not to place restrictions on it. The software soon spread through file transfer protocol (ftp) sites on the Internet, where hundreds of enthusiastic users (mainly at universities) helped to improve Linux, adding features and writing drivers to enable it to work with more kinds of hardware.

Even more so than Stallman's GNU project, Linux became the best-known example of the success of the open-source model of software development and distribution. Because the source code was freely available, programmers could develop new utilities or features and send them to Torvalds, who served as the gatekeeper and decided when to incorporate them and when it was time to release a new version of the Linux kernel.

By the mid-1990s, the free and reliable Linux had become the operating system of choice for many website servers and developers. A number of companies created their own added-value Linux distributions, including such names as Red Hat, Debian, and Caldera. Meanwhile, many large companies such as Corel, Informix, and Sun began to offer Linux distributions and products.

The combination of the essentially free Linux operating system and Java, a language developed by JAMES GOSLING primarily to run programs on Web servers and browsers, threatened to be the one-two punch that knocked out Microsoft Windows, the heavyweight champion of PC software. Linux had undeniable populist appeal: It was democratic and communitarian, not corporate. The main drawback for Linux as a PC operating system was the vast number of software packages (including the key office software developed by Microsoft) that ran only on Windows, and the fact that Linux was somewhat harder for new users to set up.

Meanwhile Torvalds, who still worked at the University of Helsinki as a researcher, faced an ever-increasing burden of coordinating Linux development and deciding when to release successive versions. As more companies sprang up to market software for Linux, they offered Torvalds very attractive salaries, but he did not want to be locked into one particular Linux package. Torvalds also found that he was in great demand as a speaker at computer conferences, particularly groups of Linux developers and users.

Instead, in 1997 Torvalds moved to California's Silicon Valley where he became a key software engineer at Transmeta, a company that makes Crusoe, a processor designed for mobile computing. However, Torvalds continues to keep strong ties to the Linux community.

As Torvalds told *PC Magazine* he believes that with the growth in free and open source operating software "the business model for software will evolve away from selling standard blocks of software such as Windows toward a future where software is customized to meet specific needs." As of 2002, Linux has thus not made serious inroads into the PC market, although a number of developments may eventually make Linux more competitive even there. These include improved installation programs and "Lindows," a Linux package that can run Windows programs through emulation. Meanwhile, the penguin, the friendly symbol for Linux adopted by Torvalds, waddles forward confidently into the future.

Torvalds has received numerous awards for his work on Linux, which reflect its widespread populist appeal. He was the *PC Magazine* 1999 Person of the Year. In 2002, *Internet Magazine* proclaimed him one of the top 10 "gods of the Internet." The Takeda Foundation in Japan named Torvalds and open source pioneer Richard Stallman as corecipients of its 2001 award, and *Reader's Digest* made him its 2000 "European of the Year." Torvalds told his own story in his autobiography, *Just for Fun*.

Further Reading

Bartholomew, Doug. "Lord of the Penguins." *Industry Week* 249 (February 7, 2000): 52.

Dibona, Chris. *Open Sources: Voices from the Open Source Revolution*. Sebastopol, Calif.: O'Reilly, 1999.

Moody, Glenn. *Rebel Code: Linux and the Open Source Revolution*. New York: Allen Lane, 2001.

Torvalds, Linus. *Just for Fun: The Story of an Accidental Revolutionary*. New York: HarperBusiness, 2001.

Turing, Alan

(1912–1954)
British
Mathematician, Computer Scientist

At the dawn of the computer age, Alan Mathison Turing's startling range of original thought led to the creation of many branches of computer science, ranging from the fundamental theory of computability to the question of what might constitute true artificial intelligence.

Turing was born in London on June 23, 1912. His father worked in the Indian (colonial) Civil Service, and his mother came from a family that had produced a number of distinguished scientists. Because his parents were often away, Turing was raised mainly by relatives until he was of school age, and then went as a boarding student to various schools and finally going to Sherborne, a college preparatory school.

As a youth, Turing showed great interest and aptitude in both physical science and mathematics, although he tended to neglect other subjects. One of his math teachers further observed that Turing "spends a good deal of time apparently in investigations of advanced mathematics to the neglect of elementary work."

When he entered King's College, Cambridge, in 1931, Turing's mind was absorbed by Einstein's relativity and the new theory of quantum mechanics, subjects that few of the most advanced scientific minds could grasp. At this time, he also encountered the work of mathematician JOHN VON NEUMANN, a many-faceted mathematical genius who would also become a great computer pioneer. Meanwhile, Turing pursued the study of probability and wrote a well-regarded thesis on the Central Limit Theorem.

Turing's interest then turned to one of the most perplexing unsolved problems of contemporary mathematics. Kurt Gödel had shown that in any system of mathematics there will be some assertion that can be neither proved nor disproved. But another great mathematician, David Hilbert, had asked whether there was a way to tell whether any particular mathematical assertion was provable.

Instead of pursuing conventional mathematical strategies to tackle this problem, Turing reimagined the problem by creating the Turing Machine, an abstract "computer" that performs only two kinds of operations: writing or not writing a symbol on its imaginary tape, and possibly moving one space on the tape to the left or right. Turing showed that from this simple set of states, any type of calculation could be constructed. His 1936 paper "On Computable Numbers," together with ALONZO CHURCH's more traditional logical approach, defined the theory of computability. After publishing this paper, Turing then came to America, studied at Princeton University, and received his Ph.D. in 1938.

Turing did not remain in the abstract realm, however, but began to think about how actual machines could perform sequences of logical operations. When World War II erupted, Turing returned to Britain and went into service with the government's secret code-breaking facility at Bletchley Park. He was able to combine his previous work on probability and his new insights into computing devices, such as the early special-purpose computer COLOSSUS, to help analyze cryptosystems, such as the German Enigma cipher machine, and to design specialized code-breaking machines.

Alan Turing "invented" an imaginary computer and showed how it could be used to determine what mathematical problems were computable. During World War II, Turing helped the Allies build automated systems to crack German U-boat codes. Later, Turing raised many of the key questions that would shape the field of artificial intelligence. *(Photo © Photographer, Science Source/Photo Researchers)*

As the war drew to an end, Turing's imagination brought together what he had seen of the possibilities of automatic computation, and particularly the faster machines that would be possible by harnessing electronics rather than electromechanical relays. In 1946, after he had moved to the National Physical Laboratory in Teddington, England, he received a government grant to build the Automatic Computing Engine, or ACE. This machine's design incorporated such advanced programming concepts as the storing of all instructions in the form of programs in memory without the mechanical setup steps required for machines such as the ENIAC. Another of Turing's important ideas was that programs could modify themselves by treating their own instructions just like other data in memory. However, the engineering of the advanced memory system ran into problems and delays, and Turing left the project in 1948 (it would be completed in 1950). Turing also continued his interest in pure mathematics and developed a new interest in a completely different field, biochemistry.

Turing's last and perhaps greatest impact would come in the new field of artificial intelligence. Working at the University of Manchester as director of programming for its Mark 1 computer, Turing devised a concept that became known as the Turing test. In its best-known variation, the test involves a human being communicating via a Teletype with an unknown party that might be either another person or a computer. If a computer at the other end is sufficiently able to respond in a humanlike way, it may fool the human into thinking it is another person. This achievement could in turn be considered strong evidence that the computer is truly intelligent. Since Turing's 1950 article, "Computing Machinery and Intelligence," computer programs such as ELIZA and Web "chatterbots" have been able to temporarily fool people they encounter, but no computer program has yet been able to pass the Turing test when subjected to extensive, probing questions by a knowledgeable person.

Turing also had a keen interest in chess and the possibility of programming a machine to challenge human players. Although he did not finish his chess program, it demonstrated some relevant algorithms for choosing moves, and led to later work by CLAUDE E. SHANNON, ALLEN NEWELL, and other researchers—and ultimately to Deep Blue, the computer that defeated world chess champion Garry Kasparov in 1997.

However, the master code breaker Turing himself held a secret that was very dangerous in his time and place: He was gay. In 1952, the socially awkward Turing stumbled into a set of circumstances that led to his being arrested for homosexual activity, which was illegal and heavily punished at the time. The effect of his trial and forced medical "treatment" suggested that the coroner was correct in determining Turing's death from cyanide poisoning on June 7, 1954, a suicide.

Alan Turing's many contributions to computer science were honored by his being elected a Fellow of the British Royal Society in 1951 and by the creation of the prestigious Turing Award by the Association for Computing Machinery, given every year since 1966 for outstanding contributions to computer science.

In recent years, Turing's fascinating and tragic life has been the subject of several autobiographies and even the stage play and TV film *Breaking the Code*.

Further Reading

Henderson, Harry. *Modern Mathematicians*. New York: Facts On File, 1996.

Herken, R. *The Universal Turing Machine*. 2nd ed. London: Oxford University Press, 1988.

Hodges, A. *Alan Turing: The Enigma*. New York: Simon & Schuster, 1983. Reprint, New York: Walker, 2000.

Turing, Alan M. "Computing Machinery and Intelligence." *Mind* 49 (1950): 433–460.

———. "On Computable Numbers, with an Application to the Entscheidungsproblem." *Proceedings of the London Mathematical Society* 2, no. 42 (1936–37): 230–265.

———. "Proposed Electronic Calculator." In A. M. *Turing's ACE Report of 1946 and Other Papers*. B. E. Carpenter and R. W. Doran, eds. Charles Babbage Institute Reprint Series in the History of Computing, vol. 10. Cambridge, Mass.: MIT Press, 1986.

The Alan Turing Home Page. Available on-line. URL: http://www.turing.org.uk/turing. Updated on October 13, 2002.

Turkle, Sherry

(1948–)
American
Scientist, Writer

The computer has had a tremendous impact on business and many other aspects of daily life. Many analysts and pundits have sought to understand the ongoing impact of developments such as the personal computer, the Internet, e-commerce, and mobile communications. However, relatively few researchers have looked at how computer networking may be affecting people psychologically. Sherry Turkle is a pioneer in this study who has raised provocative questions about cyberspace and the self.

Turkle was born Sherry Zimmerman on June 18, 1948, in Brooklyn, New York. When she was born, her biological father had left, and she received a new last name, Turkle, from her stepfather. Because she was referred to as Zimmerman at school but Turkle at home, she first experienced what would become a central theme of her later work—the question of multiple identities.

After graduating from Abraham Lincoln High School as valedictorian in 1965, she enrolled in Radcliffe College (which later became part of Harvard University) in Cambridge, Massachusetts. However, when she was a junior her mother died, she quarreled with her stepfather, and dropped out of Radcliffe because she was no longer able to keep up with her studies.

Turkle then went to France, where she worked as a live-in house cleaner for about a year and a half. The late 1960s in France was a time of both social and intellectual turmoil. A new movement called post-structuralism was offering a radical critique of modern institutions. Turkle became fascinated by its ideas and attended seminars by such key figures as Michel Foucault and Roland Barthes. Turkle, who had been somewhat shy, found that she was tapping into an alternate identity. As she later told author Pamela McCorduck, "I just knew that when I was in France—hey! I felt in touch with another aspect of my self. And I saw my job as bringing that through into English-speaking Sherry."

Post-structuralism and postmodernism see identity as something constructed by society or by the individual, not something inherent. The new philosophers spoke of people as having multiple identities between which they could move fluidly. Being able to try on a new identity for herself, Turkle could see the applicability of these ideas to her own life and explore how they might also explain the changes that were sweeping through society.

The year 1968 was marked by social protest and civil unrest in many countries. In the United States, the movement against the Vietnam War was reaching its peak. In Czechoslovakia, an uprising against the Soviet client state failed, and in France, a large uprising of students and workers eventually also failed to achieve significant reform. Turkle pondered the aftereffects of that failed revolution. Many French intellectuals turned inward, embracing the psychoanalytical theories of Freud that had been generally regarded with suspicion earlier. She considered the relationship between social conflict and the adoption of ideas.

Turkle decided to return to the United States to resume her studies. In 1970, she received an A.B. degree in social studies, summa cum laude, at Radcliffe. After working for a year with the University of Chicago's Committee on Social Thought, she enrolled in Harvard, receiving an M.A. degree in sociology in 1973. She went on to receive her doctorate in sociology and personality psychology in 1976. The title of her dissertation was "Psychoanalysis and Society: The Emergence of French Freud." In it, she showed how many French people had responded to severe social stress by using the ideas of Freud and the postmodern Freudian Jacques Lacan to create what amounted to a new identity. Turkle expanded her findings in her first book, *Psychoanalytic Politics: Freud's French Revolution*, published in 1978.

After getting her Harvard Ph.D., Turkle accepted a position as an assistant professor of sociology at the nearby Massachusetts Institute of Technology (MIT). Here she found a culture as exotic as that of the French intellectuals, but seemingly very different. In encountering the MIT hackers who would later be described in Steven Levy's book, she "came upon students who talked about their minds as machines. I became intrigued by the way in which they were using this computer language to talk about their minds, including their emotions, whereas I used a psychoanalytic one."

For many of the MIT computer students, the mind was just another computer, albeit a complicated one. An emotional overload required "clearing the buffer," and troubling relationships should be "debugged." Fascinated, Turkle began to function as an anthropologist, taking notes on the language and behavior of the computer students. In her second book, *The Second Self: Computers and the Human Spirit* (1984), Turkle said that the computer for its users is not an inanimate lump of metal and plastic, but an "evocative object" that offers images and experiences and draws out emotions. She also explained that the computer could satisfy deep psychological needs, particularly by offering a detailed but structured "world" (as in a video game) that could be mastered, leading to a sense of power and security.

Although the computer culture of the time was largely masculine, Turkle also observed that this evocative nature of technology could also allow for a "soft approach," based on relationship rather than rigorous logic. Of course, this "feminine" approach usually met with rejection. As Turkle told an interviewer, "Being a girl comes with a message about technology—'Don't touch it, you'll get a shock.'" Turkle believed that this message had to be changed if girls were not to be left behind in the emerging computer culture.

Turkle had already observed how computer activities (especially games) often led users to assume new identities, but she had mainly studied stand-alone computer use. In the 1990s, however, on-line services and the Internet in particular increasingly meant that computer users were interacting with other users over networks. In her 1995 book *Life on the Screen*, Turkle takes readers inside the fascinating world of the MUD, or multi-user dungeon. In this fantasy world created by descriptive text, users could assume any identity they wished. An insecure teenage boy could become a mighty warrior—or perhaps a seductive woman. A woman, by assuming a male identity, might find it easier to be assertive and would avoid the sexism and harassment often directed at females.

Cyberspace offers promises, perils, and potential, according to Turkle. As with other media, the computer world can become a source of unhealthy escapism, but it can also give people practice in using social skills in a relatively safe environment—although there can be difficulty in transferring skills learned on-line to a face-to-face environment.

Although the computer has provided a new medium for the play of human identity, Turkle has pointed out that the question of identity (and the reality of multiple identities) is inherent in the modern (or postmodern?) world. As she noted in an interview for the MIT *Technology Review*, "We live an increasingly multi-roled existence. A woman may wake up as a lover, have breakfast as a mother, and drive to work as a lawyer. . . . So even without computer networks, people are challenged to think about their identities in terms of multiplicity."

Increasingly, people no longer log in to a computer, do some work, then log out. The connection is always on, and with laptops, notebook computers, and palm computers the connection is available everywhere. High-capacity (broadband) networks also make it possible to deliver video and sound. Looking to the future, Turkle suggested to an interviewer for *Fortune* magazine that

> broadband is an intensive and extensive computational presence in which the computer doesn't become just an instrumental machine to get things done—to do things for us—but a subjective machine that changes us, does things to us. It opens up the computer as a partner in daily life in a way that it hasn't been before.

Turkle suggests that the boundary between cyberspace and so-called real life is vanishing, and thus "It's going to create a kind of crisis about the simulated and the real. The notion of what it is to live in a culture of simulation—how much of that counts as real experience and how much of that is discounted—is going to become more and more in the forefront of what people think and talk about."

Turkle married artificial intelligence pioneer and educator SEYMOUR PAPERT; the marriage ended in divorce. She continues today at MIT as a professor of the sociology of science and as director of the MIT Initiative on Technology and Self. She has received a number of fellowships, including a Rockefeller Foundation Fellowship (1980), a Guggenheim Fellowship (1981), Fellow of the American Association for the Advancement of Science (1992), and World Economic Forum Fellow (2002).

In 1984 Turkle was selected as Woman of the Year by *Ms.* magazine, and she has made a number of other lists of influential persons, such as the "Top 50 Cyber Elite" of *Time Digital* (1997) and *Time Magazine's* Innovators of the Internet (2000).

Further Reading

"Sherry Turkle" [Home page] Available on-line. URL: http://web.mit.edu/sturkle/www. Downloaded on November 3, 2002.

Turkle, Sherry. *Life on the Screen: Identity in the Age of the Internet*. New York: Simon and Schuster, 1995.

———.*The Second Self: Computers and the Human Spirit*. New York: Simon and Schuster, 1984.

———."Session with the Cybershrink." *Technology Review* 99 (February–March 1996): 41ff.

W

⊠ Wang, An
(1920–1990)
Chinese/American
Inventor, Entrepreneur

There was a time not long ago when "word processor" meant an office machine, not a program running on a personal computer (PC). From the 1970s through the early 1980s, An Wang's Wang Laboratories dominated the word processor market. Wang also invented the magnetic memory "core," an important form of mainframe memory until the advent of the memory chip.

An Wang, whose name means "Peaceful King," was born in Shanghai, China, on February 7, 1920. The family sometimes also lived in the village of Kun San, about 30 miles from Shanghai, where his father taught English. Young Wang started school in third grade when he was six years old, because the school had no earlier grades.

When Wang was only 16, he entered the Chiao Tung University, the Shanghai technical school that has been called "the MIT of China." By then, however, Japan was invading China, Shanghai was being bombed, and the countryside was plunging into turmoil. Wang tried to concentrate on his electrical engineering studies, as well as improving the English skills that his father had imparted. One way he practiced English was by translating popular science articles from American magazines to Chinese and publishing them in a digest. Meanwhile, the fighting came closer to the university, although it was not overrun.

In 1940, Wang received his bachelor's degree in electrical engineering and began to work as a teaching assistant at the university. However, in 1941 he volunteered to help the Chinese army design and build radio equipment for use by troops in the field. It is there that Wang learned to be what he described as a "seat of the pants engineer"—scrounging parts to build generators and fleeing constant Japanese bombing raids.

After the war, the Nationalist Chinese government sponsored a program to send engineers to the United States to learn about the latest technology. When Wang arrived in the United States in June 1945, he soon decided to take a step further and enroll at Harvard University. Because he already had practical experience, he was rushed quickly through the program, receiving a master's degree in 1946. In 1947, he entered Harvard's doctoral program in applied physics. By then, however, China was again embroiled in war, this time between the Nationalist and Communist forces. In his memoir *Lessons* (1986) Wang explained his decision not to return to China. "I . . . knew myself well enough to know that I could not thrive under a totalitarian Communist system. I had long been independent, and I wanted to continue to make my own decisions about my life."

In 1948, Wang received his Ph.D. in engineering and applied physics. He also received a research fellowship at the Harvard Computer Laboratory, which would prove to be a crucial turning point in Wang's career. Wang worked under HOWARD AIKEN, who had designed the Harvard Mark I, a huge programmable calculator that used thousands of electromechanical relays. Wang's assignment was to devise a storage (memory) system for this machine that did not rely on slow mechanical motion.

Magnetism seemed to be a good possibility, since data stored as little bits of magnetic material need not be continually refreshed. It was easy enough to magnetize something, creating a positive or negative magnetic "flux" that could stand for the binary 1 and 0. The problem was that reading the flux to determine whether it was a 1 or 0 reversed it, destroying the data. Wang spent weeks wracking his brain trying to come up with a solution to the dilemma. One afternoon, while walking through Harvard Yard, the solution suddenly came to him (as recounted in *Current Biography Yearbook 1987*): "It did not matter whether or not I destroyed the information while reading it. . . . I could simply rewrite the data immediately afterward."

Having solved that problem, Wang still had to create the memory device itself. After trying a nickel-iron alloy trade-named Deltamax he finally decided on ferrite, an iron composite. The ferrite was shaped into tiny doughnut-like rings and arranged in a three-dimensional lattice. Wang's patented "ferrite core memory" would become the standard fast memory device for mainframe computers in the 1950s and 1960s.

In 1951, Wang left Harvard and started his own company, Wang Laboratories, to manufacture magnetic core memories and hopefully to develop other new technologies. Wang was creative in finding applications for the devices—one of the earliest was the electronic scoreboard at New York City's Shea Stadium.

Wang Laboratories grew slowly, until in 1956 Wang was able to sell his memory patents to IBM for $400,000. Wang used the money to expand the company's production to include electronic counters, controls for automatic machine tools, and data coding devices. By the early 1960s, Wang had also designed LINASEC, an electronic typesetting system that became popular with newspapers, and LOCI, a desktop scientific calculator. This and Wang's later calculators gave scientists and engineers a relatively inexpensive alternative for calculations too complex for a simple calculator but which did not require an expensive computer. Wang Laboratories' sales grew steadily from $2.5 million in 1965 to $6.9 million in 1967, the year the company went public as one of the year's hottest stocks.

Wang anticipated that the calculator market would soon be flooded with cheaper models, so he decided to start building minicomputers. The System 2200, introduced in 1972, was sold to businesses for the attractive price of $8,000 and ran software written in the relatively easy-to-use BASIC programming language.

Wang believed that business data processing had three key aspects—calculation, text processing, and data processing. Leaving the calculators to others, Wang decided to focus on text processing. After all, the creation and revision of documents accounts for the bulk of activity in any office. Thus, the same year the System 2200 came out Wang introduced the Model 1200 Word Processing System. In doing so, Wang was challenging IBM, whose typewriters and related equipment had long dominated the office market.

While the first Wang word processor used a typewriter equipped with magnetic tape cassettes to store text, the next model, the WPS, used a cathode-ray tube, allowing text to be created and edited on screen. This was the birth of modern word processing. Wang's word processors and office computers sold well and Wang opened a large new factory in Lowell, Massachusetts, revitalizing that city of old factories and mills.

In 1979, Wang launched a new family of machines called the Office Information System. These systems integrated word processing and database functions, anticipating the modern office application suite. Wang then added facilities for processing voice, images, and video and for networking users together. The goal was to integrate an office's entire document, data processing, and communications needs into a single system that could be scaled upward to meet a company's growing needs.

By 1981, Wang seemed to be well on the way to achieving his business goals. The company was earning $100 million a year and had been growing at an annual rate of 55 percent for the preceding five years. However, Wang's attempt to achieve lasting domination of the office information systems market ran into formidable challenges.

A number of very large companies, including Hewlett-Packard and Digital Equipment Corporation (DEC), began to offer systems that competed with at least a portion of Wang's offerings. But the ultimate downfall of Wang Laboratories turned out to be the IBM PC, which debuted in 1982. At first, the PC with its limited memory and storage capacity could not match the performance of Wang's word processing and information systems. However, as Wang struggled to make its advanced technology (especially networking) work smoothly and to convince businesses to buy equipment that cost many thousands of dollars, the PC like the proverbial tortoise began to overtake Wang's hare. In 1985, Wang Laboratories' phenomenal growth finally shuddered to a halt: Earnings dropped precipitously and the company laid off 1,600 workers.

By the late 1980s, the PC's processor, display, memory, and hard drive had all been upgraded. Perhaps more important, unlike those committed to the dedicated Wang equipment, PC users could buy whatever software they wished. A new generation of software such as WordPerfect, dBase, and Lotus 1-2-3 gave the PC open-ended capabilities. In the following decade, software suites such as Microsoft Office would add the integration that had been such an attractive feature of the Wang system.

Wang, who had partly retired from the company, returned in an attempt to turn Wang Laboratories around. He streamlined operations and landed a huge $480 million contract to supply minicomputers to the U.S. Air Force. Wang also developed an improved Wang Integrated Office Solution. In 1986, he named his son Fred Wang as chief executive officer. Unfortunately, the turnaround was only temporary, and by the end of the decade the company's fortunes were again in decline. An Wang died of esophageal cancer on March 24, 1990.

He had reached the first rank both as inventor and entrepreneur, perhaps by following his own observation that "success is more a function of consistent common sense than it is of genius." He received the U.S. government's Medal of Liberty in 1986, and in 1988 he was elected to the National Inventors Hall of Fame. Wang also donated generously to many philanthropic causes, including a performing arts center in Boston, a clinic at Massachusetts General Hospital, and his own Wang Institute of Graduate Studies, specializing in training software engineers.

Further Reading

Bensene, Rick. "Wang Laboratories: From Custom Systems to Computers." Available on-line. URL: http://www.geocities.com/oldcalculators/d-wang-custom. html. Posted on October, 2001

Kenney, Charles C. *Riding the Runaway Horse: The Rise and Decline of Wang Laboratories*. Boston: Little, Brown, 1992.

Moritz, Charles, ed. "Wang, An." *Current Biography Yearbook, 1987*. New York: H. W. Wilson, 1987, pp. 586–590.

Wang, An, and Eugene Linden. *Lessons: An Autobiography*. New York: Addison-Wesley, 1986.

Watson, Thomas J., Sr.
(1874–1956)
American
Entrepreneur

For more than a generation, the word most closely associated with "computer" in the public mind had three letters: IBM. Just as Bell became the iconic name for telephone and Ford for the automobile, the products of International Business Machines dominated the office through most of the 20th century. The business leader who took the company from typewriters and tabulators to the mainframe computer was Thomas J. Watson Sr.

Watson was born February 17, 1874, in Campbell, New York. His father was a successful lumber dealer and the family also ran a farm. Watson did well in school, attending first the Addison (New York) Academy, and then the School of Commerce in Elmira, New York. Although Watson's father wanted him to study law, young Watson was determined to make his own way. Watson's first job after graduation in 1892 was as a salesman selling pianos, organs, and sewing machines from a wagon near Painted Post, New York.

In 1898, Watson entered the corporate world at National Cash Register (NCR), having been hired by a reluctant manager after persistent visits to his office. It was a tough year to start a corporate career: the country was in the midst of an economic depression and it was difficult to generate many sales. Nevertheless, Watson soon impressed his superiors with his continued persistence and energy, which by 1898 made him the top salesman in the company's Buffalo office. Watson would later attribute his success to following "certain tried and true homilies," and he would later use slogans to focus the attention of his employees at IBM.

Watson continued to move up through the managerial and executive ranks, becoming general sales manager at NCR's home office in Dayton, Ohio, in 1903. Watson's aggressive pursuit of sales and targeting of the competition eventually took him to the edge of ethics and legality. NCR's executives asked him to do something about the companies that were selling secondhand cash registers and thus undercutting NCR's business. Watson responded by creating and running a used cash-register company that was supposedly independent but in reality was secretly funded by NCR. The company successfully undercut its competitors, but in 1910 NCR's main competitor, the American Cash Register Company, sued NCR for violating the Sherman Antitrust Act, and the government joined in the suit. On February 13, 1913, NCR and a number of its executives, including Watson, were convicted of antitrust violations. Watson was fined $5,000 and sentenced to one year in prison.

An appeals court overturned Watson's conviction in 1915, but by then he had left NCR. The company had wanted him to sign a consent decree admitting illegal practices. Watson had steadfastly refused, claiming that he had done nothing illegal. Watson was then fired. Later he would say that it was then that he vowed to form a company even larger than the mighty "Cash," as NCR was informally called.

Watson was not out of work for long. He was offered the job of general manager at a company called Computing-Tabulating-Recording, or C-T-R. This company manufactured items such as time clocks, scales, and punch-card tabulators (using technology that had been developed in the late 19th century by HERMAN HOLLERITH).

Watson marshaled the sales and managerial skills he had mastered at NRC to build C-T-R's business. He decided to focus on the tabulating machine part of the business, seeing it as having the most growth potential. Refining practices common at NCR, Watson emphasized the motivation and training of the sales force. He gave each salesman a defined territory and a sales quota. Salesmen who made their quotas were given a special recognition as members of the "100 Percent Club." A dress code and the required abstention from alcoholic beverages resulted in a disciplined sales force and a reputation for conservatism that would later characterize IBM's corporate culture.

Overall, Watson's management philosophy was that the company needed to be sales driven. All other divisions, including research and development and manufacturing, were expected to understand what was needed to promote sales. Slogans such as "Sell and Serve," "Aim High," and especially "Think," were used in offices and company meetings.

The fruits of Watson's priorities and practices soon became evident. C-T-R's sales rose from $4.2 million in 1914 to $8.3 million in 1917, largely accounted for by a growing market share in the tabulating equipment area. In 1924, Watson became chief executive officer of C-T-R. That same year, he changed the company's name to something that better reflected the scope of his ambitions: International Business Machines, or IBM.

Under Watson's leadership, IBM expanded its product line, particularly card sorters and tabulators, which grew in capabilities and became essential to many large corporations and government offices. By the eve of World War II, IBM's annual sales were approaching $25 million. The *New York Times* referred to Watson as "an industrial giant" and later the *Saturday Evening Post* attributed IBM's success to Watson's "ingenuity in creating new markets, perfecting of educational-sales technique, and stubborn strength for hard work."

Although Watson was also criticized for being a foe of labor unions he did support President Franklin D. Roosevelt's New Deal economic programs, although he stayed out of direct involvement with politics. Watson also promoted the "International" part of the company's name through a high-profile exhibit at the 1939 World's Fair in New York and the slogan "world peace through world trade."

Watson had shown some interest in computation (the product line included some basic electromechanical calculators) but during the war years he became more aware of the potential of large-scale automatic computation. He became an important backer of the research of HOWARD AIKEN at Harvard, who built the Mark I, a huge programmable calculator (or early computer) that used relays for its operation. Watson became upset when the Mark I was unveiled at Harvard because he thought that Aiken and the university had taken most of the credit while downplaying IBM's important role in the funding and research.

After the war, Watson and other IBM executives debated over whether the company should produce electronic digital computers. IBM did produce a small machine, the Selective Sequence Electronic Calculator (SSEC), but the usually aggressive and innovative Watson seemed reluctant to commit the company to computers on a large scale. However, it is also true that no one in the late 1940s really knew whether there was a viable market for more than a few of the huge, unwieldy vacuum tube computers.

However, the urgings of his son THOMAS WATSON JR., and the success of ENIAC's successor Univac, built by J. PRESPER ECKERT and JOHN MAUCHLY with assistance from JOHN VON NEUMANN, spurred IBM to respond. Watson backed his son's instincts in committing the company wholeheartedly to the computer age. In 1952, Watson made his son the company president while retaining the chairmanship of the board. Thomas Watson Sr. died on June 19, 1955. He had left a company with a worldwide reputation and annual revenues of $700 million.

Watson had considerable interest in the arts, to which he contributed generously. He also contributed to innovations in the medical field, including arthritis research and the development of an eye bank for corneal transplants. Watson received the U.S. Medal of Merit for meritorious service during World War II. He received numerous awards from other countries as well.

Further Reading

"Facts, Factoids, Folklore." *Datamation* 32 (January 1, 1986): 42ff.

Rodgers, W. *THINK: A Biography of the Watsons and IBM*. New York: Stein and Day, 1969.

Sobel, Robert. *Thomas Watson, Sr., IBM and the Computer Revolution*. Frederick, Md.: Beard Books, 2000.

Watson, T. J., Jr. and P. Petre. *Father, Son, and Company: My Life at IBM*. New York: Bantam, 1990.

Watson, Thomas, Jr.
(1914–1993)
American
Entrepreneur

Thomas J. Watson Sr. (at right) turned a tabulator company into the office machine giant IBM. His son Thomas Watson Jr. (at left) then took the company into the computer age in the early 1950s. It would dominate the market for mainframe computers for three decades or more. *(IBM Corporate Archives)*

In the course of three decades, THOMAS J. WATSON SR. had built the small Computing-Tabulating-Recording company (CTR) into the giant International Business Machines (IBM), taking it to the dawn of the computer age. His son Thomas Watson Jr. took the reins in the early 1950s and built IBM's computer business until its mainframes dominated business and government data processing.

Thomas Watson Jr. was born on January 8, 1914, in Dayton, Ohio. He was the eldest son of Thomas Watson Sr., the chief executive office (CEO) of IBM. The boy was exposed to his father's business from a very young age: At five years of age, he was given a tour of the IBM factory. When he was nine, Watson Jr. joined his father on an inspection tour of the company's European plants. As Watson would later tell the *New York Herald Tribune* on the verge of his taking over the reins of IBM: "I take real pride in being a great man's son. My father has set a fine example for me. Because he is who he is, I realize that nearly every act of mine will be scrutinized very closely."

However, that was in retrospect. In actuality, the young Watson responded to the nearly overwhelming presence of his father by rebellion: He became known as "Terrible Tommy Watson," and often got into trouble for pranks. Alternately, he became depressed and lacked self-confidence.

After attending various private secondary schools, at which he was an indifferent student, Watson Jr. enrolled in Brown University in Providence, Rhode Island, graduating in 1937 with a B.A. degree, again without any academic distinction. That same year, he joined IBM as a junior salesman. His father kept a scrupulous hands-off policy, making it clear that there would be no favoritism. Watson Jr. was assigned to one of the toughest sales districts, the Wall Street area in New York. Well aware that all eyes (including his father's) were on him, Watson Jr. buckled down and not only met his sales quota but exceeded it by 150 percent.

During World War II, Watson Jr. enlisted as a private in the U.S. Army Air Force. He became a pilot and flew on difficult ferry missions between Alaska and Russia and later flew staff missions between Russia and the Middle East. By the time of his discharge in 1946, he had reached the rank of lieutenant colonel and had earned a U.S. Air Medal.

Upon return to IBM, Watson Jr. was rapidly groomed to step into the responsibilities gradually being relinquished by his aging father. Watson Jr. started as vice president in charge of sales and was appointed to the board of directors four months later. In 1949, he became executive vice president, and in 1952 he replaced his father as IBM's president, with his father becoming chairman of the board.

IBM had made its success on electromechanical technology—typewriters, tabulators, and calculators of increasing sophistication. World War II had seen the birth of a much faster and more versatile information-processing technology—the electronic digital computer pioneered by researchers such as J. PRESPER ECKERT and JOHN MAUCHLY, who had built the giant vacuum tube computer ENIAC at the University of Pennsylvania. Now Eckert and Mauchly had started a company to produce an improved electronic computer, Univac.

But while small companies were willing to build on a radical new technology, IBM was in a more difficult position. A company that has built such a commanding position on one technology cannot easily shift to another while maintaining its business, and Watson Sr. was reluctant to make the investment needed for new factories, particularly when it was not clear whether the market for the large, expensive machines would amount to more than a few dozen.

However, by 1955 *Time* was reporting that "the mechanical cogs and gears have given way to electronic circuits, cathode-ray tubes, and transistors." It was largely the result of the firm conviction of Watson Jr. that computers were the wave of the future, as well as the need he felt to match and, if possible, exceed the achievements of the father he admired.

Watson Sr. died on June 19, 1955, and the following year Watson Jr. became IBM's CEO. At that time, the company had grown to 72,500 employees and had annual revenue of $892 million. With Watson Jr. in control, the company developed its IBM 702 (primarily an accounting machine), the 650 (designed for smaller businesses) and the 701 (a scientific computer). The threat from Sperry-Rand's Univac subsided. IBM also extended its research and introduced innovations such as the first disk drive (with the IBM 305 RAMAC), which had a capacity of five megabytes.

In the early 1960s, IBM's product line became fully transistorized with the 1400 series. The IBM/system/360, announced in 1964, redefined the mainframe computer. It was designed to be scalable, so that users could start with a small system and add larger models without any hardware or software becoming incompatible. This era also marked the introduction of the first integrated circuits, the next step in miniaturization after the transistor.

As a manager, Watson Jr. inherited many useful practices from his father, but had innovations of his own. One was the "penalty box" by which an executive who had made a serious mistake would be temporarily moved from the fast track to a post where he had time to think things over. But although he drove people hard and could be harsh, Watson Jr. did not want meek compliance. He once noted, "I never hesitated to promote someone I didn't like. The comfortable assistant—the nice guy you like to go fishing with—is a great pitfall. I looked for those sharp, scratchy, harsh, almost unpleasant guys who see and tell you about things as they really are."

With IBM's growing dominance came increased legal threats of antitrust action. In 1952, Watson Jr. had settled a case with the government over his father's objections. In 1969, the government filed a more serious case in which it claimed that IBM dominated the computer industry to such an extent that the company should be broken up. IBM had always sold everything as a package—hardware, software, and service. This "bundling" had made it very difficult for competitors to get a toehold in the IBM-compatible software market. Under pressure from the government, Watson agreed to "unbundle" the hardware and software components and sell them separately. The case would drag on until 1981 when the Reagan administration dropped it.

In 1970, Watson Jr. had a heart attack. Although he recovered, the experience made him decide to take an earlier retirement to reduce health-threatening stress. Watson did retain some input into the company's executive board, however. Meanwhile, he indulged his passion for sailing, making long voyages, including one to Greenland and the Arctic Circle.

As a successful business leader, Watson Jr. was also in great demand for government service. In the Kennedy administration, he had served on a number of advisory commissions. President Johnson asked Watson Jr. to be his secretary of commerce, but he declined. In 1979, Watson did take the post of U.S. ambassador to Moscow. He entered at a time of high tension between the United States and the Soviet Union as a result of the latter's invasion of Afghanistan and disputes over the deployment of new missile systems. During the 1980s, Watson Jr. became an advocate for arms reduction. He also watched and occasionally offered advice as IBM launched the IBM PC, the machine that would define a whole new computer market.

In addition to sailing, Watson Jr. enjoyed flying small planes (including jets and a helicopter), and collecting scrimshaw and antique cars. He also took the time to write his autobiography, *Father, Son and Company: My Life at IBM and Beyond*.

Thomas Watson Jr. died on December 31, 1993, from complications from a stroke. He received many awards from trade groups in recognition of his business leadership and ranks high in many lists of the 20th century's greatest executives.

Further Reading

"The Businessman of the Century: Henry Ford, Alfred P. Sloan, Tom Watson, Jr., Bill Gates." *Fortune*, November 22, 1999, pp. 108ff.

Rodgers, W. *THINK: A Biography of the Watsons and IBM*. New York: Stein and Day, 1969.

Watson, T. J., Jr., and P. Petre. *Father, Son, and Company: My Life at IBM*. New York: Bantam, 1990.

Weizenbaum, Joseph
(1923–)
German/American
Computer Scientist

Since its inception in the early 1950s, research in artificial intelligence (AI) has led to dramatic achievements and frustrating failures (or partial successes). Although researchers have had to accept that the challenge of creating computer programs that exhibit true humanlike intelligence is more difficult than they had imagined, most have continued to be basically optimistic about the effort and sanguine about its potential results. However, one researcher, Joseph Weizenbaum, after writing one of the most famous programs in the history of AI, became one of the most persistent and cogent critics of the AI project itself.

Weizenbaum was born on January 8, 1923, in Berlin to Jewish parents. In 1934, he enrolled in a Berlin preparatory school, but after two semesters he was dismissed because of recently passed Nazi racial laws. In 1936, the Weizenbaum family, increasingly fearful about the future in Germany, emigrated to the United States.

In 1941, Weizenbaum enrolled in Wayne State University in Detroit, Michigan. The following year, he enlisted in the U.S. Army Air Corps. After the war, he resumed his study of mathematics. While working as a research assistant, Weizenbaum had the opportunity to help design and build an early digital computer, and although he received his master's degree in mathematics in 1950, he would spend his career in the computer field.

From 1955 to 1963, Weizenbaum worked for General Electric's Computer Development Laboratory as a systems engineer. During this time, he would oversee the design and implementation of the first integrated computerized banking system, for Bank of America.

In 1963, Weizenbaum returned to academia, joining the faculty at the Massachusetts Institute of Technology (MIT), which had one of the nation's foremost programs in artificial intelligence research. He contributed to the development of the time-sharing computer system at MIT and early computer networks, but the work for which he would be most remembered started with his interest in natural language processing. This was the effort to get computers to communicate in ordinary language and ideally, to carry on meaningful conversations with human beings.

In 1966, Weizenbaum created a remarkable program called ELIZA (named for the character in George Bernard Shaw's play *Pygmalion* who is taught "proper English" by Professor Henry Higgins). ELIZA was remarkable not for its complexity or for innovative use of AI techniques, but for the way it used a few simple procedures to convey the impression that it was a true artificial intelligence—or indeed, a real person.

ELIZA worked by identifying key words and basic sentence structure and then "mirroring" the user's statements back, much in the manner of the nondirective psychotherapy that Carl Rogers had made popular. A sample snippet of human-ELIZA dialog might look like this (ELIZA's words are in caps).

> Men are all alike.
> IN WHAT WAY?
> They're always bugging us about something or other.
> CAN YOU THINK OF A SPECIFIC EXAMPLE?
> Well, my boyfriend made me come here.
> YOUR BOYFRIEND MADE YOU COME HERE
> He says I'm depressed much of the time.
> I AM SORRY TO HEAR YOU ARE DEPRESSED
> It's true. I am unhappy
> DO YOU THINK COMING HERE WILL HELP YOU NOT TO BE UNHAPPY
> I need some help, that much seems certain.
> WHAT WOULD IT MEAN TO YOU IF YOU GOT SOME HELP
> Perhaps I could learn to get along with my mother.
> TELL ME MORE ABOUT YOUR FAMILY

According to legend, one night Weizenbaum had absent-mindedly gone home without shutting down ELIZA on his terminal. Around 4 A.M. a graduate student went looking for Weizenbaum (whom he thought was still logged onto the computer). Having run into problems with his program, the student wanted to ask Weizenbaum to log off so he would have access to the whole computer, which might speed up his debugging effort.

When the student arrived there was no one in Weizenbaum's office, but the terminal was on and a cursor was blinking. Thinking Weizenbaum had logged on from home, the student typed, "Joe, please log off."

The interaction then proceeded something like this:

> WHY DO YOU SAY JOE LOG OFF?
> I need the computer for an AI demo tomorrow, Joe
> WHAT MAKES YOU BELIEVE THAT?
> Joe, I'm serious. I've got a demo here tomorrow and I need the computer.
> PERHAPS THE PEOPLE YOU HANG AROUND WITH COULD HAVE CAUSED THIS.

The student grew increasingly angry at what seemed to be Weizenbaum's obnoxious behavior, and finally called him at home. It took awhile for Weizenbaum to realize that the student had been talking to ELIZA. ELIZA had inadvertently passed the test that ALAN TURING had posed, in which a human user tries to determine whether he or she is talking to another human being or a computer.

Weizenbaum's 1976 book *Computer Power and Human Reason* is a collection of essays that both explain the achievements of AI pioneers and point out their limitations. In particular, Weizenbaum became dismayed at the ease with which many people responded to ELIZA as though it were human. This posed the possibility that computer technology would be allowed to create shallow substitutes for a human intelligence that science was no longer really trying to understand. If, as Weizenbaum observes, "the computer programmer is creator of universes for which he alone is responsible. . . . Universes of almost unlimited complexity. . ." then indeed the computer scientist must take responsibility for his or her creations. This is the challenge that Weizenbaum believes has not been taken seriously enough.

As the 1960s progressed, the United States plunged into the Vietnam War and racial tension crackled in the streets of major cities. Weizenbaum became increasingly concerned that technology was being used for warlike and oppressive purposes. Later, he recalled that

> The knowledge of behavior of German academics during the Hitler time weighed on me very heavily. I was born in Germany, I couldn't relax and sit by and watch the university in which I now participated behaving in the same way. I had to become engaged in social and political questions.

As an activist, Weizenbaum campaigned against what he saw as the misuse of technology for military purposes, such as missiles and missile defense systems. He was founder of a group called Computer Professionals Against the ABM (antiballistic missile). In a 1986 article, he wrote that:

> It is a prosaic truth that none of the weapon systems which today threaten murder on a genocidal scale, and whose design, manufacture and sale condemns countless people, especially children, to poverty and starvation, that none of these devices could be developed without the earnest, even enthusiastic cooperation of computer professionals. It cannot go on without us! Without us the arms race, especially the qualitative arms race, could not advance another step.

Although most pundits consider the computer to be a source of revolutionary innovation, Weizenbaum has suggested that it has actually functioned as a conservative force. He gives the example of banking, which he had helped automate in the 1950s. Weizenbaum asks

> Now if it had not been for the computer, if the computer had not been invented, what would the banks have had to do? They might have had to decentralize, or they might have had to regionalize in some way. In other words, it might have been necessary to introduce a social invention, as opposed to the technical invention.

Weizenbaum does not consider himself to be a Luddite, however, and he is not without recognition of the potential good that can come from computer technology:

> Perhaps the computer, as well as many other of our machines and techniques, can yet be transformed, following our

> own authentically revolutionary transformation, into instruments to enable us to live harmoniously with nature and with one another. But one prerequisite will first have to be met: there must be another transformation of man. And it must be one that restores a balance between human knowledge, human aspirations, and an appreciation of human dignity such that man may become worthy of living in nature.

As a practical matter, he says, "The goal is to give to the computer those tasks which it can best do and leave to man that which requires (or seems to require) his judgment."

During the 1970s and 1980s, Weizenbaum not only taught at MIT but also lectured or served as a visiting professor at a number of institutions, including the Center for Advanced Studies in the Behavioral Sciences at Stanford (1972–73), Harvard University (1973–74), and the Technical University of Berlin and the University of Hamburg.

In 1988, Weizenbaum retired from MIT. That same year, he received the Norbert Wiener Award for Professional and Social Responsibility from Computer Professionals for Social Responsibility (CPSR). In 1991, he was given the Namur Award of the International Federation for Information Processing. He has also received European honors, such as the Humboldt Prize from the Alexander von Humboldt Foundation in Germany.

Further Reading

Ben-Aaron, Diana. "Weizenbaum Examines Computers [and] Society." *The Tech* (Massachusetts Institute of Technology) vol. 105, April 9, 1985. Available on-line. URL: http://the-tech.mit.edu/V105/N16/weisen.16n.html.

ELIZA. Available on-line. URL: http://www.uwec.edu/jerzdg/orr/articles/IF/canon/Eliza.htm. Updated on March 6, 2001.

Weizenbaum, Joseph. *Computer Power and Human Reason* San Francisco: W. H. Freeman, 1976.

———. "ELIZA—A Computer Program for the Study of Natural Language Communication Between Man and Machine." *Communications of the* ACM, 9 (January 1966): 35–36. Available on-line. URL: http://i5.nyu.edu/~mm64/x52.9265/january1966.html. Downloaded on December 9, 2002.

Wiener, Norbert

(1894–1964)
American
Mathematician, Computer Scientist

Although computers often appear to be abstract entities, they exist in, and interact with, the physical world. Indeed, today many "hidden" computers control the heat and air conditioning in homes, help steer cars, and for that matter, assemble parts for those cars in the factory. Norbert Wiener developed the theory of cybernetics, or the process of communication and control in both machines and living things. His work has had an important impact on philosophy and on design principles.

Wiener was born on November 26, 1894, in Columbia, Missouri. His father was a professor of Slavic languages at Harvard University, and spurred an interest in communication which the boy combined with an avid pursuit of mathematics and science (particularly biology). A child prodigy, Wiener started reading at age three. His father force-fed him knowledge far ahead of his years. He demanded perfect answers to every question or problem and became enraged when the boy made mistakes. Later, in his autobiography *Ex-Prodigy* (1953), Wiener would reveal how his father had made him feel a sense of inadequacy that, together with his isolation from other children, stunted Wiener's social development.

Wiener's academic development proceeded at a dizzying pace, however. He entered Tufts

University at age 11 and earned his B.A. degree in mathematics in 1909 at the age of 14, after concluding that his lack of manual dexterity made biological work too frustrating. He earned his M.A. degree in mathematics from Harvard only three years later, and his Harvard Ph.D. in mathematical logic just a year later, in 1913, when he was still only 18. Wiener then traveled to Europe on a fellowship, where he met leading mathematicians such as Bertrand Russell, G. H. Hardy, Alfred North Whitehead, and David Hilbert. When America entered World War I, Wiener served at Aberdeen Proving Ground in Maryland, where he designed artillery firing tables.

Norbert Wiener developed the science of cybernetics, or automatic control systems. To do so, he drew on physiology, neurology, mathematics, and electronics. In this book *The Human Use of Human Beings,* he argued for bringing humanist values to decisions about how to employ technology. *(Photo © Bettmann/CORBIS)*

After the war, Wiener was appointed as an instructor in mathematics at the Massachusetts Institute of Technology (MIT), where he would serve until his retirement in 1960. He continued to travel widely, serving as a Guggenheim Fellow at Copenhagen and Göttingen in 1926 and a visiting lecturer at Cambridge (1931–32) and Tsing-Hua University in Beijing (1935–36). Wiener's scientific interests proved to be as wide as his travels, including research into stochastic and random processes (such as the Brownian motion of microscopic particles) where he sought more general mathematical tools for the analysis of irregularity. His work took him into the front rank of world mathematics.

During the 1930s, Wiener began to work more closely with MIT electrical engineers who were building mechanical analog computers. He learned about feedback controls and servomechanisms that enabled machines to respond to forces in the environment.

When World War II came, Wiener did secret military research with an engineer, Julian Bigelow, on antiaircraft gun control mechanisms, including methods for predicting the future position of an aircraft based upon limited and possibly erroneous information.

Wiener became particularly interested in the feedback loop—the process by which an adjustment is made on the basis of information (such as from radar) to a predicted new position, a new reading is taken and a new adjustment made, and so on. (He had first encountered these concepts at MIT with his friend and colleague Harold Hazen.) The use of "negative feedback" made it possible to design systems that would progressively adjust themselves, such as by intercepting a target. More generally, it suggested mechanisms by which a machine (perhaps a robot) could progressively work toward a goal.

Wiener's continuing interest in biology led him always to relate what he was learning about control and feedback mechanisms to the behavior of living organisms. He had followed the work of Arturo Rosenbleuth, a Mexican physiologist who was studying neurological conditions that appeared to result from excessive or inaccurate feedback. (Unlike the helpful negative feedback, positive feedback in effect amplifies errors and sends a system swinging out of control.)

By the end of World War II, Wiener, Rosenbleuth, the neuropsychiatrist Warren

McCulloch, and the logician Walter Pitts were working together toward a mathematical description of neurological processes such as the firing of neurons in the brain. This research, which started out with the relatively simple analogy of electromechanical relays (as in the telephone system), would eventually result in the development of neural network theory, which would be further advanced by MARVIN MINSKY. More generally, these scientists and others such as JOHN VON NEUMANN had begun to develop a new discipline for which Wiener in 1947 gave the name *cybernetics*. Derived from a Greek word referring to the steersman of a ship, it suggests the control of a system in response to its environment.

The field of cybernetics attempted to draw from many sources, including biology, neurology, logic, and what would later become robotics and computer science. Wiener's 1948 book *Cybernetics, or Control and Communication in the Animal and the Machine* was as much philosophical as scientific, suggesting that cybernetic principles could be applied not only to scientific research and engineering but also to the better governance of society. (On a more practical level, Wiener also worked with Jerome Wiesner on designing prosthetics to replace missing limbs.)

Although Wiener did not work much directly with computers, the ideas of cybernetics would indirectly influence the new disciplines of artificial intelligence (AI) and robotics. However, in his 1950 book *The Human Use of Human Beings*, Wiener warned against the possible misuse of computers to rigidly control or regiment people, as was the experience in Stalin's Soviet Union. Wiener became increasingly involved in writing these and other popular works to bring his ideas to a general audience.

Wiener received the National Medal of Technology from President Johnson in 1964. The accompanying citation praised his "marvelously versatile contributions, profoundly original, ranging within pure and applied mathematics, and penetrating boldly into the engineering and biological sciences." He died on March 18, 1964, in Stockholm, Sweden.

Further Reading

Heims, Steve J. *John von Neumann and Norbert Wiener: From Mathematics to the Technologies of Life and Death*. Cambridge, Mass.: MIT Press, 1980.

Rosenblith, Walter, and Jerome Wiesner. "A Memoir: From Philosophy to Mathematics to Biology." Available on-line. URL: http://ic.media.mit.edu/JBW/ARTICLES/WIENER/WIENER1.HTM. Downloaded on November 3, 2002.

Wiener, Norbert. *Cybernetics, or Control and Communication in the Animal and Machine*. Cambridge, Mass.: MIT Press, 1950. 2nd ed., 1961.

———. *The Human Use of Human Beings: Cybernetics and Society*. Boston: Houghton Mifflin, 1950. 2nd ed., New York: Avon Books, 1970.

———. *Invention: The Care and Feeding of Ideas*. Cambridge, Mass.: MIT Press, 1993.

Wirth, Niklaus

(1934–)
Swiss/American
Computer Scientist

By the 1970s, computer programmers were writing extensive and increasingly complicated programs in languages such as FORTRAN (for science) and COBOL (for business). While serviceable, these languages were limited in their ability to define data and control structures and to organize large programs into manageable pieces. Niklaus Wirth created new programming languages such as Pascal that helped change the way computer scientists and programmers thought about their work. His work influenced later languages and ways of organizing program resources.

Wirth was born on February 15, 1934, in Winterthur, Switzerland. His father was a professor of geography. He received a bachelor's degree in electrical engineering at the Swiss Federal Institute of Technology (ETH) in 1958, then earned his M.S. degree at Canada's Laval University in 1960. He then went to the University of California, Berkeley, where he received his Ph.D. in 1963 and taught in the newly founded computer science department at nearby Stanford University.

By then, Wirth had become involved with computer science and the design of programming languages. He helped implement a compiler for the IBM 704 for a language called NELIAC, a version of Algol. This work gave Wirth familiarity with the practical problems of implementing languages with regard to specific hardware, and it also motivated him to look more closely at language design. He noted that since computer hardware could be designed using modules with specific connections for the flow of electrical signals, software should also consist of modules of code with defined connections, not a tangled mess of control "jumps" and data being accessed from all over the program.

Wirth returned to the ETH in Zurich in 1968, where he was appointed a full professor of computer science. He had been part of an effort to improve Algol. Although Algol offered better program structures than earlier languages such as FORTRAN, the committee revising the language had become bogged down in adding many new features to the language that would become Algol-68.

Wirth believed that adding several ways to do the same thing did not improve a language but simply made it harder to understand and less reliable. Between 1968 and 1970, Wirth therefore crafted a new language, named Pascal after the 17th-century mathematician who had built an early calculating machine.

The Pascal language required that data be properly defined and allowed users to define new types of data, such as records (similar to those used in databases). It provided all the necessary control structures such as loops and decision statements. Following the new thinking about structured programming (which had been introduced as a concept by EDSGER DIJKSTRA), Pascal retained the GOTO statement (considered "unsafe" because it made programs confusing and hard to debug) but discouraged its use.

Pascal became the most popular language for teaching programming. By the 1980s, versions such as UCSD Pascal and later, Borland's Turbo Pascal, developed by PHILIPPE KAHN, were bringing the benefits of structured programming to desktop computer users. Meanwhile, Wirth was working on a new language, Modula-2. As the name suggested, the language featured modules, packages of program code that could be linked to programs to extend their data types and functions. Wirth also designed a computer workstation called Lilith. This powerful machine not only ran Modula-2; its operating system, device drivers, and all other facilities were also implemented in Modula-2 and could be seamlessly integrated, essentially removing the distinction between operating system and application programs. Wirth also helped design Modula-3, an object-oriented extension of Modula-2, as well as another language called Oberon, which was originally intended to run in embedded systems (computers installed as control systems in other devices). Wirth also gave attention to multiprogramming (the design of program that could perform many tasks simultaneously, coordinating the exchange of data between them).

Looking back at the development of object-oriented programming (OOP), the next paradigm that captured the attention of computer scientists and developers after structured programming, Wirth has noted that OOP is not new. Its ideas (such as encapsulation of data) are largely implicit in structured procedural programming, even if it shifted the emphasis to binding functions into objects and allowing new objects to extend (inherit from) earlier ones. But he believes the fundamentals of good programming have not changed in 30 years. In a 1997 interview, Wirth noted that "the woes of software engineering are not due to lack of tools, or proper management, but largely due to lack of sufficient technical competence. A good designer must rely on experience, on precise, logic thinking; and on pedantic exactness. No magic will do."

Besides being noted for the clarity of his ideas, Wirth could also demonstrate a wry sense of humor. When asked how his last name should be pronounced, Wirth referred to a concept in programming in which a variable can be either passed to a procedure by reference (address) or by value (the actual number stored). Wirth said that his last name should be pronounced "Virt" if by reference, but "Worth" if by value!

Wirth has received numerous honors, including the Association for Computing Machinery Turing Award (1984) and the Institute of Electrical and Electronics Engineers Computer Pioneer Award (1987).

Further Reading

Pescio, Carlo. "A Few Words with Niklaus Wirth." *Software Development* 5, no. 6 (June 1997). Available on-line. URL: http://www.eptacom.net/pubblicazioni/pub_eng/wirth.html. Downloaded on December 9, 2002.

Wirth, Niklaus. *Algorithms + Data Structures = Programs*. Englewood Cliffs, N.J.: Prentice Hall, 1976.

———. *Project Oberon: The Design of an Operating System and Compiler*. Reading, Mass.: Addison-Wesley, 1992.

———. "Recollections about the Development of Pascal." In Bergin, Thomas J., and Richard G. Gibson, eds. *History of Programming Languages-II*. New York: ACM Press; Reading, Mass.: Addison-Wesley, 1996, pp. 97–111.

———. *Systematic Programming: an Introduction*. Englewood Cliffs, N.J.: Addison-Wesley, 1973.

Wirth, Niklaus, and Kathy Jensen. *PASCAL User Manual and Report*. 4th ed. New York: Springer-Verlag, 1991.

Wozniak, Steve

(1950–)
American
Inventor, Electrical Engineer

Steven Wozniak's last name may *sound* like a computer, but what he is best known for is designing one. Besides designing the simple but elegant Apple II personal computer, Wozniak cofounded Apple Computer Corporation and was a key innovator in its early (pre-Macintosh) years.

Born on August 11, 1950, in San Jose, California, Wozniak grew up to be a classic electronics whiz. This was perhaps not surprising; Wozniak's father was a designer of high-tech satellite guidance systems at the Lockheed Missiles and Space facility in nearby Sunnyvale. He was always at hand to help his son explore the principles of electronics.

Young Wozniak built his own transistor radio. He also built a working electronic calculator when he was 13, winning the local science fair. (However, courses other than mathematics and science bored Wozniak, and he got poor grades in them.)

By the time he graduated from Homestead High School, Wozniak knew that he wanted to be a computer engineer. He tried community college but soon quit to work with a local computer company. Although he then enrolled in the University of California, Berkeley (UC Berkeley), to study electronic engineering and computer science, he dropped out in 1971 to go to work again, this time as an engineer at Hewlett-Packard, at that time one of the most successful companies in the young Silicon Valley.

By the mid-1970s, Wozniak was in the midst of a technical revolution in which hobbyists explored the possibilities of the newly available microprocessor, or "computer on a chip." A regular attendee at meetings of the Homebrew Computer Club, Wozniak and other enthusiasts were excited when the MITS Altair, the first

At a time when microcomputers usually came as kits requiring patience and skill with a soldering gun, Steve Wozniak designed the sleek, easy-to-use Apple II. Hooked up to a cassette recorder and a TV or monitor, the friendly machine named for a fruit demonstrated that the desktop computer could be a tool for everyone. Wozniak later produced rock festivals and today promotes new wireless technology. *(Courtesy of Alan Luckow)*

complete microcomputer kit, came on the market in 1975. The Altair, however, had a tiny amount of memory, had to be programmed by toggling switches to input hexadecimal codes (rather like ENIAC), and had very primitive input/output capabilities. Wozniak decided to build a computer that would be much easier to use—and more useful.

Wozniak's prototype machine, the Apple I, had a keyboard and could be connected to a TV screen to provide a video display. He demonstrated it at the Homebrew Computer Club and among the interested spectators was his younger friend STEVE JOBS. Jobs had a more entrepreneurial interest than Wozniak, and spurred him to set up a business to manufacture and sell the machines. Together they founded Apple Computer in June 1976. Their "factory" was Jobs's parents' garage, and the first machines were assembled by hand.

Wozniak designed most of the key parts of the Apple, including its video display and later, its floppy disk interface, which is considered a model of elegant engineering to this day. He also created the built-in operating system and BASIC interpreter, which were stored in read-only memory (ROM) chips so the computer could function as soon as it was turned on.

In 1981, just as the Apple II was reaching the peak of its success, Wozniak was almost killed in a plane crash. He took a sabbatical from Apple to recover, get married, and return to UC Berkeley (under an assumed name!) to finish his B.S. degree in electrical engineering and computer science.

Wozniak's life changes affected him in other ways. As Apple grew and became embroiled in the problems and internal warfare endemic to many large companies, "Woz" sold large amounts of his Apple stock and gave the money to Apple employees whom he thought had not been properly rewarded for their work. He watched as the company's clumsily designed Apple III failed in the marketplace. Wozniak did not want to be a manager, remarking to an interviewer in 1994 that "I was meant to design computers, not hire and fire people." He left Apple for good in 1985 and founded Cloud Nine, an ultimately unsuccessful company that designed remote control and "smart appliance" hardware intended to put everything at a homeowner's fingertips.

Showing an exuberant energy that ranged beyond the technical into the social and artistic areas, Wozniak produced two rock festivals that lost $25 million, which he paid out of his own money. He would be quoted as saying "I'd rather be liked than rich."

A theme for Wozniak's life and work might be found in a speech he gave to a Macintosh convention in 2000. He said, "I really wanted technology to be fun. The best technology should be for people." He then paused, and in a softer voice, added, "to play games."

In early 2002, Wozniak announced a new venture, Wheels of Zeus (wOz), a company that would develop a variety of "intelligent" wireless devices to take advantage of modern communications technology and GPS (global positioning system) technology.

Since the 1990s, Wozniak has organized a number of charitable and educational programs, including cooperative activities with people in the former Soviet Union. He particularly enjoys classroom teaching, bringing the excitement of technology to young people. In 1985, Wozniak received the National Medal of Technology.

Further Reading

Cringely, Robert X. *Accidental Empires: How the Boys of Silicon Valley Make Their Millions, Battle Foreign Competition, and Still Can't Get a Date*. Reading, Mass.: Addison-Wesley, 1992.

Freiberger, Paul, and Michael Swaine. *Fire in the Valley: the Making of the Personal Computer*. 2nd ed. New York: McGraw-Hill, 1999.

Kendall, Martha E. *Steve Wozniak, Inventor of the Apple Computer*. 2nd rev. ed. Los Gatos, Calif.: Highland Publishing, 2001.

Srivastava, Manish. "Steven Wozniak." Available on-line. URL: http://ei.cs.vt.edu/~history/WOZNIAK.HTM. Downloaded on January 28, 2003.

Woz.org. Available on-line. URL: www.woz.org. Updated on September 27, 2002.

Y

Yang, Jerry (Chih-Yuan Yang)
(1968–)
Chinese/American
Entrepreneur, Inventor

The generation born in the 1960s and 1970s grew up in a world where computers were largely taken for granted. While hardware innovation continued, the new frontier for the 1990s was the Internet and particularly the World Wide Web, which had been developed by TIM BERNERS-LEE. Like most frontiers, however, the Web lacked much in the way of maps for travelers. Jerry Yang and his partner, David Filo, cofounded Yahoo!, the company and website that provided the best-known portal and guide to the burgeoning world of the Web.

Chih-Yuan Yang was born on November 6, 1968, in Taiwan (the Republic of China). His father died when he was only two years old. His mother, a teacher of English and drama, found her son to be bright and inquisitive. (He started to read Chinese characters when he was only three.)

When Yang was 10, the family moved to the United States. At that time, Yang received an Americanized first name, Jerry. Although Yang was made fun of at first because he knew no English, he was lightning fast in math and after three years was also in an advanced English class.

Yang went to Stanford University to study electrical engineering, earning simultaneous bachelor's and master's degrees in 1990. Embarking on his doctoral studies, Yang met David Filo, with whom he taught in a program in Japan. Yang and Filo were supposed to be working on systems for computer-aided design of electronic circuits, but because their adviser had gone on sabbatical, Yang and Filo were left mainly on their own.

One of their pastimes was surfing the Web with their Netscape browser. However, Yang in particular was frustrated by the lack of an organized way to approach the growing variety of resources on the Web. Deciding to do something about the problem, in 1994 Yang created a website called "Jerry's Guide to the World Wide Web." On it he started listing what he considered to be the best websites, divided into various categories. The list (to which Filo also started contributing) proved to be amazingly popular, registering a million "hits" (people accessing the page) a day. The site was so popular, in fact, that the Stanford system administrator complained that it was tying up their computer.

Yang and Filo were finally getting close to completing their doctoral work, but they decided that what had started as a hobby was more important than their Ph.Ds. They decided to move their list site to a commercial Web server and turn it into a business. While Filo coded an expanded program to better organize the material on the site, Yang tried to find ways to use the site to earn money. One possibility was selling advertising, but no one knew whether people would pay to advertise on the Web, a medium that many businesspeople barely knew existed. Another possibility was to have companies pay a small royalty for each hit their website received by being linked through Yang and Filo's site.

Early in 1995, Yang and Filo went on a hunt for a venture capitalist willing to invest in their ideas. At first they had no luck, but then they met Mike Moritz of Sequoia Capital, who had invested in many successful high-tech companies such as Apple, Cisco Systems, and Oracle. Moritz later to author Robert Reid recalled that his first impression of the two graduate students' operation was not very promising:

> [They] were sitting in this cube with the shades drawn tight, the Sun [Web] servers geneating a ferocious amount of heat, the answering machine going on and off every couple minutes, golf clubs stashed against the walls, pizza cartons on the floor, and unwashed clothes strewn around. It was every mother's idea of the bedroom she wished her sons never had.

Nevertheless, Mortiz was impressed by the young entrepreneurs' intelligence and energy. He agreed to invest $1 million in their venture. However, they needed a catchier name than "Jerry and Dave's List." They decided on Yahoo! (complete with exclamation point). They claimed that the name stood for "Yet Another Hierarchic Officious Oracle." (The hierarchic part referred to the organization of the listing into various levels of topics and subtopics.)

The Yahoo! site proved to be immensely popular. Previously, Web users seeking specific information had only search engines that were sometimes useful but often spewed out hundreds of irrelevant listings. With Yahoo, however, users could browse categories of information and find sites that had been reviewed for suitability by Yahoo! employees.

Yahoo!'s popularity translated to marketability: In August 1995, the company sold advertising to five businesses for $60,000 apiece. Revenues grew further as Yang and Tim Koogle (who eventually came the company's chief executive officer) began to forge relationships with "content providers" such as the Reuters news service. These relationships gave the linked companies access to millions of Web users while in turn making Yahoo! a more attractive place to visit.

Besides adding news, weather, stock quotes, and other features, Yahoo! also drew in users through another of Yang's initiatives—the ability of users to create their own custom Yahoo! page with the news and other features they desired. When users took advantage of this feature they made Yahoo! their browser's default page, ensuring that they would begin each Web session through the portal of Yahoo!

By the beginning of 2000, Yahoo! was getting 190 million visitors a month and earning nearly $1 billion in annual revenue. Despite his new personal wealth, Yang's lifestyle remained modest and his image remained one that young people easily identified with. Yang often expressed his astonishment at the business giant that had come out of a few Web pages. He told an interviewer that "I'm like a kid in a candy store—except the candy store is the size of an airport."

However, the 21st century has brought difficult challenges to Internet-based companies in general and Yahoo! in particular. Companies such as America Online (AOL) and Microsoft (with its MSN network) were creating Web information portals of their own. Another threat was that a media conglomerate such as Time-Warner (which later merged with AOL) would use its vast existing resources of content to provide on-line offerings that Yahoo! could not match. Against this background there was also the bursting of the Internet stock bubble, and the general market decline of 2001–02, all of which dried up much of the advertising revenue flowing to the Web. In the first quarter of 2001, Yahoo!'s earnings plunged 42 percent.

One possibility was that Yahoo! would try to acquire a media company to strengthen its content, or even buy the remarkably successful on-line auction company eBay. While Koogle favored such acquisitions, Yang refused, wishing to stick with what he saw as Yahoo!'s core business as a Web information integrator. Yang has also tried to distance himself from the corporate infighting and focus on presenting a positive image of Yahoo! to the outside world.

Yahoo! still has considerable financial resources and has better prospects than most dot-coms. Whatever happens, Yang has achieved remarkable stature as a business innovator and one of the pioneers of the age of e-commerce. Yang received the 1998 *PC Magazine* "People of the Year" award.

Further Reading

Elgin, Ben, et al. "Inside Yahoo!" *Business Week,* May 21, 2001, pp. 114ff.

Reid, Robert H. *Architects of the Web: 1,000 Days That Built the Future of Business.* New York: John Wiley, 1997.

Schlender, Brent. "How a Virtuoso Plays the Web." *Fortune,* March 6, 2000, pp. F-79ff.

Z

Zuse, Konrad

(1910–1995)
German
Engineer, Inventor

Great inventions seldom have a single parent. Although popular history credits Alexander Graham Bell with the telephone, the almost forgotten Elisha Gray invented the device at almost the same time. And although ENIAC, designed by J. PRESPER ECKERT and JOHN MAUCHLY, is widely considered to be the first practical electronic digital computer, another American inventor, JOHN VINCENT ATANASOFF, built a smaller machine on somewhat different principles that also has a claim to being "first." Least known of all is Konrad Zuse, perhaps because he did most of his work in a nation soon to be engulfed in the war it had begun.

Zuse was born on June 22, 1910, in Berlin, Germany, but grew up in East Prussia. He attended the Braunsberg High School, which specialized in the classics (Greek and Latin). However, Zuse had become fascinated with engineering, and in 1927 went to the Technical University in Berlin to study civil engineering, receiving his degree in 1935.

Zuse went to work at the Henschel aircraft plant in Berlin, where he was assigned to study the stress on aircraft parts during flight. At the time, such calculations were carried out by going through a series of steps on a form over and over again, plugging in the data and calculating by hand or using an electromechanical calculator. Like other inventors before him, Zuse began to wonder whether he could build a machine that could carry out these repetitive steps automatically.

Zuse was unaware of the nearly forgotten work of CHARLES BABBAGE and that of other inventors in America and Britain who were beginning to think along the same lines. He also had little background in electrical engineering, though later he would note that this lack of experience also meant a lack of preconceptions, thus enabling him to try new ideas.

With financial help from his parents (and the loan of their living room), Zuse began to assemble his first machine from scrounged parts. His first machine, the Z1, was completed in 1938. The machine used slotted metal plates with holes and pins that could slide to carry out binary addition and other operations (in using the simpler binary system rather than decimal, Zuse was departing from other calculator designers). Such memory as was available was provided in similar fashion.

The Z1 had trouble storing and retrieving numbers and never worked well. Undeterred, Zuse began to develop a new machine that used electromechanical telephone relays (a ubiquitous component that was also favored by HOWARD AIKEN). The new machine worked much better, and Zuse successfully demonstrated it at the German Aerodynamics Research Institute in 1939.

With World War II underway, Zuse was able to obtain funding for his Z3, which was able to carry out automatic sequences from instructions (Zuse used discarded movie film instead of punched tape). The relatively compact (closet-sized) machine used 22-bit words and had 600 relays in the calculating unit and 1,800 for the memory. It also had an easy-to-use keyboard that made many operations available at the touch of a button, but the machine could not do branching or looping like modern computers.

The Z3 was destroyed in a bombing raid in 1944. However, Zuse also designed the specialized, simplified S1, which was used to design the guidance system for the V-1, an early "cruise missile" that was launched in large numbers against targets in Britain and western Europe toward the end of the war.

Meanwhile, Zuse used spare time from his military duties at the Henschel aircraft company to work on the Z4, which was completed in 1949. This machine was more fully programmable and was roughly comparable to Howard Aiken's Mark I. What amazed American and British observers was that Zuse had designed the machine almost single-handedly, without either the ideas or resources available to designers such as Aiken.

By that time, however, Zuse's electromechanical technology had been surpassed by the fully electronic vacuum tube computers such as ENIAC and its successors. (Zuse had considered vacuum tubes but had rejected them, believing that their inherent unreliability and the large numbers needed would make them impracticable for a large-scale machine.) During the 1950s and 1960s, Zuse ran a computer company called ZUSE KG, which eventually produced electronic vacuum tube computers.

Zuse's most interesting contribution to computer science was not his hardware but a programming language called Plankalkül, or "programming calculus." Although the language was never implemented, it was far ahead of its time in many ways. It started with the radically simple concept of grouping individual bits to form whatever data structures were desired. It also included program modules that could operate on input variables and store their results in output variables. Programs were written using a notation similar to mathematical matrices.

Zuse labored in obscurity even within the computer science fraternity. However, toward the end of his life his work began to be publicized. He received numerous honorary degrees from European universities as well as awards and memberships in scientific and engineering academies. Zuse also took up abstract painting in his later years. He died on December 18, 1995.

Further Reading

Bauer, F. L., and H. Wössner. "The Plankalkül of Konrad Zuse: A Forerunner of Today's Programming Languages." *Communications of the ACM* 15 (1972): 678–685.

Lee, J. A. N. *Computer Pioneers*. Los Alamitos, Calif.: IEEE Computer Society Press, 1995.

Zuse, Konrad. *The Computer—My Life*. New York: Springer-Verlag, 1993.

Entries by Field

Computer Science

Backus, John
Baran, Paul
Bartik, Jean
Bell, Chester Gordon
Berners-Lee, Tim
Brooks, Frederick P.
Cerf, Vinton
Codd, Edgar F.
Corbató, Fernando
Davies, Donald Watts
Dertouzos, Michael
Diffie, Bailey Whitfield
Dijkstra, Edsger
Feigenbaum, Edward
Gelernter, David Hillel
Gosling, James
Hamming, Richard Wesley
Hillis, W. Daniel
Hopper, Grace Murray
Joy, Bill
Kahn, Robert
Kay, Alan C.
Kemeny, John G.
Kernighan, Brian
Kildall, Gary
Kleinrock, Leonard
Knuth, Donald E.
Lanier, Jaron
Lenat, Douglas B.
Licklider, J. C. R.
Lovelace, Ada
Maes, Pattie
McCarthy, John
Minsky, Marvin
Nelson, Ted
von Neumann, John
Newell, Allen
Nygaard, Kristen
Papert, Seymour
Perlis, Alan J.
Postel, Jonathan B.
Rabin, Michael O.
Rees, Mina Spiegel
Ritchie, Dennis
Roberts, Lawrence
Sammet, Jean E.
Samuel, Arthur
Shannon, Claude E.
Simon, Herbert A.
Stallman, Richard
Stroustrup, Bjarne
Sutherland, Ivan
Thompson, Kenneth
Tomlinson, Ray
Turing, Alan
Weizenbaum, Joseph
Wiener, Norbert
Wirth, Niklaus

Engineering

Baran, Paul
Bush, Vannevar
Cray, Seymour
Drexler, K. Eric
Eckert, J. Presper
Engelbart, Douglas
Felsenstein, Lee
Forrester, Jay W.
Grove, Andrew S.
Hewlett, William Redington
Kahn, Robert
Kilburn, Thomas M.
Kilby, Jack
Kleinrock, Leonard
Lenat, Douglas B.
Mauchly, John
Metcalfe, Robert M.
Noyce, Robert
Olsen, Kenneth H.
Packard, David
Torres Quevedo, Leonardo
Wozniak, Steve
Zuse, Konrad

Entrepreneurism

Amdahl, Gene M.
Andreessen, Marc
Bezos, Jeffrey P.
Bushnell, Nolan
Case, Steve
Dell, Michael
Ellison, Larry

INVENTION

MATHEMATICS

PROGRAMMING

WRITING

CHRONOLOGY

1836 Charles Babbage conceives of the Analytical Engine.

1843 Ada Lovelace explains the work of Charles Babbage.

1847 George Boole introduces his "Algebra of Logic."

1886 William Burroughs founds the American Arithmometer Company.

1890 Herman Hollerith's tabulator aids the U.S. Census.

1893 Leonardo Torres Quevedo demonstrates an equation-solving machine.

1924 Thomas Watson Sr. changes his company's name to IBM.

1927 Vannevar Bush develops the Differential Analyzer.

1936 Alonzo Church and Alan Turing publish theories of computability.

1939 John Atanasoff demonstrates the ABC computer.

William Hewlett and David Packard start Hewlett-Packard.

1941 George Stibitz builds the Complex Number Calculator.

1942 Hermann Goldstine becomes head of ENIAC project.

1944 Howard Aiken's Mark I calculator is installed at Harvard.

1946 J. Presper Eckert and John Mauchly complete ENIAC.

Adele Goldstine documents ENIAC and trains programmers.

John von Neumann describes the modern stored program computer.

1947 Wallace J. Eckert builds SSEC electronic calculator for IBM.

Norbert Wiener coins the term *cybernetics*.

1948 Claude Shannon revolutionizes communications theory.

Jean Bartik begings to develop programs for BINAC and Univac. ca. late 1940s

Richard Hamming develops error-correcting codes.

Elizabeth Holberton develops symbolic programming commands.

1949 Mina Rees begins to direct mathematical research for the U.S. Navy.

1950 Claude Shannon describes algorithms for computer chess.

Alan Turing proposes a test for determining artificial intelligence.

1951 Thomas Kilburn develops larger, faster electronic and magnetic memory.

Marvin Minsky builds SNARC neural network computer.

An Wang starts to market magnetic core memories.

1952 Univac predicts the results of the presidential election.

Thomas Watson Jr. takes charge at IBM, backs computer development.

1953 Jay Forrester introduces magnetic "core" memory in the Whirlwind.

1956 John McCarthy organizes Dartmouth conference on artificial intelligence (AI) research.

Newell, Simon, and Shaw demonstrate automated theorem-proving.

1957 Jim Backus's FORTRAN language is released.

Grace Hopper's Flow-Matic compiles English-like commands.

Alan Perlis develops programming and computer science courses.

1958 John McCarthy begins to develop LISP language.

1959 Jack Kilby and Robert Noyce patent different versions of the integrated circuit.

1960 Gordon Bell and Kenneth Olsen introduce the PDP-1 minicomputer.

Kenneth Olsen's Digital Equipment Corporation introduces the PDP-1 minicomputer.

1961 Fernando Corbató begins implementing time-sharing at the Massachusetts Institute of Technology (MIT).

Arthur Samuel's checkers program beats a champion.

1962 Joseph Licklider becomes head of ARPA computer research.

1963 Ivan Sutherland develops interactive Sketchpad program.

1964 Paul Baran publishes "On Distributed Communications."

The IBM System/360 series is developed by Gene Amdahl and Frederick Brooks.

John Kemeny and Thomas Kurtz develop BASIC at Dartmouth.

1965 Edward Feigenabum invents the expert system.

Moore's law predicts rapid doubling of processor power.

Theodor Nelson coins the term *hypertext*.

Seymour Papert founds the MIT artificial intelligence laboratory

1966 Lawrence Roberts put in charge of designing ARPANET.

Joseph Weizenbaum's ELIZA program bemuses computer users.

1967 Donald Davies and Lawrence Roberts further develop packet-switching.

Kristen Nygaard's Simula 67 introduces object-oriented programming.

Seymour Papert develops the LOGO language at MIT.

1968 Edsger Dijkstra urges structured programming in his paper "GOTO Considered Harmful."

Douglas Engelbart demonstrates the mouse and graphical interface.

Donald Knuth publishes first volume of *The Art of Computer Programming*.

Gordon Moore, Robert Noyce, and Andrew Grove found Intel Corporation.

Niklaus Wirth begins to develop Pascal.

1969 The ARPANET is successfully tested by Leonard Kleinrock.

1970 Edgar Codd introduces the relational database model.

Alan Kay begins to develop object-oriented Smalltalk language.

Marvin Minsky and John McCarthy found Stanford AI Laboratory.

Ken Thompson and Dennis Ritchie begin to design UNIX.

1971 Ray Tomlinson invents e-mail.

1972 Dennis Ritchie begins to develop the C language.

Nolan Bushnell introduces Pong, which becomes a runaway seller.

Hubert Dreyfus critiques AI in *What Computers Can't Do*.

Lee Felsenstein's Community Memory becomes the first computer bulletin board.

An Wang markets the first dedicated word processing system.

1973 Vinton Cerf and Robert Kahn announce the TCP/IP network protocol.

Gary Kildall develops CP/M operating system.

1974 Jean Sammet becomes first woman president of the Association for Computing Machinery (ACM).

1975 Michael Rabin devises a new way to test for prime numbers.

1976 Seymour Cray's Cray 1 redefines supercomputing.
Whitfield Diffie and Martin Hellman reveal public key cryptography.
Brian Kernighan publishes *Software Tools*.
Raymond Kurzweil demonstrates a reading machine for the blind.
Robert Metcalfe introduces Ethernet.

1977 Steve Wozniak's Apple II is released.

1979 Daniel Bricklin's VisiCalc spreadsheet becomes a hit for the Apple II.
Larry Ellison's Oracle begins to develop relational databases.

1980 Bill Gates scores a coup in selling the DOS operating system to IBM.

1981 The IBM PC, designed by Philip Estridge, enters the market.

1982 Esther Dyson's *Release 1.0* becomes a leading computer industry journal.
Gordon Eubanks enters utility software market with Symantec.
Bill Joy and Scott McNealy found Sun Microsystems.
Jonathan Postel outlines the Internet domain name server organization.

1983 Philippe Kahn releases Turbo Pascal 1.0.
Mitchell Kapor releases Lotus 1-2-3 spreadsheet.
Jaron Lanier begins commercial virtual reality research.
Richard Stallman begins the GNU project for open-source UNIX.
Bjarne Stroustrup begins to develop C++

1984 William Gibson's novel *Neuromancer* popularizes cyberspace.
Steve Jobs markets the Apple Macintosh.
Douglas Lenat begins to compile a huge knowledge base called Cyc.
Sherry Turkle explores computer culture in *The Second Self*.

1985 Chris Crawford's *Balance of Power* game mirrors the cold war.

1986 K. Eric Drexler's *Engines of Creation* introduces nanotechnology.
W. Daniel Hillis builds the first Connection Machine.
Marvin Minsky proposes theory of multiple intelligences.

1987 Andrew Grove becomes chief executive officer of Intel Corporation.

1988 Dell Computer goes public after growing direct personal-computer sales.
Clifford Stoll begins to track a mysterious hacker at a government lab.

1991 Tim Berners-Lee's World Wide Web appears on the Internet.
Steve Case's America Online reaches out to home users.
Eric Raymond publishes *The New Hacker's Dictionary*.
Linus Torvalds releases the first version of Linux.

1992 The Internet Society is founded.

1993 David Gelernter is badly injured by the Unabomber.
Howard Rheingold's *The Virtual Community* explores on-line life.

1994 Marc Andreessen and Jim Clark found Netscape.

1995 Michael Dertouzos writes *What Will Be*, his book about a computerized future.
James Gosling's Java language is introduced.
Pattie Maes begins commercial development of "software agents."
Jerry Yang and David Filo's Yahoo! becomes popular on the Web.

1997 Jeff Bezos's Amazon.com goes public.

1998 Pierre Omidyar's eBay goes public and has a banner year.

2001 Michael Rabin proposes an unbreakable randomized code.

2002 Shawn Fanning's Napster files for bankruptcy after losing in court.

GLOSSARY

algorithm A step-by-step procedure for solving a problem.

analog computer A computer that uses natural forces or relationships to solve problems. Its values are continuously variable, rather than discrete as in digital computers.

applet A small Java program that runs in a user's Web browser.

ARPA (Advanced Research Projects Agency) An agency of the U.S. Department of Defense which, starting in the mid-1960s, funded computer research projects such as ARPANET, ancestor of the Internet.

artificial intelligence (AI) The attempt to create computer programs, systems, or robots that exhibit behavior characteristic of human intelligence, such as the ability to adapt to changing conditions and to learn.

Backus-Naur Form (BNF) A standard notation for describing the grammatical structure of computer languages.

BASIC (Beginners' All-purpose Symbolic Instruction Code) An easy-to-use, interactive computer language developed by John Kemeny and Thomas Kurtz at Dartmouth College in the mid-1960s and later used on many personal computers.

broadband An Internet connection that can transport large quantities of data, as over a cable or DSL line. This makes it practical to receive media such as streaming video or sound.

browser A program that connects to and displays Web pages.

bulletin board system (BBS) A dial-up system (first developed in the late 1970s) that allowed users to post messages and share files.

C A sparse but powerful computer language developed in the early 1970s by Dennis Ritchie and often used for systems programming.

C++ A version of C developed in the early 1980s by Bjarne Stroustrup. It includes object-oriented features such as classes for better organization of programs. It has largely supplanted C for most applications.

character recognition The identification of alphabetic characters from printed matter and their translation to character codes stored in the computer.

COBOL (COmmon Business-Oriented Language) A programming language developed in the late 1950s, based considerably on previous work by Grace Hopper. COBOL became the preferred language for business programming with mainframe computers for more than two decades.

cognitive psychology The study of the process of cognition (thinking) in human beings. The field has both contributed to and benefited from research in artificial intelligence.

Compiler A program that takes instructions written in a higher-level (symbolic) computer language and translates them into low-level machine codes that can be executed by the processor.

computer science The study of the design, programming, and operation of computers. It includes topics such as algorithms, data structures, and computer language grammar.

core Computer memory consisting of a lattice of tiny doughnut-shaped ferrite (iron) rings that can be magnetized in two different ways to indicate a 1 or 0 in the data. Core was the primary kind of fast random-access memory for computers from the mid-1950s until it was supplanted by memory chips.

cybernetics A term coined by Norbert Wiener, derived from a Greek word meaning the steersman of a boat. It refers to the science of automatic control of machines. It has largely been subsumed into computer science.

cyberspace A term popularized in the science fiction novel *Neuromancer* (1984) by William Gibson. It refers to the world perceived by computer users as they interact with computer programs and with one another. Virtual reality is an intense form of cyberspace that immerses the user in graphics and sound.

database An organized collection of files containing information. Information is usually broken down into records, each containing the information about one "thing" (such as a customer). Records in turn are broken into fields which contain particular items of information, such as a phone number or zip code.

data glove A glove that contains sensors and transmitters such that its position can be tracked by a computer system. It allows the user to interact with a virtual reality simulation, such as by picking up virtual objects.

data mining The analysis of data in databases to extract patterns or other information that can be used for purposes such as marketing.

data structure A way of representing or organizing a particular kind of information in a computer. For example, an integer (whole number) might be stored as two or four 8-bit bytes in memory.

differential analyzer A form of analog computer developed by Vannevar Bush in the 1930s. It could solve sets of equations using several variables.

digital computer A computer that stores information as specific digits (numbers), usually using the binary system (1 and 0). Digital computers have largely replaced analog computers, which stored data as continuous quantities based on natural forces or relationships.

distributed In computer design, refers to storing data or programs in a number of different locations (or different computers). Distributed systems can offer greater flexibility and the ability to work around failures.

DNS (Domain Name System) The system that sets up a correspondence between names such as "stanford.edu" or "well.com" and the numbers used to route messages around the Internet. The part of an Internet address following the "dot" is a domain, and domains correspond to different types of sites, such as .edu (educational institutions) or .com (businesses).

dot-com A business that is heavily involved with the Internet and World Wide Web, such as an on-line bookstore.

e-commerce Short for "electronic commerce," it is the buying and selling of goods and services on-line.

electromechanical A system using electricity and/or magnetism together with mechanical parts, such as relays or switches. Some electromechanical computers were built in the late 1930s and early 1940s but they were soon replaced by electronic computers.

electronic Involving the control of the flow of electrons with no mechanical parts. The first electronic control device was the vacuum tube. The transistor replaced the vacuum tube in the 1950s, and in the 1960s the equivalent of thousands of transistors were built into tiny integrated circuits.

expert system A program that uses a set of rules (a knowledge base) to perform analyses or make decisions. The rules are usually developed in consultation with humans who are experts in performing the task that the software is intended to implement.

file-sharing system A program (such as the ill-fated Napster) that lets users make files (such as MP3 music files) readily available to other users.

file system A system for organizing data into files on a storage device such as a hard disk. Files are usually grouped into directories or folders. The file system also provides programs with a way to read, write, or copy files.

floating point A method of keeping track of the location of the decimal point (and thus the magnitude) of a number.

FORTRAN (FORmula TRANslation) A programming language designed by John Backus and his colleagues in the mid-1950s. Because of its ability to handle mathematical expressions, it became the most widely used language for scientific and engineering applications, although it began to decline in the 1970s with the popularity of C.

frame In artificial intelligence, a way of organizing "common sense" knowledge so that computer programs can work with it. For example, a frame might include information about the steps involved in a visit to a restaurant (being seated, ordering food, eating, paying for the meal, and so on).

ftp (file transfer protocol) A program designed for copying files between computers in a UNIX network. (Versions are now available for Windows and other operating systems.)

graphical user interface (GUI) A way of controlling a computer using visual cues (menus and icons) and a pointing device, usually a mouse. The first modern GUI was developed in the 1970s by Alan Kay and others at the Xerox Palo Alto Research Center. It was then adopted by Steve Jobs for the Macintosh (1984) and later by Microsoft Windows.

HTML (hypertext mark-up language) A system of codes originally devised by Tim Berners-Lee for formatting and otherwise controlling text, graphics, and links on a Web page.

HTTP (hypertext transport protocol) Also developed by Berners-Lee as part of the World Wide Web system, HTTP is a standard set of rules for connecting Web servers and browser so that Web pages can be fetched and displayed.

hypertext The embedding of links in text documents that allow the reader to go to other parts of the document or to other documents. Although the idea of hypertext was outlined by Vannevar Bush in the 1940s, the term itself was coined by Ted Nelson in the mid-1960s. Today the World Wide Web is the most familiar example of hypertext.

integrated circuit A circuit whose components are embedded into a single "chip" such as of silicon. Today the equivalent of millions of transistors can be embedded in a chip the size of a thumbnail.

Internet The worldwide connection of many millions of computers and networks that use a standard system for transferring data.

Java A language (similar to C++) designed by James Gosling in the mid-1990s. It is primarily used for applications related to the World Wide Web.

knowledge base An organized collection of rules that describe a subject (such as protein chemistry) in such a way that an expert system can use it to make decisions or recommendations, perform analyses, and so on.

knowledge engineering The design of knowledge bases and expert systems.

Lambda calculus A system for manipulating mathematical functions devised by Alonzo Church. It was a key to the design of LISP and other functional programming languages.

Linux A freely available version of the UNIX operating system. Developed by Linus Torvalds in the early 1990s, today Linux is very popular for Web servers and workstations, although it has not made much headway against Windows on the desktop.

LISP (LISt Processing Language) A language designed by John McCarthy in the late 1950s primarily for artificial intelligence research. Unlike most languages that are designed for specifying procedures, LISP is built around lists of symbols and functions that manipulate them. This makes it easy to write programs that can modify themselves.

LOGO A language designed by Seymour Papert in the late 1960s to teach computer science concepts, particularly to young people. Although functionally similar to LISP, LOGO uses simpler syntax and features a graphic "turtle" that can be manipulated on the screen or even driven in the form of a small wheeled robot.

mainframe A large computer (named for the big cabinet that held the central processing unit). Originally, virtually all computers were mainframes, but in the 1960s when the smaller minicomputer came along, the term *mainframe* was adopted to distinguish the larger machines.

microprocessor An integrated circuit chip that contains a complete arithmetic and logic processing unit. It can thus serve as the CPU (central processing unit) of a computer. The availability of microprocessors in the mid-1970s spurred the creation of the personal computer (PC).

minicomputer A type of computer built starting in the early 1960s. It is smaller than a mainframe but larger than today's personal computers. Minicomputers generally handled data in small chunks but made computing power considerably more affordable to smaller schools and businesses.

nanotechnology The direct manipulation of molecules or even single atoms. Nanotechnologists such as K. Eric Drexler hope to build tiny machines that could act as computer components or even robots for a variety of purposes.

neural network A system first developed by researchers such as Marvin Minsky in the 1950s. It mimics the operation of the brain, in which individual nodes act like neurons and respond to stimuli by sending signals. "Correct" responses can be reinforced, allowing the system to gradually "learn" a task.

object-oriented programming (OOP) A way of organizing programs into objects that have defined characteristics and capabilities. The program then runs by creating suitable objects based on a definition called a class, and then interacting with them. Object-oriented programming became popular starting in the 1980s with the development of languages such as Smalltalk and particularly C++.

open source The distribution of software along with its program codes, allowing users to freely modify or extend the program. Generally users are required to include the modified source if they in turn distribute or sell the program. This alternative to proprietary software was publicized by Richard Stallman; today the most widely used open-source product is probably Linux.

operating system (OS) The software that controls the basic operation of a computer, including access to files, connection to printers or other devices, and processing of user commands. Applications programs call upon the operating system when they want to perform a task such as opening a file.

packet switching Developed by Leonard Kleinrock and others during the 1960s, packet-switching is the breaking up of messages into small separately addressed chunks (packets) for transmission over the network. The ability to use many alternate routes for packets makes it possible to get around failures or outages as well as to use the connections most efficiently.

parallel processing The use of more than one processor, either in a single computer or in a group of computers. Special programming languages can be used to assign various program tasks to different processors.

parity A system of error detection where each group of data bits is set so that it always has an even (or odd) number of 1 bits. As a result, if the data arrives with an odd number of ones when it should be even (or vice versa), the system knows that an error has occurred.

pipe A facility for connecting the output of one program to the input of another. Made popular by UNIX, pipes make it possible to use several simple utility programs in succession to perform a more elaborate task.

portability The ability of a program or operating system to be easily adapted to many different models of computer.

portal A website, such as Yahoo!, that offers access to many kinds of information or services in one place.

punch card A card with columns of spaces that can be punched out to indicate various data items. Punch cards were first used for automatic data processing by Herman Hollerith in the 1890s. They were widely used with mainframe computers until the 1970s, when they were largely replaced by magnetic tape and disk drives.

RAM (random access memory) A form of memory in which the computer can fetch any desired bit of data directly without having to go through intervening information (as with a tape drive). Early forms of RAM include cathode-ray tubes (CRTs), magnetic drums, and magnetic "core." Today RAM is in the form of memory chips.

relational database A database system in which data from different files can be connected by referring to a field (such as a customer number) that they have in common. This allows for more efficient and better-defined organization of data.

relay An electromagnetic switch that is triggered by an electrical impulse. Relays were used extensively in the telephone system starting in the 1930s and were adopted to early computers before the advent of electronics (vacuum tubes), which is about 1,000 times faster.

scalable Of a computer system, the ability to smoothly and easily add capacity while keeping existing hardware and software. One of the IBM System/360's selling points was its scalability.

semiconductor A material such as germanium or treated silicon that is neither a good conductor nor a good insulator, but whose conductivity can be controlled, thus controlling the flow of electrons. The semiconductor is the basis of the transistor and thus of modern computer components.

software agent A program that helps a user perform tasks based on general instructions, much in the way that a human, such a travel agent, might do.

speech synthesis The generation of spoken words by the proper combination of sound elements (phonemes).

SQL (structured query language) A standard system for specifying data and operations to be performed using a relational database. For example: SELECT * WHERE COST > 50.00 would extract all records where the Cost field exceeded $50.00.

stored program concept Early computers such as ENIAC did not store instructions but simply read them from cards, then executed and discarded them. This made it difficult to change programs or perform repetition (looping). In 1945, John von Neumann formalized the idea that computers should store their instructions in memory where they could be referred to repeatedly, making looping and self-modifying programs much easier.

structured programming A movement starting in the 1970s that emphasized the proper control of program flow (with loops and the like), the avoidance of haphazard GOTO "jumps" and the grouping of code into procedures or subroutines. This movement was reflected in the writing of Edsger Dijkstra and the design of languages such as C (by Dennis Ritchie) and particularly Pascal (by Niklaus Wirth).

subroutine A defined portion of a program that performs a specified task and then returns control to the main program. This allows large programs to be broken into more manageable pieces. The structured programming movement led to further refinement in the form of procedures and functions that accepted only data of a specified type and that did not allow internal variables to be changed from other parts of the program.

supercomputer A somewhat nebulous term referring to a computer of unusual speed and power, often built for experimental purposes or for particular applications (such as aircraft design or nuclear physics) that required as much processing power as possible. Seymour Cray built a series of famous supercomputers in the 1970s and 1980s.

tabulator A device that counts data from punched cards, sorts cards, and so on. Tabulators were first built by Herman Hollerith in the late 19th century, and IBM sold increasingly elaborate punch card systems in the 1920s and 1930s.

TCP/IP (transmission control protocol/internet protocol) The basic rules for routing data around the Internet and for managing transmission. TCP/IP was developed by Vinton Cerf and Robert Kahn and announced in 1973.

time-sharing A system that allows many users to use the same computer. The computer rapidly switches from one user or program to the next, giving each a small amount of execution time. If the system capacity is not strained, each user experiences virtually instant response. Time-sharing was a valuable innovation in the 1960s and 1970s, when computers were relatively expensive. However, today each user generally has his or her own computer.

transistor A device that allows the control of an electron flow, turning it on or off, amplifying it, and so on. The transistor replaced the vacuum tube during the 1950s because it is more compact and uses much less power.

UNIX A popular operating system developed by Kenneth Thompson and Dennis Ritchie in the early 1970s. It is characterized by having a "kernel" containing core functions (such as the file system and device control) and allowing for a variety of user command processors (called shells). UNIX also uses "pipes" to allow programs to be connected to one another. During the 1970s, UNIX users developed many additional utilities, and UNIX became the operating system of choice for campuses and research laboratories.

virtual community The experience of ongoing relationship between participants in an on-line system. As with physical communities, members of virtual communities can form friendships, conduct feuds, and work together on projects. The WELL (Whole Earth 'Lectronic Link) became a famous virtual community starting in the mid-1980s.

virtual machine A system that interprets instructions in a general language (such as Java) for a particular hardware or operating system environment.

virtual reality The use of realistic graphics and sound and interactive devices (such as data gloves) to give the user the experience of being immersed in a world generated by the computer. Virtual reality has been used for military and medical training, entertainment, and other applications.

voice recognition The conversion of speech to text by analyzing its sound elements (phonemes) and identifying the words.

workstation A computer specialized for tasks such as computer-aided design or graphics. Workstations are generally more powerful than ordinary PCs, although the distinction has diminished in recent years.

World Wide Web The system of pages of information and media linked by hypertext and accessible over the Internet. The Web was created by Tim Berners-Lee in the early 1990s and became popular in the mid-1990s with the advent of graphical Web browsers such as Mosaic and Netscape.

BIBLIOGRAPHY

Bach, M. J. *The Design of the Unix Operating System*. Upper Saddle River, N.J.: Prentice Hall, 1986.

Berlinksi, David. *The Advent of the Algorithm: The Idea That Rules the World*. New York: Harcourt, 2000.

Biermann, Alan W. *Great Ideas in Computer Science: A Gentle Introduction*. 2nd ed. Cambridge, Mass.: MIT Press, 1997.

Brookshear, J. Glenn. *Computer Science: An Overview*. 6th ed. Reading, Mass.: Addison-Wesley, 2000.

Cortada, James W. *Historical Dictionary of Data Processing: Biographies*. New York: Greenwood Press, 1987.

Freiberger, Paul, and Michael Swaine. *Fire in the Valley: The Making of the Personal Computer*. New York: McGraw-Hill, 1999.

Goldstine, Hermann. *The Computer from Pascal to von Neumann*. Princeton, N.J.: Princeton University Press, 1972.

Greenia, Mark W. *History of Computing: An Encyclopedia of the People and Machines that Made Computer History*. Revised CD Ed. Lexikon Services, 2001.

Hafner, Katie, and Matthew Lyon. *Where Wizards Stay Up Late: The Origins of the Internet*. New York: Simon and Schuster, 1996.

Hillis, Daniel W. *The Pattern on the Stone: The Simple Ideas that Make Computers Work*. New York: Basic Books, 1998.

Kernighan, B. W., and R. Pike. *The Unix Programming Environment*. Upper Saddle River, N.J.: Prentice Hall, 1984.

Kidder, Tracy. *The Soul of a New Machine*. New York: Modern Library, 1997.

Knuth, Donald E. *The Art of Computer Programming*. Vols. 1–3. 3rd ed. Reading, Mass.: Addison-Wesley, 1998.

Lee, J. A. N. *Computer Pioneers*. Los Alamitos, Calif.: IEEE Computer Science Press, 1995.

Levy, Steven. *Hackers: Heroes of the Computer Revolution*. Updated ed. New York: Penguin, 2001.

Malone, Michael S. *The Microprocessor: A Biography*. New York: Springer-Verlag, 1995.

McCartney, Scott. *ENIAC: The Triumphs and Tragedies of the World's First Computer*. New York: Berkeley Books, 1999.

Raymond, Eric. *The New Hacker's Dictionary*. 3rd ed. Cambridge, Mass.: MIT Press, 1996.

Shasha, Dennis, and Cathy Lazere. *Out of Their Minds: The Lives and Discoveries of 15 Great Computer Scientists*. New York: Springer-Verlag, 1997.

Slater, Robert. *Portraits in Silicon*. Cambridge, Mass.: MIT Press, 1987.

Spencer, Donald D. *The Timetable of Computing*. Ormond Beach, Fla.: Camelot Publishing, 1999.

White, Ron. *How Computers Work*. Millenium ed. Indianapolis, Ind.: Que, 1999.

Winslow, Ward. *The Making of Silicon Valley: A 100-Year Renaissance*. Palo Alto, Calif.: Santa Clara Valley Historical Association, 1996.

INDEX

Note: Page numbers in **boldface** indicate main topics. Page numbers in *italic* refer to illustrations.

B

C

D

E

H

I

J

K

L

M

N

O

P